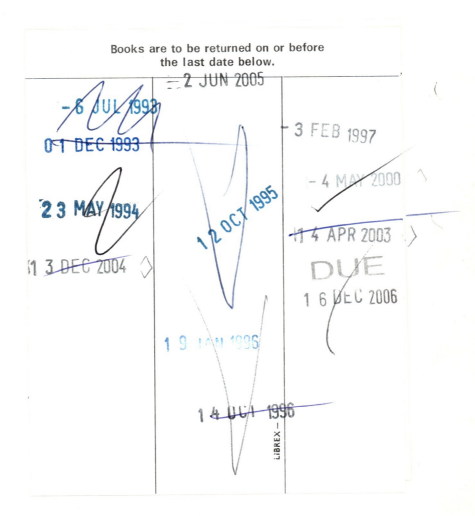

BIOTECHNOLOGY IN FOOD PROCESSING

BIOTECHNOLOGY IN FOOD PROCESSING

Edited by

Susan K. Harlander and Theodore P. Labuza

Department of Food Science and Nutrition
University of Minnesota
St. Paul, Minnesota

np

NOYES PUBLICATIONS
Park Ridge, New Jersey, U.S.A.

Library of Congress Catalog Card Number: 86-5242
ISBN: 0-8155-1073-X
Printed in the United States

Published in the United States of America by
Noyes Publications
Mill Road, Park Ridge, New Jersey 07656

10 9 8 7 6 5 4 3 2 1

Library of Congress Cataloging-in-Publication Data

Biotechnology in food processing.

Includes bibliographies and index.
1. Biotechnology. 2. Food industry and trade.
I. Harlander, Susan K. II. Labuza, T.P.
TP248.2.B5537 1986 664 86-5242
ISBN 0-8155-1073-X

Foreword

Biotechnology 1945-1985

Joshua Lederberg

Dedication

Dedicated to the memory of Edward L. Tatum (1909-1975)

The editors requested, or at least acquiesced in the suggestion, that I look back over the 40 years of my own active involvement in biotechnology research. My historical focus will be on three of the publications that underlay much of the theoretical insight and experimental methodology used in biotechnology today: Beadle and Tatum (1941); Avery, MacLeod and McCarty (1944) and Lederberg and Tatum (1946). They concern biochemical genetics *(Neurospora)*, DNA mediated transformation (pneumococcus) and conjugal gene exchange (*E. coli* K12), respectively. Think back to, or for most of you try to imagine, a time when genetic research had not yet reached *E. coli* K12 as an experimental object; when the very notion of genes in bacteria was problematical. Table 1 summarizes some of the milestones of physiological genetics before 1945.

Our discipline really does begin in the middle of the last century. We had Gregor J. Mendel with the first account of genes as segregating units in carefully designed crosses of garden plants. We had Friedrich Miescher with the first isolation of a material that we now know as nucleic acid. Within a few years we had the pure culture of microorganisms and their role in disease, putrefaction and fermentation clearly established. Mendel's work lay dormant for 35 years to be suddenly rediscovered by three groups of investigators (de Vries, Correns, and Tschermak) simultaneously in 1900: the beginning of modern genetics.

Within two years Archibald Garrod, an internist, was looking at a trait called alcaptonuria, distinguished by the blackening of urine. He studied the pedigrees and with Bateson's help inferred that this attribute followed the recently rediscovered Mendel's Laws. Over the next decade he established several clear examples of "inborn errors of metabolism:" genetic blocks of a metabolic pathway. Garrod's work was only dimly recognized by mainstream genetics. It did appear in textbooks of biochemistry and medicine. It will come up again because some of the very concepts that Garrod used in his explanation of these diseases in man were to be redeveloped in the modern era of biochemical genetics.

Table 1. History of Ideas Leading to Crossing Bacteria, *(E. coli* K12)

	Genetics
1865	Mendel—first account of genes
1900	deVries, Correns, Tschermak—rediscovery
1902	Garrod—inborn errors of metabolism
1904	Blakeslee—*Mucor*
1928	Dodge—*Neurospora* life cycle
1941	Beadle & Tatum—*Neurospora* biochemical mutants
1942	F.J. Ryan—*Neurospora* biochemical mutants at Columbia U.
	Biochemistry
1865	Miescher—nucleic acids
1920	Levine—nucleotides
1926	Sumner—crystalline urease: protein
1930	Northrop—crystalline pepsin: protein
1935	Stanley—crystalline TMV: protein plus RNA
1944	Avery, MacLeod, McCarty—DNA has genetic activity in pneumococcus
	Microbiology
1676	van Leeuwenhoek—bacteria
1695	van Leeuwenhoek—protozoa copulating
1870's	Pasteur, Koch, Cohn—*Schizomycetes*
1928	Griffith—pneumococcus transformation
1930's	*Salmonella* serotypes
1943	Luria & Delbruck—population statistics of bacterial mutations
1944	USNH—malaria life cycle
1945	Dubos—"The Bacterial Cell" contra "Traditional" medical school teaching
1945-6	Lederberg (Columbia, Yale)—*Escherichia coli* K12

One early venture into microbial genetics did flourish briefly between 1900 and 1941. Albert Blakeslee wrote his doctoral thesis at Harvard University on the sexual cycle of the *Mucorales* in 1904. His work with a few species of *Mucor* and *Phycomyces* established a segregation of sex determining factors in these fungi, and he was obviously influenced, like Garrod, by the resurgence of Mendelism. However after he completed his doctoral work he could not get a job working on the genetics of fungi; it had no obvious agricultural or any other applied significance. Instead he was able to get work at the Agricultural Experiment Station in Connecticut and began his work in plant breeding and cytogenetics where he did make many important contributions. So fungal genetics lay fallow for several decades until other sources of interest in fungi, their life cycle and their application to genetics could be revived.

In 1928, B.O. Dodge dearly loved one of these fungi that had been originally isolated in Java growing on moldy peanuts, and then other species growing on bread—the red bread mold *Neurospora.* He found that *Neurospora* was uncommonly easy to handle and to carry through its complete sexual cycle. He worked out the segregation of sex in this particular organism and found that whereas the zygospores in *Phycomyces* do not germinate very efficiently, the ascospores in *Neurospora* do. It ended up being an ideal organism for Mendelian genetics.

George Beadle and Edward Tatum come into our story out of their efforts to discover genetic markers with which to probe the relationship of genes to development. Among the earliest mutants discovered in *Drosophila* were albino or other pigment anomalies in eye color. It was obvious that there was a chemical difference in white eyes versus red eyes, and some effort was made in the extraction of these pigments. Around 1935, Boris Ephrussi and George Beadle discovered diffusible substances which were accumulating in the larval body of some mutants and could restore the color of implants of eye primordia from other mutants; what we would now call a simple complementation test. They called this diffusible substance the V+ hormone. We would now call this "hormone" a metabolic intermediate in a blocked chemical pathway. Beadle then went to Stanford University in 1937 to continue these studies and advertised for a biochemist to help him in the conduct of this work. Tatum had done his doctoral work at the University of Wisconsin at Madison. Shortly thereafter with Harland Wood he was the first to show that propionic bacteria had a growth factor requirement for thiamine, a vitamin previously known to be important in the metabolism of mammals and of yeast. From that experience he became thoroughly imbued with a sense of comparative biochemistry; namely that many metabolic systems would be found to be very similar in a wide variety of organisms. He thus felt unrestrained in his choice of experimental material.

At Stanford University, this sense was soon to be demonstrated with a vengeance in his work on *Neurospora.* His immediate task was to isolate the blocked intermediate, the V+ hormone, out of the fruit flies. Isolating it by the gram was a formidable task, for they had few modern methods like chromotographic isolation. Then, at the very last minute, after having done all this backbreaking work, they were scooped by Butenandt. He had tested a substance that had been found in dog urine and believed to be an anabolite of tryptophane, namely kynurenine. It functioned in the bioassay test exactly like the elusive V+ hormone. That easy win just leapfrogged all over the various laborious isolations that Tatum had done throughout this time. This chastening experience motivated them to seek better experimental material for biochemical study. Guided in part by Tatum's prior background in microbial nutrition it occurred to Beadle that they could reverse their research strategy. Instead of painfully pursuing the biochemistry of the mutants one happened to find in

Drosophila, why not pick an organism where one could look for mutants blocked in specific nutritional pathways? In other words, by mutational block, create additional examples of what nature had provided in the biochemical requirements; e.g., for thiamine in the propionibacterium. It worked like a charm within a few months. The project involved irradiating *Neurospora*, crossing the irradiated spores to cultures of the opposite mating type, isolating single ascospore cultures and testing them for growth on basal medium. It took as few as 299 manual isolations of single spores, which is a couple of days' work once you get the hang of it, before they had a culture which had an induced requirement for pyridoxine. The growth requirement segregated in Mendelian fashion; and the modern era of experimental biochemical genetics had begun.

Another milestone appeared in 1928, a casual observation outside of any contemporary paradigm. Fred Griffith was working on the very practical problems of the serological classification of the organisms of pneumonia. He pondered what was different about type 1, type 2, type 3, and so on which were recognized by the serological reagents. His experiment, the speculative roots of which were not clearly explained in his published paper, was to put killed cells of one type into a mouse. He then inoculated the mouse with live cells of another type and came out with live cells of the first type; i.e., of the killed organisms. So he postulated a transforming principle that on some occasion would transfer an attribute from those dead bacterial cells to the ones that were still living. This work, published in the Journal of Hygiene, hardly reached the attention of geneticists. In any case there was no framework with which to understand its biological significance and no single geneticist in those days was working on bacteria as test organisms.

O.T. Avery, a biochemist and microbiologist at The Rockefeller Institute, was very much concerned with pneumonia for the same reasons as was Griffith. When Griffith's work was first announced he was quite skeptical. However, some of his postdoctoral associates were able to repeat Griffith's published observation. We certainly have to credit Avery and the leadership of The Rockefeller Institute for devoting a large part of the energy and resources of his laboratory from 1930 to 1944 to doing what he did best—the fractionation, isolation and characterization of chemical entities; the extraction of the material out of those killed pneumococcal cells that could cause the transformation of types. Much to Avery's astonishment, this material turned out not to be protein (which the theory of the time would have predicted) but DNA. It is perfectly correct to say that the DNA revolution began with this finding by Avery, MacLeod and McCarty: the first operational assay for the biological specificity of a preparation of DNA.

Let us jump now to 1945. Beadle and Tatum have published a series of papers on *Neurospora* out of Stanford University. Avery and his group have published their report that the transforming factor in the pneumococcus is DNA. I find myself, a medical student, working with Francis J. Ryan in the Department of Zoology at Columbia University. In fact I had begun working for him in my undergraduate sophomore year, 1942, when Ryan returned from his postdoctoral fellowship with Beadle and Tatum. He was one of the first people to bring *Neurospora* out of the Stanford Laboratory. Of course in the department we were all full of the excitement of this new world of biochemical genetics.

It is hard to convey the meteoric impact of Avery's work. Well before it was formally published, we heard about it from Alfred Mirsky who traveled frequently between The Rockefeller Institute and Columbia University. The inspiration that this was going to be the key to what we would now call molecular biology was transparent to everybody in the department. I say this with such vehemence because others have denied that Avery's work was well understood. Perhaps it was not well understood in other quarters, but in a place that has the metabolism of knowledge that New York does, the atmosphere was really quite electrified. This had to be the opening of the door to a great new era, and we had to find every conceivable way to try to exploit this breakthrough.

There were problems on all sides; e.g., whether the transforming factor was pure DNA in the first place. At one level that didn't matter too much, for the course had already been clearly set and in short order it would be settled as to whether the gene was DNA or protein, or both, or something else. If the pneumococcal transformation factor was really a gene, a unit of Mendelian heredity, how could we verify that fact, when one knew nothing else of the genetics of bacteria?

The first effort was to try to transform *Neurospora* with DNA containing extracts. Ryan had taught me the technology of the auxotrophic mutants that would not grow in a simple synthetic medium without some supplement. So it was not difficult to imagine an experimental design that would greatly facilitate looking for transformation in this organism; namely to seek a few transformed cells (prototrophs) that would grow without the supplement. If you could get DNA transfer in *Neurospora*, one could answer all the questions that separated pneumococcus from *Drosophila.* Well it didn't work! But I won't take too much discredit for that, for it is only in the last three or four years that systems have been developed to enable *Neurospora* to be transformed, and they do require very arcane handling of the organism to get DNA into it.

That was discouraging, but at least it had laid out the experimental methodology by which one could efficiently look for genetic transfer in a microorganism. So to ring the changes, if we can't transform *Neurospora* maybe we can develop a broader genetics of easily handled bacteria like *E. coli.*

Tatum, having a very strong bacteriological background himself, had started to generalize on the study of metabolic pathways from *Neurospora* to *E. coli,* and in late 1944 he published a report on auxotrophic mutants of the same variety as had been found in *Neurospora.* This was an analogy between *E. coli* and *Neurospora* that would at least encourage one to look further into attempts to dissect its genetic structure. But there were no known crosses, no transformations, no other methods of dealing with the relationships of genes to one another, and certainly no way to identify them as DNA. Based on these premises, there were two further lines of experiment. One that I did in a rather desultory way was to try to transform *E. coli,* but it was to take many years to concoct the witches' brew that makes it possible (calcium treatment and so forth).

Another possibility emerged out of the need, this desperate need, to do something genetically significant with *E. coli* in order to justify the Avery experiment and its generalization to genetics. If you could show that bacteria could be crossed, a bridge between Avery's experiment and the rest of the world of genetics would be completed. So that led to a reexamination of whether *E. coli* could be crossed.

The experimental design recorded in my notebook dated July 15, 1945 was "Diplophase in bacteria can be selected for by using two different mutant strains of *E. coli* and growing them in continuously renewed minimal medium." In practice it sufficed to plate washed cells from a mixed culture onto minimal agar.

Francis Ryan encouraged me to continue these experiments with Ed Tatum. He had just left Stanford (where biology departments in those days did not have that much use for biochemists) and he was to start a new department of microbiology at Yale University. He had then just published a second research paper which explained that he had taken some of the single step mutants in *E. coli*, mentioned before, and subjected them to a second round of radiation to yield double mutants.

These double mutants were exactly what I had been looking for to prevent "adaptation or reverse mutation." The design was: if there were two genetic blocks in each culture, then the statistical probability of both undergoing reverse mutation in a single selection cycle would be reduced to a negligible quantity. We were getting reversion rates of 10^{-6} or 10^{-7} per cell—enough to interfere with the selection experiment. One could expect double mutants to be occurring at 10^{-12} or 10^{-14} and that is a sufficiently small number to give acceptable blanks.

Tatum very graciously accepted my proposal and invited me to join him in New Haven. I arrived in New Haven in March 1946, started assembling the mutants, made a bunch more, and did the first actual crossing experiments on the 2nd of June. Soon I had done about a dozen of them and was therefore quite convinced that the experiment would work and be reproducible. The Cold Spring Harbor Symposium in 1946 was coming up in early July. We had no plans to present the material after just one month of experimentation; but this was the first international conference of microbial genetics. Everyone in the world was there and almost everyone was bewailing the lack of a sexual process in bacteria. So we decided to present the first results even with that very limited number of experiments.

We were fortunate to be able to get the first confrontation in front of nearly every expert in the world all at once, under the discipline of their mutual critical outlook.

Things certainly have changed from then to now. A comparison of the earlier linkage map published for *E. coli* to a contemporary one shows a fantastic increase in the number of gene loci. Back then we knew nothing about F^+ or F^- (male and female). Presently there are well over a thousand loci mapped in *E. coli*, close to a third of the total expected to be there.

May I now take up a question that has dogged me for many, many years? Why did it take until 1946 for successful crossing in bacteria? The experiment I've described to you is so simple that many high school students have done it. Many of the concepts needed to conduct it were available in the early 1900's.

Why wasn't it done then? Why did it take 40 years?

To answer this question, let us review some older history. Microbiology began with the first microscopic visualization of microbes at the hands and eyes of Leeuwenhoek in 1676. He continued his observations and in 1695 he reported on a wide variety of animalcules, larger than the tiny dancing dots, the bacteria. These larger animals were the protozoa. He gave an eloquent description of the copulation of protozoa that made it clear that they indeed had a sexual process. That protozoa were sexual while bacteria were not was then an attribution of the scale of nature that was firmly engrained with the very initiation of microbiology. This was reaffirmed with the development of pure culture methodology by Koch and Cohn. This concept of bacteria as organisms that breed true to type and do not divide by any other process than fission became crystalized in 1875 in the first formal taxonomy of bacteria. Ferdinand Cohn called them the *Schizomycetes*, the fission fungi. The myth of bacterial asexuality was thus engrained in the very class names of the organisms in question. The compartmentalization of teaching, of thought, reflected in the separation of microbiology as a medical specialty with little interest in the fundamental biology of these organisms, the contrary with respect to genetics, is at least part of the explanation for bacterial genetics having a delayed start.

Today, the intersection of applied research with fundamental study is being pursued on a broad interdisciplinary foundation. As I look back over the most important and revolutionary advances in science, I observe that many of them have come from the intersection of the world of practice (medical practice, engineering needs, biotechnological applications, natural history) with the cutting edge of existing knowledge. In a firmly established discipline the canon of experimental design is to narrow down the relevant variables and to control all the appropriate inputs to allow a reproducible and controllable result. That is just fine as long as your theoretical outlook suffices to embrace all the relevant variables. In that world of practice one can no longer choose the constraints for experimental convenience. The real world invariably intrudes operational variables you hadn't thought about, had no way of controlling, and many new discoveries have come from that sphere. For example, I am confident that efforts to optimize biotechnical organisms for hyperproduction will uncover new principles of gene regulation. In this, and in many other ways, I am confident that research with applied orientation, such as is being celebrated at this symposium, will also surely nourish the most revolutionary findings when one looks beyond the most proximal practical goals. In complementary fashion, the most exciting advances in technology have resulted from pure science in totally unrelated fields. When Ed Tatum was pondering where to begin his definitive work after his postdoctoral fellowship, his mentors urged him to go into the microbiology of butter. He could hardly have made a more important contribution to food processing biotechnology than what he did, beginning with the eye pigments of the fruit fly.

Rockefeller University
New York, New York
February 1986

Joshua Lederberg

BIBLIOGRAPHY

1. Beadle, G.W. and Tatum, E.L. Genetic control of biochemical reactions in *Neurospora*. Proc. Nat. Acad. Sci. 27:499–506 (1941).
2. Avery, O.T., MacLeod, C.M., and McCarty, M. Studies on the chemical nature of the substance inducing transformation of pneumococcal types. J. Exptl. Med. 79:137–158 (1944).
3. Lederberg, J. and Tatum, E.L. Gene recombinations in *Escherichia coli*. Nature 158:558 (1946).

Preface

Broadly defined, biotechnology is the collection of industrial processes that involve the use of biological systems. In some instances these processes involve recombinant DNA technology, in which the hereditary apparatus of an animal, plant or bacterial cell is altered such that it can produce more or different chemicals or perform new functions. However, biotechnology is not limited to genetic engineering, as it also encompasses enzyme and protein engineering, plant and animal tissue culture technology, biosensors for biological monitoring, and bioprocess and fermentation technology.

Biotechnology as an integrated multidisciplinary field will have a profound impact on a number of areas, including pharmaceuticals for human and veterinary health care, the chemical and energy industries, agriculture and the food industry, and the environment. Biotechnology will have a dramatic effect on the food processing industry, creating exciting opportunities in new product development and differentiation, cost reduction, and creation of novel processing methods. The challenge for food processors will be to determine how, when, and in what areas biotechnology will impact their industry.

Although numerous symposia dealing with biotechnology have been offered, none has adequately addressed the unique needs of the food processing industry. In response to this need for a realistic assessment of the current status and potential impact of biotechnology in the food industry, the Department of Food Science and Nutrition at the University of Minnesota sponsored a widely attended symposium "Biotechnology in the Food Processing Industry," on October 7-9, 1985, to explore the economic, technological and regulatory impact of evolving areas of high technology on food and food related industries. This volume represents the proceedings of the symposium.

The success of the symposium was made possible by the support, cooperation and enthusiasm of many institutions and individuals. We are grateful for the support of our cosponsors, the Agricultural Experiment Station at the University of Minnesota and the Cooperative State Research Service of the

U.S. Department of Agriculture. Contributions from the following individuals and companies are also gratefully acknowledged: Bayer AG/Miles Laboratories; Charlotte E. Biester; General Mills, Inc.; Hershey Foods Corporation; Kraft, Inc.; Schwan's Sales Enterprises, Inc.; The Pillsbury Company; The Procter and Gamble Company; and, The Quaker Oats Company.

We gratefully acknowledge the valuable consultation and expertise of the Local Organizing and Scientific Advisory Committees, and of the Office of Special Programs at the University of Minnesota. We also express special thanks to Y.C. Albert Hong for technical assistance in editing and typesetting the manuscripts in this volume.

Ultimately the success of the symposium was the result of the cooperation, imagination and enthusiasm of the speakers, convenors and authors whose efforts we now sincerely acknowledge.

St. Paul, Minnesota
February, 1986

Susan K. Harlander
Theodore P. Labuza

Contributors

Warren Alm
Kraft, Inc.
Glenview, IL

Robert Buchanan
Food Safety Laboratory
U.S. Department of Agriculture
Philadelphia, PA

Theodore Cayle
Kraft, Inc.
Glenview, IL

Bruce Chassy
National Institutes of Health
Bethesda, MD

Robert Dinwoodie
Kraft, Inc.
Glenview, IL

David Easson, Jr.
Department of Applied
Biological Sciences
MIT
Cambridge, MA

David Evans
DNA Plant Technology
Cinnaminson, NJ

Renee Fitts
Department of Applied
Biological Sciences
MIT
Cambridge, MA

Susan Gordon
Calgene
Davis, CA

Roy Grabner
Monsanto Company
St. Louis, MO

Susan K. Harlander
Department of Food Science
and Nutrition
University of Minnesota
St. Paul, MN

David Jackson
Genex Corporation
Gaithersburg, MD

Spiros Jamas
Department of Applied
Biological Sciences
MIT
Cambridge, MA

Rena Jones
Labatt Brewing Company
London, Ontario, Canada

Sarah Kiang
Kraft, Inc.
Glenview, IL

Theodore P. Labuza
Department of Food Science
and Nutrition
University of Minnesota
St. Paul, MN

Russell Larson
Kraft, Inc.
Glenview, IL

Joshua Lederberg
Rockefeller University
New York, NY

Samir Ma'ayeh
Kraft, Inc.
Glenview, IL

Jeremy Mathers
Kraft, Inc.
Glenview, IL

Larry McKay
Department of Food Science
and Nutrition
University of Minnesota
St. Paul, MN

Stephen McNamara
Hyman, Phelps and McNamara, P.C.
Washington, D.C.

David Mehnert
Kraft, Inc.
Glenview, IL

Raymond Moshy
International Plant Research
Institute
San Carlos, CA

Saul Neidleman
Cetus Corporation
Emeryville, CA

Nanette Newell
San Francisco, CA

Maura Raines
Kraft, Inc.
Glenview, IL

ChoKyun Rha
Department of Applied
Biological Sciences
MIT
Cambridge, MA

John Roland
Kraft, Inc.
Glenview, IL

Richard Ronk
Center of Food Safety and
Applied Nutrition, FDA
Washington, D.C.

Igne Russell
Labatt Brewing Company
London, Ontario, Canada

Richard Saunders
Kraft, Inc.
Glenview, IL

Peter Senior
Imperial Chemical Industries
Billingham, United Kingdom

Michael Sfat
BIO-Technical Resources, Inc.
Manitowac, WI

William Sharp
DNA Plant Technology
Cinnaminson, NJ

Sharon Shoemaker
Genencor, Inc.
South San Francisco, CA

Anthony Sinskey
Department of Applied Biological Sciences
MIT
Cambridge, MA

Graham Stewart
Research and Quality Control
Labatt Brewing Company
London, Ontario, Canada

Nayan Trivedi
Research and Development—Fermentation
Universal Foods Corporation
Milwaukee, WI

Ronald Wetzel
Genentech, Inc.
South San Francisco, CA

David Wheat
Arthur D. Little, Inc.
Cambridge, MA

Contents

Chapter 1

Biotechnology: Its Potential Impact on Traditional Food Processing

Raymond Moshy

INTRODUCTION

Biotechnology, broadly defined, includes any technique that uses live organisms (or parts of organisms) to make or modify products, to improve plants or animals, or to develop microorganisms for specific uses.

The subject of the role of biotechnology in food processing is very broad. The chapters in this monograph deal with several food processing applications. However, these represent but a small segment of the total food processing industry. The chapters on enzymology, protein engineering, modified polysaccharides, ingredients from plant cell cultures, and production of ascorbic acid all can be characterized as dealing with bioconversions. This chapter will specifically address the history of biotechnology and the overall impact of plant biotechnology on food processing.

A Brief History of DNA

The story of genetics begins in the middle 1800's. Gregor Mendel, a priest and only son of a peasant family, began the study of gene expression in part of what is now Czechoslovakia. Mendel entered a monastery in 1843, where he did the work that launched modern genetics. Most of Mendel's work was done with pea plants because they grew easily and rapidly; and they fertilize each other. More importantly, peas have obvious and reproducible characteristics; such as flower location, height, color, pod shape and the shape of the pea itself.

Mendel mated purebred pea plants and observed the offspring. He would take two sets of inbred pea plants which had the genetically pure distinctive single traits, such as wrinkled versus smooth peas. Upon crossing, plants of the first generation produced identical fruits; the pods all contained smooth peas. The obvious question was, "What happened to the heredity factor responsible for the "wrinkle" trait?" Had it been lost or destroyed? When plants of this first generation were allowed to fertilize each other, some of their progeny had wrinkled peas. The proportion of plants with smooth peas and wrinkled peas in the second generation was about three to one. Obviously it had not dropped out, but only dropped back. Mendel perceived that the factors which determine hereditary traits are discrete entities, which we now call genes. He further perceived that these entities can disappear in the first progeny of a cross because they are not expressing themselves, but could reappear, or reexpress the trait in a subsequent generation.

Mendel's work published in 1866, received little recognition by the scientific community and remained in the shadows for nearly fifty years until rediscovered in 1900 by DeVries in Holland, Correng in Germany and Von Tischernak in Austria.

The pollen and ova in pea plants get only one of the pair of genes, either the (W) or the (w) representing smooth and wrinkle respectively. When the male or female cells fuse to form a new plant, each of the offspring has (W) paired with (w) which is the only possible combination. To Mendel, it followed then that in these hybrid plants one type of the wrinkle or smooth trait determining gene was dominant (W) and one was recessive (w). Since the (W) gene controlling smoothness of shape is dominant, all the hybrid plants have smooth peas. However, when these hybrid plants, in turn, produce their pollen and ova, each pollen grain will be either (w) or (W); and each ovum will be either (w) or (W). When the genes pair to form the next generation, and if there are enough plants to provide an adequate statistical sample, a quarter of them will be pure dominant (W,W), a quarter mixed, dominant recessive (W,w), another quarter mixed (w,W), and the final quarter, which is made up of wrinkled peas, is pure recessive (w,w). The pairs containing at least one dominant (W) gene, make up three quarters of the plants, all produce smooth peas because the (W) is dominant.

In the early 1900's, Thomas Hunt Morgan of Columbia University was working with the common fruit fly (*Drosophila melanogaster*). Morgan knew that chromosomes were involved in heredity and that they were observable under the microscope in the cell nuclei. He showed that genes were in the chromosomes and that Mendel's trait determining factors were part of these visible chromosomes in the cells. He also demonstrated that the genes must be arranged in a linear order on the chromosomes like a string of pearls. Morgan ascertained the order of genes by crossing fruit flies with genetically distinct traits, such as wing shape or eye color. The frequency at which these traits reappeared in the progeny were determined and he soon learned that some of the traits appeared more frequently in offspring than could be accounted for by the random reassortment that Mendel had found. The traits had to be linked, perhaps as a consequence of genes that were literally connected one to the other.

Since chromosomes appeared to be the place where the genes actually were, Morgan concluded that a linkage between genes implied they were on the same chromosome.

The next step was the crossing of flies that had different linked traits. The result was different pairs of linked traits in the progeny with differing frequencies. By using the frequency of reappearance of the different pairs of traits in offspring, Morgan determined that the farther apart the genes of the linked pair were on the chromosome, the more frequent was the reappearance of the trait in the offspring. This implied that genes had to be arranged in a linear manner on the chromosome. This method for determining the position of genes on the chromosome has become widely applicable in genetics. Morgan won a Nobel Prize in 1933 for his discoveries relating to the hereditary functions of the chromosomes.

Dr. Morgan's student, Hermann Muller, began a study of what caused mutations in genes. It was shown that certain factors would cause a change in one gene but not affect the many thousands of others in the same cell. A physical force, such as X rays, could penetrate the cell and perhaps have a point effect upon a single gene. Muller irradiated hundreds of fruit flies with X rays and crossed them with nonradiated fruit flies with the result of enormous numbers of mutations. The mutated flies were flat, indented, had bulging eyes, different colors, no eyes, no hair, no antennae, truncated cave wings, downturn wings, upturn wings or no wings at all. These mutated flies compared with "normal" inherited changes by normal mutations differed apparently only in number. The effect of X rays was analogous to the effects of natural mutagens but manyfold more potent. The importance of this work is that it shed light on natural mutations and it provided an exceptionally powerful mutation tool for studying genes.

Natural mutations make possible the continuing diversification that leads to the innumerable species of our planet. Mutations occurring naturally are more likely to be destructive than they are to be constructive. Evolution effected only by naturally occurring mutation is unpredictable, slow, and restricted to a single line of descent. The exchange of genes between individuals simplifies the process and speeds up the amount and scope of diversity. When two cells randomly meet and fuse to form a single large cell, containing the contents of both, the process lacks discrimination. When the fused cell divides, if all of the DNA from the parent cells is divided equally into identical progeny, there is no mixing of DNA and there are no new combinations. A higher evolutionary step involves the introduction of specificity into the process of cellular fusion where, rather than a random fusion, a fusion occurs in which one cell is complementary to another. In this type of fusion, there is the beginning of a primitive form of sexual transfer in which one cell donates DNA to another cell. Subsequent recombination of genes inside the recipient ensures that some genes from the donor are included in the chromosomes of the offspring. This is called gene transfer. Gene transfer is a useful promoter of diversity because the DNA is from living organisms and contains genes of predictable value unlike the unpredictable nature of naturally occurring mutation.

The discovery of gene transfer in bacteria by Dr. Joshua Lederberg laid the groundwork for the incredible science of bacterial genetics. In 1945, Dr. Lederberg postulated that if laboratory-prepared DNA could be introduced into bacteria and recombined, then bacteria must have some biochemical mechanism for recombining their genes. The question was whether bacteria could release DNA into the growth medium, which in turn would be taken up by other bacteria and thus result in a recombination which could be detected by a genetic change in the progeny. Lederberg used double mutants. He reasoned that if each bacterial parent were defective in two different genes, the chance of either parent spontaneously reverting to normal would be exceedingly small. The chance of one mutation reverting to normal is about one in a million while the chance of two mutations reverting to normalcy is one in a trillion.

Lederberg took about a million cells of a population defective in two genes, A and B, and a million cells of a population defective in gene C and D. One strain had the genetic makeup A-, B-, C+, D+ with the A and B being the defective genes and the C and D being the normal genes. This strain grew only when given compounds a and b. The other strain was made up of A+, B+, C-, D-, with genes A and B being normal and permitting the strain to make normal products a and b. Genes C and D are defective so that the organism must have c and d in their medium to grow. After mating had occurred, samples were transferred to a medium lacking a, b, c, d which allows only normal cells to grow. Some colonies did appear and grow, and they were of normal gene makeup, i.e., A+, B+, C+, D+. This could only have come about as a consequence of an exchange of genes between the double mutant strains of bacteria. Lederberg called this process conjugation, which is the sex-like process involved in the fusion of two bacteria and the mixing of their genes. This then formed the basis for modern bacterial genetics.

In 1953, James Watson and Francis Crick described the structure of DNA. Watson and Crick's revelation of the DNA structure has often been equated with Darwin's "Origin of Species" in terms of its impact on science and society. Although this judgement might be premature, it is safe to say that it has been the most important single event in biology since the Darwin period. Watson and Crick's research was helped greatly because a great deal of information was on hand already. The chemistry of DNA was already worked out, and DNA was the right molecule to look at to unlock the secrets of the gene. DNA always has just four nucleotides containing the bases adenine, thymine, guanine, and cytosine.

These nucleotides were known to be linked together through their phosphate sugars. In 1948, Erwin Chargaff of Columbia University, wrote, while examining the chemical composition of DNA "I saw before me in dark contours the beginning of a grammar of biology .... I started from the conviction that, if different species exhibited different biological activities, there should also exist chemically demonstrable differences between (their) DNA's." In 1950, Chargaff published his work which showed that the relative proportions of the four nucleotides differed considerably in DNA samples from humans, pigs, sheep, oxen, bacteria and yeast. He also discovered that the amount of adenine always equaled the amount of thymine and the amount of guanine always equaled the amount of cytosine. Chargaff's rules of nucleotide composition are one of the most important sets of observations which contributed to the final solution of the structure of DNA.

The DNA double helix consists of intertwined spirals joined together by physical bonds to form a long thin twisted ladder. This unique structure has the means of conferring a unifying chemical principle upon all living things and has the ability to vary that unifying thread to an incredible number of combinations and recombinations using but a few chemical identities. The nucleotide bases form each rung of the ladder, and the sequence of base pairs form an exact code that dictates to each cell in an organism the products that it can make and what it can do during its life cycle. The sugar phosphate moieties make up the side pieces. As the ladder unwinds and is pulled apart, each strand maintains its half of the bases and serves as the template for the production of a new complementary ladder. The sequence in which the four nucleotide bases occur on the strand of DNA determines the order in which some 20 amino acids are put together to result in different proteins. All living organisms share this same genetic code and the same nucleotides and the same amino acids, resulting in an unbelievable diversity despite the similarity. There are several thousand genes with several million base pairs involved in the life processes of a bacterium. These numbers multiply by orders of magnitude in the higher plants and animals.

In 1973, Cohen and Boyer extracted a piece of genetic information from the DNA of one organism and inserted it into another with its message intact. This action was the creation of recombinant DNA technology and the birth of the science of genetic engineering. Recombinant DNA is the technique many think of first when speaking of biotechnology. As important as it is, it is not the only technology that goes to make up biotechnology. The manipulation of tissue, cells and genes is included in the definition of biotechnology, as are the new techniques of microculture for the selection of cell variants, cell fusion and the regeneration of whole plants from single cells, embryo recovery and transfer. In agricultural research, these powerful techniques, coupled with advanced breeding systems, are providing new capabilities in understanding and manipulating the basic biochemical processes in plants. For the first time, the family of techniques, including recombinant DNA, cell biology, tissue culture and plant breeding are being focused as one, on the development of new and improved agricultural products.

Agricultural Genetics

In a sense, whole plant (as well as microbial and animal) genetic engineering has been used for centuries to improve the properties of plants for human use. The entire field of agriculture as well as that of industrial fermentation have their bases in whole plant or microorganism genetic engineering. In the food processing industry, microorganisms are selected and improved for industrial fermentation use in making alcoholic beverages, amino acids, antibiotics, baked goods and cheese. In agriculture, plants are selected and bred for improved productivity, disease resistance, and quality. New genotypes of superior adaptation to environments, whether static or changing, are generated by the plant breeder typically by manipulating gene frequencies in populations. The physical basis of inheritance, the understanding of what is in fact occurring when two parents are crossed, had not been understood until Mendel's work was rediscovered and the conclusions applied in the early 1900's. Plant breeders now make their crosses with somewhat more precision, and knowledge of modern genetics permits them to manipulate selected plant genomes to produce predetermined improved varieties. These improved varieties have been responsible for the extraordinary developments in agricultural output in the last five decades. Farm productivity in the U.S. has increased almost 300%, while total farm acreage has actually declined slightly. Hybrid corn, which was developed in the 1930's, doubled U.S. corn yield and since then the yield has again been increased from 21 to 91 bushels per acre.

However, there are two drawbacks to even the most advanced breeding techniques. These are the limited genetic diversity used in most breeding programs and the time required to develop a desired strain. Genetic engineering offers the promise of reshuffling genes to produce recombinations never before available, resulting in a greatly increased range and source of genetic diversity for crop improvement. While Mendelian genetics has gone a long way in making classical breeding a more precise science, there is still a great deal of trial and error. In combining two complete genomes by a sexual cross, the outcome, more than likely, will be a surprise. A trait may be controlled by a single gene, or by a gene sequence, but the many thousands of other genes often times obscure the results and complicate the job.

Theoretically, the genetic engineer can obtain a gene of choice from a plant and insert it into another plant without the complications of extraneous genes. Theoretically, the gene transfer resulting in an improved variety can be accomplished in the equivalent of one generation. Conversely, conventional plant breeding might take repeated backcrosses of anywhere from five to ten generations (and years) to rid the strain of the undesirable genes. The promise of recombinant DNA technology is to speed up this process and direct the transfer exactly. The reality is that an enormous amount of work remains to be done, beginning with an all out assault on overcoming the basic knowledge voids in plant biochemistry. Advances in transformation techniques will not proceed rapidly until we know which genes we want to transfer in plants. Assuming that the rapid progress in plant transformation continues and more effective vectors and promoters are developed and knowledge of gene regulation grows, we have the problem that very few useful genes have been identified and cloned. This difficulty is compounded by the fact that few useful traits in plants are the consequence of a single gene, such as the case for carbohydrate metabolism.

More immediately, plant cell and plant tissue culture techniques are already providing a significant amount of genetic diversity for a number of crops in laboratories throughout the world. Further, plant cell culture techniques are well along in the research phase and some are moving into the development phase in the production of phytoproducts. Phytoproducts are plant ingredients produced by tissue culture techniques.

BIOTECHNOLOGY: ITS POTENTIAL IMPACT ON TRADITIONAL FOOD PROCESSING

Formulation is a key component of food product development and food technology. The myriad food products found in our supermarkets attest to the advanced state of product formulation in the food industry. Simplistically, formulated-processed food products are the result of proper mixing followed by some type of processing. Process engineering is the second component of food technology.The food scientist starts with one or two basic commodity type materials, such as a cereal grain (corn, wheat, rice) which is then modified either by a processing or bioconversion procedure resulting in a cereal mass. Flavors, sweeteners, thickeners, colorants or antioxidants might then be added. The next step is some form of process engineering. For example, the mixture might be extrusion cooked, baked, puffed or toasted, and dried, and a new ready-to-eat cereal product is rolled out. The final step is packaging which will keep the product in as high a quality as possible.

Application of biotechnology might well be the means to add another component to food product development. One key aspect that has been missing has been the ability to custom produce basic food raw materials with predetermined functional properties to achieve better formulation, processing, and stability after packaging. Although laudable progress has been realized in enhancing yield and productivity of raw agricultural ingredients, the development of agricultural commodities with properties tailored to meet end-product requirements lags significantly behind the advances in formulation, process development, and packaging. For decades, food scientists, engineers, and packaging scientists have been trained in the science and art of manipulating food ingredients, in the disciplines of processing of food ingredients and in package engineering. However, little or no attention has been paid to the virtually untapped potential of the manipulation of the basic agricultural raw material to achieve desired goals.

Typically first stage agricultural raw materials, whether they be orange juice concentrate, corn meal, tomato paste or wheat flour are viewed as relatively uniform nonfunctional commodities. If they lack texture, they are texturized; if they lack stability, stabilizers are added; if they become rancid, an antioxidant is added and the package is modified. Technologists have not been trained to ask, nor conditioned to look demandingly at the raw material. Why is there no wheat variety that results in a flour that does not stale? Why is there no variety of peas that when canned will give quality equivalent to that of frozen peas or when blanched and frozen, will give quality equivalent to that of fresh peas? Rather, much effort goes into processing and packaging these raw materials in a manner that enhances or at least retains what functionality they do have, and processing and packaging that minimally degrades the organoleptic characteristics. There has been no concerted effort to develop varieties of agricultural materials so designed that, upon processing organoleptically and functionally superior products are produced. Thus, biotechnology offers an approach to customize raw materials and thus provides the food scientist and food product developer with a new powerful tool.

Functional Attribute Crops

On a conceptual basis, agricultural food raw materials are made up of the four macro components: water, protein, carbohydrate, and fat, plus other minor chemical constituents. These chemicals are converted into food products. If one could relate the physicochemical characteristics of one or more of these components to functionality and if one could then relate these physicochemical characteristics to their biosynthesis, then that particular plant raw material could begin to take on the character of a functional attribute crop. A functional attribute crop is a term coined for a raw commodity with a specific attribute or function, which has added value to the food processor or the consumer and end-user. Such attributes would not be present in existing crops.

Corn: A corn hybrid yielding a starch with a high amylose content and low gelatinization temperature, or a sweet corn hybrid devoid of lipoxygenase and not requiring blanching prior to freezing would be examples of functional attribute crops. It is important to first discuss how one develops a new corn hybrid with special characteristics. Classical plant breeding is a long, tedious and demanding exercise in which the plant breeder seeks to introgress specific genetic characteristics. Those plants which retain the desirable characteristics are selected as they are gradually being freed of undesirable traits by repeated backcrossing techniques. It is like a mammoth juggling act since the target characteristic, like a high-amylose, low gelatinization temperature starch in a corn hybrid, must be accompanied by all of the requisite agronomic characteristics of yield, disease and insect resistance, and uniformity.

There are several steps in the typical procedure used for developing new inbred lines, which, when finally crossed, will result in a superior proprietary hybrid. The corn plant has a male organ, the tassel, which is the source of pollen, and in the ear below is the silk, which is the female organ. Just before the emergence of the silk from the ear, the ear is bagged to protect it from unwanted and random pollination. Sometime later when the tassel is ready to drop pollen, it is also bagged. The pollen is shaken into the bag and can then be dusted on the silks of another plant to complete the cross. Care is taken so that pollen is transferred only to the silks of specific plants. The pollinated ear is then covered with a pollinating bag. The pollinating bags are secured around the ear and the kernels are permitted to develop as the ear matures. Pollination with the pollen from the same plant is called selfing, while that from a sister plant is sibbing. These are the steps taken in developing an inbred line. When two inbred lines are crossed the result is an F_1 hybrid, hopefully with gratifying heterotic vigor and the wonderful sought-after functional attributes.

Teosinte, found in South America where there are many native varieties of corn, is thought to be the primitive original ancestor of corn. It is a perennial crop, which might be a desirable trait for the present annual corns grown in the U.S. There are many other South American varieties available which could serve as rich source material for introducing new characteristics into commercial U.S. corn lines. The difficulty is that without an assist from biotechnology, classical breeding techniques would be overwhelmed if one attempted to access specific genetic characteristics buried among the millions of genes represented in the large corn population. This probably explains the limited genetic background in commercial corn in the U.S. The result is a relative uniform commodity corn which enters into the food processing stream as an inert carrier upon which the food scientists or food engineers superimpose formulation and process manipulations. Sweet corn in the U.S. is bred and raised for its organoleptic characteristics at the expense of field yield. Taste, appearance and shelf life are the primary requirements whether it be for the fresh market or for frozen cob or cut kernel products. Ohio 43 is an important inbred parent of hybrid field corn in the U.S., with many isogenic lines representing identical genetic backgrounds but differing from each other by single gene mutations.

There are probably two dozen mutants in maize that are known to affect starch synthesis. There are many as yet uncharacterized mutants which also play key roles. Knowledge of the biochemical pathways which lead to a specific carbohydrate end-product reveals the enzyme systems involved, which in turn can lead to the gene sequences responsible for the expression of these enzymes. It is known that the endosperm mutants dull (*du*), sugary-2 (*su-2*), waxy (*wx*) and amylose-extender (*ae*) all effect the relative proportions of amylopectin and amylose (1,2). The waxy endosperms produce no amylopectin; and the other mutations all increase the amylose content, with *ae* being more potent than either *du* or *su-2*. The sugary-1 (*su-1*) mutant effects the production of the water-soluble phytoglycogen which contributes to the creamy texture of sweet corn. There are two shrunken mutants, *sh-1* and *sh-2* and a brittle-2 (*bu-2*) mutant which limit reducing sugars and sucrose availablility for starch synthesis, and, as a consequence, those endosperms will have an increased level of those sugars thus enhancing the sweetness taste impact. Unfortunately, the fine structure biochemistry and the biochemical interactions of these processes are only slightly understood. Nevertheless, tracking these genes using molecular probes in a crossing program provides a useful but still rather primitive means to orchestrate the development of corn hybrids with predetermined carbohydrate composition and functionality.

It is possible to have double, triple or quadruple gene mutations in corn. These materials when coupled with molecular or genetic probe techniques offer a substantial assist in the acceleration and precision of a breeding program. Breeders when selecting progeny from crosses made with parents with known mutations such as these, must go through several generations in the belief that the genes being selected for are in the F_1, F_2 and F_3 progenies because being recessive traits they will not be phenotypically detectable in the first few generations. However, with the use of molecular probes, i.e., with radioactively labeled genes, such as sugary-1 or amylose extender, genes in the F_2 population can be detected at the very earliest plantlet stage, so that only those plantlets will be retained and grown out and all others discarded. This can save the time of growing two or three generations in the field. It should be noted that the gene mutants discussed here affected the carbohydrate synthesis cycles, which in turn, affect sweetness, creaminess and texture.

Shelf life is another trait that can be selected for. Lipoxygenase is the enzyme system which catalyzes the oxidation of unsaturated fatty acids (3). The reaction of oxygen with linolenic acid produces aldehydes such as cis-3-hexenal and cis-nonadienal which have typical rancidity notes (4,5,6). These carbonyls may, in turn, be reduced to the corresponding alcohols and also isomerized to the trans form by the actions of ADH and isomerase, respectively. Blanching in water at 95-100°C for about 10 minutes is required to inactivate lipoxygenase and consequently minimize the development of oxidative rancidity off-flavors during frozen storage. However, blanching has a marked deleterious effect on the texture of the reconstituted frozen cob corn, making it too pasty and soft. The solution is to develop a mutant which does not produce lipoxygenase. This might be accomplished by supplementing the classical breeding and selection approach with lipoxygenase probes. Soybean lipoxygenases have been cloned. If they were to cross hybridize with corn lipoxygenase, they then could be used as molecular probes. If not, corn lipoxygenase would have to be isolated and used as the probe.

Thus two new functional attributes of sweet corn are: 1) enhanced sweetness and creaminess by manipulation of some of the carbohydrate determining genes; and, 2) improvement of the texture and storage life of the frozen sweet corn by deleting lipoxygenase. These are improvements which have obvious consumer and economic benefits.

Wheat: There is greater technical difficulty in hybridizing wheat which is matched by the complexity of the economics of hybrid wheat (7). Wheat is a self pollinating crop. The male and female organs are in close proximity to each other and mechanical or manual emasculation is impractical. When one considers the quantity of wheat that is used in food processing, in baked goods, pasta, sweet goods and confections, it is amazing how undifferentiated a commodity like wheat flour has become and has always been. Hybrid wheat is important economically to protect proprietary improvements. As functional attribute wheat is developed, the importance of hybridizing wheat will come to the forefront. As of today, the quantity of hybrid wheat produced in the U.S. is insignificant. The approach to hybrid wheat in the past has been to induce male sterility in wheat followed by the use of genes to restore pollen fertility, a technique called the cytoplasmic genetic method. There is still much work to be done to develop appropriate parents that have the required combining ability and also give high yields. However, improvement has been made and by 1982, at Kansas State University field trials, eight out of ten of the better yielding wheats were from hybrids. From 1983-1985, more that two thirds of the leading yield-trial wheats have been hybrids. Another optimistic factor is that several multinational corporations including Shell, Rohm and Haas, and Orsan are developing chemical gameticides or pollen suppressants. Therefore, it does appear that wheat hybrids will become a commercial reality and the proprietary protection that goes along with a hybrid appears to be in the offing.

The first improvement in wheat quality was related to inhibition of bread staling. In the 1950's, Bechtel (8,9) defined bread staling as the deleterious changes in the crumb of baked goods which were not the result of or caused by spoilage organisms. Staling is more than a simple moisture redistribution from the crumb to the crust (10). Generally, starch is the component playing the key role in bread staling, but proteins, lipids and the pentosans have also been implicated (11). The consensus is that staling is the organoleptic consequence resulting from or coincident with amylopectin retrogradation. Staling may be minimized by the addition of temperature resistant amylopectin, which helps to maintain the perception of freshness by a continual breakdown of starch after baking (12,13). Surface active agents are also used. They very likely function by forming an inclusion complex with amylopectin which prevents migration from the starch granule. Bread staling is minimal below 0°C, moderate at room temperature and most rapid at refrigerator temperature, which parallels the rate of amylopectin retrogradation. Establishing an optimum range of amylopectin content, molecular weight and molecular weight distribution which results in minimal staling of wheat would give the biotechnologist objective molecular targets.

It appears reasonable, therefore, that a similar approach to that suggested for corn, using the carbohydrate controlling gene probes could be applied to wheat. The basic assumption is that staling can be controlled by controlling the starch composition of wheat. In 1974, Maleki and Seibel (14) reported significant differences in the staling rate of bread made from flour derived from different varieties of wheat. They concluded that the ultimate answer to staling may lie in trying to find varieties of wheat that produce bread of desired qualities instead of using antistaling agents in the formula. Somewhere buried in the fine structure of the starch components of those wheat varieties lies the answer to why one variety of wheat gives bread that stales at a rate different than another. It waits to be studied.

Vegetable Oil: Another example of biotechnology applied to the macro components of plants is modification of vegetable oil. Oil palm is one plant that has seen some development in this area. Palm oil production began in earnest in the 1950's (15,16). Oil palms have a productive life of 25-30 years. By the 1980's many producing oil palms had reached their most productive stage and were rapidly declining. There is a need for replanting new stock in virtually all oil palm plantations. Beginning about 1990, in Malaysia alone, 4 to 5 million new trees per year are required to replenish about one million hectares of oil palm plantations.

Cross sections of the oil palm fruit show that they are uniform within a tree but vary considerably, from tree to tree. If one were to plant the seeds from different trees, a large degree of diversity would result. Planting materials with optimal agronomic characteristics without diversity would have obvious advantages. Oil yields also vary considerably within a group of palms. One would hope to reproduce and use only the new planting materials representing palms which are optimal in yield. Similarly, palms having disease resistance, ideal physical stature and, of interest to the food scientist, specific fatty acid composition, would be very desirable. Rather than propagating oil palms from seed, with the consequent segregating and nonuniform progeny, biotechnology permits the use of the micropropagation technique, which is a subset of tissue culture to produce clones of more uniform planting materials.

A major tissue culture laboratory in Malaysia is currently engaged in a large scale production of oil palm planting materials by micropropagation techniques. The process begins with a tiny root cutting which has been induced to form a callus. A callus is a proliferation of undifferentiated cells. The callus is first induced to differentiate. When embryogenesis has occurred, embryos can be separated and each can result in a plant which can be propagated to produce a larger plant in stages. Once large enough, the plants with the designed characteristics can be put out into the plantation.

Tomato: A last example of a functional attribute crop which could be developed by applications of biotechnology is the tomato. Two aspects under consideration are high solid tomatoes with less moisture and enhanced flavor tomatoes. This section will focus on the latter development.

In the previous discussion of lipoxygenase, the fatty acid oxidation products had a negative influence on frozen cob corn quality. Conversely lipoxygenase can have a positive effect in other biotechnically manipulatable areas (17). For example, the freshness notes might be restored or enhanced in processed tomato or strawberry products by the judicious use of lipoxygenase activity (18,19,20). There are a number of reports on the positive flavor and aroma impacts of lipoxygenase catalyzed fatty acid oxidation products. Just as one could use biotechnology tools to inactivate lipoxygenase in corn, similarly one could envision developing overproducers in crops where compounds such as cis- or trans-3-hexenal, hexenal and other short chain aldehydes and alcohols are clearly desirable. Since the tomato is a dicotyledon and more amenable to transformation, a lipoxygenase gene from another source conceivably might be inserted into the tomato genome, and if it were expressed, interesting flavor results could be anticipated.

Conclusions

In summary, biotechnology and food processing are in very early stages of interaction. Biotechnology is difficult to understand, not necessarily each of the scientific segments in isolation, but rather in the composite, made up of virtually every discipline of biology as well as molecular chemistry, biochemistry and, of course, food science. Pulling together a cohesive understanding of these tenuously related fields is difficult, particularly as information is being developed at a startling rate. Fitting the science part of the equation to an understanding of the impacts of raw material modification and economics on product development potential, not to mention the regulatory issues, puts biotechnology applications in food processing at an infant stage. There has been much interest and discussion of "Why biotechnology and food processing?" and "When biotechnology and food processing?" "How does one make the union of biotechnology and food processing?" An effective dialogue among industrial and academic biotechnologists and industrial and academic food scientists is needed for the union to work and for the food industry to reach the next level of technology.

REFERENCES

1. Nelson, O.E., In: "Progress in Biotechnology 1," (Hill, R.D., and Munck, L., eds.) Elsevier Press, The Netherlands. p. 19-28 (1984)
2. Nelson, O.E., In: "Advances in Cereal Science and Technology III" (Y. Pomeranz, ed.) AACC, St. Paul, MN. p. 41-71 (1980)
3. Gardner, H.W. and Inglett G.E. Food products from corn germ; enzyme activity and oil stability. J. Food Sci. 36:645 (1971)
4. Kazeniac, S.J. and Hall, R.M. Flavor chemistry of tomato volatiles. J. Food Sci. 35:519 (1970)
5. Arens, D., Laskawy, G. and Grosch, W. Lipoxygenase aus erben. Bildung Fluchtinger aldehwde, aus linolsaure. Z. Lebensm. Unters. Forsch. 151:162 (1973)
6. Grosch, W. and Shwenke, D. Lipoxygenase aus soyabohnen: Fluchtige aldehyde und alcohole uas linosaure. Lebensm. Wiss. Technol. 2:109 (1969)
7. Johnston, R.A. Hybrid wheat economics. Crops and Soils Magazine, p. 15-18 (1985)
8. Bechtel, W.G. and Meisner, D.F. Present status of the theory of staling. Food Tech. 5:503 (1951)
9. Bechtel, W.G. A review of bread staling research. Trans. AACC 13:108 (1955)
10. Hertz, K.D. Staling of bread: A review. Food Tech. 19:1828 (1965)
11. Kim, S.K. and D'Appolonia, B.L. Bread staling studies II. Effect of protein content and storage temperatures on the role of starch. Cereal Chem. 54:216 (1977)
12. Bechtel, W.G. Staling studies of bread made with flour fractions V. Effect of heat stable amylase and a cross linked starch. Cereal Chem. 36:368 (1959)
13. Rubenthaler, G., Finney, K.F. and Pomeranz, Y. Effect on loaf volume and bread characteristics of alpha-amylases from cereal, fungal, and bacterial sources. Food Tech. 4:671 (1965)
14. Maleki, M. and Seibel, W. Staling of bread, its measurement and retardation. In: "Proc. IV. International Congress Food Sci. and Technol." Vol. II, p. 277-282 (1974)
15. Wood, B.J. and Beattie, T.E. Processing and marketing of palm oil. Planter, p. 379 (1981)
16. Wood, B.J. Technical developments in oil palm production in Malaysia. Planter, p. 361 (1981)
17. Drawert, F. and Berger, R.G. Uber die biogenese von aromastoffen bei Pflanzen and fruchten. XVII. A. Naturforsch Sect. C: Biosci 37(10):849 (1982)
18. Grenli, H. and Wild, J. Enzymatic flavor regeneration in processed foods. In: "Proceedings of the Fourth Congr. Food Sci. and Tech." Madrid Spain (1976)
19. Sha, F.M., Salunkhe, D.K. and Olsen, L.E. Effects of ripening processes on chemistry of tomato volatiles. Am. Soc. Hortic Sci. 94:171 (1969)
20. Drawert, F. and Berger, K. Possibilities of the biotechnological production of aroma substances by plant tissues cultures. In: "Flavour '81 Third Weurman Symposium". (Schreier, P. ed.) de Gruyter, Berlin. p. 509 (1981)

Chapter 2

Regulatory Issues in the Food Biotechnology Arena

Stephen McNamara

INTRODUCTION

New techniques of biotechnology present awesome possibilities for change of a great many aspects of our world. As stated recently by the Executive Office of the President of the United States:

> "Only forty years ago, DNA was discovered to be the repository of genetic information. This discovery has been followed by an explosion in our understanding and ability to manipulate the gene as manifest [sic] by the new commercial biotechnology which has introduced a new and profound dimension into the field of classical genetics. Today, new techniques for manipulating genetic information offer exciting advances, as remarkable as the discovery of antibiotics or the computer chip.
>
> While some techniques of biotechnology are not new — the use of yeast in baking and brewing began around 6000 B.C. — the most recently developed techniques are far more sophisticated. Modern biotechnology promises to benefit many fields of human endeavor . . . [It] will alleviate many problems of disease and pollution and increase the supply of food, energy, and raw materials" (1).

New technologies, however, also present new safety concerns, and the products of modern biotechnology inevitably will be subjected to regulation by the Federal Government for the purpose of assuring safety.

This chapter reviews the existing requirements of the Federal Food, Drug, and Cosmetic Act (FDC Act) (2) that apply to the introduction of a new food substance, and considers how these requirements may be applied by the United States Food and Drug Administration (FDA) to products of modern biotechnology.

REGULATION OF NEW FOOD SUBSTANCES UNDER THE FEDERAL FOOD, DRUG, AND COSMETIC ACT

New food substances are subject to regulation by the FDA pursuant to the food additive provisions of the FDC Act. In general, that Act establishes two pertinent classifications: 1) substances that are "generally recognized as safe" (GRAS) for their intended use in food, and 2) substances that are not generally recognized as safe (not GRAS) for their intended use in food, which are termed "food additives" (3). Specifically:

> "the term 'food additive' means any substance the intended use of which results or may reasonably be expected to result, directly or indirectly, in its becoming a component or otherwise affecting the characteristics of any food (including any substance intended for use in producing, manufacturing packing, processing, preparing, treating, packaging, transporting, or holding food; and including any source of radiation intended for any such use), if such substance is not generally recognized, among experts qualified by scientific training and experience to evaluate its safety, as having been adequately shown through scientific procedures (or, in the case of a substance used in food prior to January 1, 1958, through either scientific procedures or experience based on common use in food) to be safe under the conditions of its intended use; except that such term does not include: 1) a pesticide chemical in or on a raw agricultural commodity; or 2) a pesticide chemical to the extent that it is intended for use or is used in the production, storage, or transportation of any raw agricultural commodity; or 3) a color additive; or 4) any substance used in accordance with a sanction or approval granted prior to September 6, 1958 pursuant to this chapter, the Poultry Products Inspection Act (21 U.S.C. 451 and the following) or the Meat Inspection Act of March 4, 1907, as amended and extended (21 U.S.C. 71 and the following); or 5) a new animal drug" (3).

In addition FDA regulations provide agency interpretation of the GRAS definition where in 21 CFR 170 (a),(b) GRAS means:

> "General recognition of safety requires common knowledge about the substance throughout the scientific community knowledgeable about the safety of substances directly or indirectly added to food. . . ." "General recognition of safety through scientific procedures shall ordinarily be based upon published studies which may be corroborated by unpublished studies and other data and information."

Hereafter, these two classes of food substances will be referred to as "GRAS substances" and "food additives."

It should be noted that, as defined in the FDC Act, "food additive" is a term of art that does not always conform to common parlance. If a food substance is GRAS, then by definition it is not a food additive. Thus, substances that are used in food as preservatives or for other technological purposes, which the ordinary consumer would consider to be classic food additives, are not food additives under the law if they are deemed to be "generally recognized as safe" for use. For example, FDA regards monosodium glutamate as GRAS and not a food addiitive (4).

Regulation of GRAS Substances

There is no requirement that a GRAS substance be approved by FDA prior to use. If a substance is GRAS, a food manufacturer is free to use it without first obtaining FDA approval or even notifying FDA of its use.

FDA has published regulations that recognize the GRAS status of numerous food substances (5). Nevertheless, FDA has explicitly acknowledged that the food ingredients listed as GRAS in FDA regulations:

> "do not include all substances that are generally recognized as safe for their intended use in food. Because of the large number of substances the intended use of which results or may reasonably be expected to result, directly or indirectly, in their becoming a component or otherwise affecting the characteristics of food, it is impracticable to list all such substances that are GRAS" (6).

If a manufacturer chooses to use a food substance that is believed to be GRAS but that FDA has not officially listed as GRAS, the manufacturer assumes the risk that some day the agency may take note of this use and disagree with the users assessment of GRAS status. If FDA should conclude that such a substance is not GRAS, the agency can undertake regulatory action to stop such marketing (7). If not GRAS, the substance is a food additive unless an exception applies as noted earlier. As discussed below, a food additive may not be used until FDA issues an authorizing food additive regulation.

Because of the uncertainties and regulatory risks attendant upon use of a new food substance without explicit prior approval by FDA, many companies are reluctant to use in their products a new substance that the agency has not either recognized as GRAS or approved by a food additive regulation, even if they believe the substance is in fact GRAS and properly may be used without FDA approval.

Regulation of Food Additives

As discussed above, generally if a food substance is not GRAS, it is by definition a food additive. The FDC act prohibits commercial use of a food additive until the FDA first has published a food additive regulation that authorizes the intended use of the substance. The Act provides that a food additive shall be "deemed to be unsafe" if it is used without an approving food additive regulation, (8) and that a food shall be "deemed to be adulterated" if it "is, or it bears or contains" an unapproved food additive (9). An exemption provides for investigational use of an unapproved food additive (10).

Obtaining the issuance of a food additive regulation can be a lengthy and difficult undertaking. To initiate the process, an interested person must file a food additive petition, which is required to provide extensive information concerning the identity of the substance and other relevant matters, including information about intended use and, especially, "Full reports of investigations made with respect to the safety of the food additive" (11). After the petition is filed, FDA publishes a notice of filing in the Federal Register, providing the name of the petitioner and a brief description of the proposal in general terms (12).

The FDC Act and FDA regulations provide that, within 90 days after the filing of the petition (or within 180 days if the time is extended by the agency, as authorized by the FDC Act), FDA either will publish in the Federal Register a food additive regulation prescribing the conditions under which the additive may be safely used, or will deny the petition (13). In practice, however, FDA action on a food additive petition routinely takes much longer than 180 days. It has been calculated that the average total time (prefiling research plus FDA review) required for approval of a direct food additive is five to seven years (14).

APPLYING THE FDC ACT TO PRODUCTS OF MODERN BIOTECHNOLOGY

OSTP Publication

The Reagan Administration has formed a Cabinet Council Working Group on Biotechnology, under the White House Cabinet Council on Natural Resources and the Environment, to study the broad range of potential uses of biotechnology and attendant regulatory issues. In the Federal Register of December 31, 1984, the Executive Office of the President, Office of Science and Technology Policy (OSTP), published the Working Group's report of its research and views, and invited public comment (15).

The Working Group's document includes an index of United States laws relating to biotechnology, sets forth the policies of the major regulatory agencies (including FDA) that are involved in reviewing research and products of biotechnology, proposes a scientific advisory mechanism for assessment of biotechnology issues, and explains how the activities of federal agencies will be coordinated with respect to biotechnology-related matters.

The introduction to the Working Group document has a notably optimistic tone. It expresses great enthusiasm for the many benefits that may be provided to society by biotechnology, including benefits to the economy, as well as a commitment to foster safe development. The document specifically states:

> "Modern biotechnology promises to benefit many fields of human endeavor by offering new services and a wide variety of products superior to those currently available because they will be more effective, convenient, safer, or more economical. Biotechnology already has successfully produced new drugs and improved existing drugs.... Exciting research is underway in agricultural applications....
>
> ...The tremendous potential of biotechnology to contribute to the nation's economy in the near term, and to fulfill society's needs and alleviate its problems in the longer term, makes it imperative that progress in biotechnology be encouraged.
>
> ...legitimate concerns about safety have also been raised.... Accordingly, it is incumbent upon the government, the business community, and the public to take responsible and timely measures to insure that the public health and the environment are protected and that societal concerns are promptly addressed....
>
> ...The fundamental purpose of the Working Group is to ensure that the regulatory process adequately considers health and environmental safety consequences of the products and processes of the new biotechnology as they move, from the research laboratory to the marketplace. The Working Group recognizes the need for a coordinated and sensible regulatory review process that will minimize the uncertainties and inefficiencies that can stifle innovation and impair the competitiveness of U.S. industry..." (16).

FDA Statement of Policy

The OSTP document includes a formal "Statement of Policy for Regulating Biotechnology Products" by the FDA (17). FDA's statement is, of course, consistent with introductory statements by the White House OSTP. FDA's statement also provides general information about the agency's policies with respect to its enforcement of the FDC Act concerning the products of modern biotechnology, including food substances.

In this document, FDA states that it "proposes no new procedures or requirements for regulated industry or individuals" and that:

> "...For food products, [the FDC Act] requires FDA preclearance of food additives including those prepared using biotechnology.... The implementing regulations for food...additive petitions and for affirming generally recognized as safe (GRAS) food substances are sufficiently comprehensive to apply to those involving new biotechnology.

...the use of a new microorganism found in a food such as yogurt could be considered a food additive. Such situations will be evaluated case-by-case..." (18).

The foregoing quotations provide just about all of the explicit guidance in the document concerning FDA intentions specifically with respect to regulation of substances that are produced by biotechnology for use as human food. The essential points appear to be that FDA does not intend at this time to establish new procedures or requirements to govern the new products of biotechnology that are intended for food use and that, instead, FDA believes its existing requirements and procedures with respect to food additives and GRAS food substances are sufficient. FDA states that it will apply the existing requirements and procedures to the products of modern biotechnology on a "case-by-case" basis. The one concrete example is the statement that a "new" microorganism in a food product "could be considered a food additive."

QUESTIONS AND COMMENTS

While FDA's statement of policy provides useful general guidance, it nevertheless leaves many important questions for future resolution "case-by-case" without presenting much detailed insight at this time to assist concerned manufacturers. Several pertinent questions and comments are reviewed below.

May a New Product of Modern Biotechnology be GRAS?

A fundamental question may immediately occur: Will all food substances that are produced by new techniques of modern biotechnology be considered food additives, and therefore require the issuance of approving food additive regulations prior to commercial use, or is it possible that a new product of modern biotechnology could immediately be regarded as GRAS and be used without the delays and expenses entailed in seeking a food additive regulation?

The FDA statement of policy, quoted previously, appears to accept the premise that some products of modern biotechnology may be GRAS. "The implementing regulations for food...additive petitions and for affirming generally recognized as safe (GRAS) food substances are sufficiently comprehensive to apply to those involving new biotechnology" (18). However, this statement also appears to suggest a need either to obtain a food additive regulation or to have FDA "affirm" a new substance as GRAS. As discussed previously, FDA has accepted the view that a food substance that is in fact GRAS need not have this status affirmed by regulation prior to use in food.

In a subsequent passage in the FDA statement of policy that applies to additives for animal feeds, FDA states that approval of a food additive petition will be required for all substances that are produced by recombinant DNA technology:

> "Substances that are used in animal feeds, other than drugs, and that are produced by recombinant DNA technology, are considered to be food additives and require approval of a food additive petition (FAP). Other products of new biotechnology may also be considered to be food additives, requiring an FAP. Animal. . .food additives produced by recombinant DNA technology must be the subject of approval even if the active substance is shown to be identical or similar to the active substance in approved products produced by conventional methods" (19).

If FDA believes that use of an animal food substance that is produced by recombinant DNA must be approved by a food additive regulation, presumably the agency would have the same view about a human food substance that is produced by recombinant DNA. Human food should require at least as much regulatory scrutiny as animal feed. Notwithstanding the quoted FDA statement, however, if a substance that is produced by recombinant DNA technology is in fact "identical" to a substance that is already recognized as GRAS, it would appear that a good argument could be made by its developer that the substance produced by means of biotechnology is also GRAS and not a food additive. FDA regulations recognizing GRAS status usually identify the subject substance and any limitations on its food use without limiting GRAS status to a particular means of manufacture.

Nevertheless, even though the text of the FDA policy statement is not entirely clear on this point, it would appear that the statement reflects an agency view that the products of modern biotechnology are likely to be required to have FDA approval, if not by a food additive regulation then by a regulation affirming GRAS status. Recent speeches by FDA Commissioner Frank E. Young, M.D., also point toward FDA's requiring that a food substance produced by modern biotechnology either be the subject of an approved food additive regulation or be affirmed as GRAS by FDA. For example, Dr. Young has stated:

> "I will give you a brief description of how we regulate. . . . As you know, foods have long been produced by biotechnology, particularly those using fermentation techniques. We believe that new foods will also be produced by genetic manipulation, such as cheeses in which new microorganisms can improve their flavor or texture.
>
> FDA routinely requires specific information on microorganisms used to produce food ingredients, because we consider them to be food additives. Since the law requires that all food additives be approved by FDA as safe for human consumption, manufacturers of additives resulting from gene manipulation

must collect supporting data to demonstrate the safety of those additives.

Manufacturers may also seek to have food ingredients placed on FDA's list of ingredients approved as being 'generally recognized as safe.' Whether produced through biotechnology or not, the ingredient must be recognized through extensive experience with it as an acceptable food substance. Products not intended to be food ingredients may also be regulated because they affect a foodstuff. For example, the introduction into a plant of a gene coding for a pesticide or growth factor may constitute adulteration of the foodstuff derived from the plant.

As I have said, however, no food products will be reviewed by FDA under special procedures for biotechnology. Rather, each food, regardless of its manufacturing process, will be assessed on a case-by-case basis" (20).

Commissioner Young has considerable personal interest and expertise in the subject of biotechnology. He has made at least six presentations on the subject of biotechnology at scientific or educational conferences in 1985 (20,21,22,23,24,25). FDA is fortunate to have at this time a Commissioner who is so well-informed on this critical matter.

If a New Product of Modern Biotechnology Is Identical to a Substance That Is Already Approved by a Food Additive Regulation, Is a New Food Additive Regulation Required?

As in the case of a product of modern biotechnology that is identical to a GRAS substance, so also with respect to a product of modern biotechnology that is identical to a substance that is already approved by a food additive regulation, one reasonably may ask why a new FDA regulation should be needed. FDA food additive regulations generally do not limit approved additives to a particular method of manufacture. If it happens that a substance that is already approved by a food additive regulation now can also be manufactured by a new method, i.e., techniques of modern biotechnology, one reasonably may question why the substance, as manufactured by the new method, is not, as a matter of law, already approved by the existing food additive regulation. This assumes, of course, that it is essentially the same substance that is already approved by regulation and that the existing regulation does not limit the method of manufacture.

In its statement about animal feeds, FDA asserts that use of a feed substance produced by recombinant DNA requires approval by a new food additive regulation "even if the active substance is shown to be identical or similar to the active substance in approved products produced by conventional methods" (19). Nevertheless, if the product of a new biotechnological method is in fact identical to the product of older "conventional" methods, and FDA previously has issued a food additive regulation that approves use of the product without restricting the method of its manufacture, FDA could have difficulty in convincing a court that an additional food additive regulation is required before the product of the new biotechnological method can be used.

Minor Differences in a Newly Produced Substance

Furthermore, even if a product of modern biotechnology is slightly different from a traditional substance that is already recognized as GRAS or approved by a food additive regulation, it still would appear to be possible for approval for the new substance to be encompassed by the existing GRAS recognition or food additive approval. Prior FDA practice does not indicate that every genetic adjustment in a food substance creates a significantly new substance that requires a new food additive regulation. American agriculture has been breeding new strains of crops for years without FDA's asserting food additive status for every new variety.

It is of interest here to note that FDA regulations provide as follows:

> "Eligibility for classification as generally recognized as safe (GRAS). . . .
>
> (f) The status of the following food ingredients will be reviewed and affirmed as GRAS or determined to be a food additive or subject to a prior sanction. . . .
>
> (1) Any substance of natural biological origin that has been widely consumed for its nutrient properties in the United States prior to January 1, 1958, without known detrimental effect, for which no health hazard is known, and which has been modified by processes first introduced into commercial use after January 1, 1958, which may reasonably be expected significantly to alter the composition of the substance.
>
> (2) Any substance of natural biological origin that has been widely consumed for its nutrient properties in the United States prior to January 1, 1958, without known detrimental effect, for which no health hazard is known, that has had significant alteration of composition by breeding or selection after January 1, 1958, where the change may be reasonably expected to alter the nutritive value or the concentration of toxic constituents" (26).

These regulations explicitly recognize that food substances may be genetically modified without necessarily losing their GRAS status. Furthermore, these regulations appear to put the burden on FDA to initiate review of the status of a GRAS substance that has been modified. The regulations do not provide that the developer of a modified substance must necessarily file a food additive petition or a petition for GRAS affirmation. Consistent with these regulations, it would appear possible that a company could produce, through new techniques of biotechnology, a modifying improvement in a food substance that is recognized as GRAS, and that the substance as modified might continue to enjoy GRAS status.

Problems of Enforcement and Imports

If FDA attempts to insist that each substance that is newly produced by modern biotechnology must be approved by a new food additive regulation or new GRAS affirmation regulation prior to marketing, even if the substance is identical or essentially similar to a familiar substance that is already approved for use by a food additive regulation or recognized as GRAS, the agency may experience significant enforcement difficulties. How is the agency to determine, efficiently, whether a substance that is present in a food product is acceptable, because it was produced by older, traditional technologies, or not acceptable, because it was produced by a new method of biotechnology that has not yet been approved by regulation? This could be especially difficult to determine with respect to ingredients in imported products, for example.

Environmental Assessment Considerations

FDA's statement of policy in the OSTP document includes a brief paragraph about "Obligations Under the National Environmental Policy Act":

> "All premarket approvals of FDA-regulated products are subject to the requirements of the National Environmental Policy Act (NEPA) as defined by the Council on Environmental Quality's regulations. . . and as further described by FDA's NEPA-implementing procedures. . . . For new products or major new uses for existing products, these procedures ordinarily require the preparation of an environmental assessment. An environmental impact statement is required if manufacture, use, or disposal of the product is anticipated to cause significant environmental impacts" (27).

In effect, environmental issues also are left to future "case-by-case" resolution. One can understand FDA's desire to wait for specific cases before deciding how to apply its responsibilities under NEPA to the products of modern biotechnology. It should at least be noted, however, that there are concerns within the food industry that theoretical environmental questions could become a significant impediment to the introduction of new products of biotechnology.

Goodwill: Official Expressions in Support of Biotechnology

One of the more reassuring aspects of the OSTP/FDA statements, from the perspective of industry, is the generally favorable tone of these texts concerning the benefits that may accrue from modern biotechnology and the need for any government supervision to operate efficiently and not to stifle innovation. These sentiments are appreciated. Nevertheless, members of industry remain concerned that, as applied in the future, requirements for food additive approval, GRAS affirmation, or environmental assessment could become significant obstacles to efficient product development and marketing. When standards and procedures are identified only in general terms, the possibility remains that the regulatory requirements as applied will not prove to be as reasonable or efficient as has been promised. As the OSTP document observes, the United States is now in the forefront of biotechnology development; no one wants to see this become a "biotechnology-lag" through well-intended but unnecessary regulatory delays.

Further FDA Action

The Regulatory Program of the United States Government, April 1, 1985 - March 31, 1986, published by the Executive Office of the President, Office of Management and Budget, reports that a "final policy statement by all agencies," responding to the comments requested in the December 1984 OSTP Federal Register publication, should be published in January 1986 (28). Perhaps the final policy statement will provide some further details in response to comments. [Editors note: On October 14, 1985, the Biotechnology Science Board which was to have been established in the HHS to coordinate biotechnology regulation was scrapped with the functions given to a new committee to be established under the Office of Science and Technology Policy's Federal Coordinating Council for Science, Engineering and Technology - FCCSET. The new group will not have oversight authority, they will attempt however to resolve disagreements over biotechnology regulation] (29).

CONGRESSIONAL VIEWS

Congressman John D. Dingell (Democrat, of Michigan), Chairman, Committee on Energy and Commerce, United States House of Representatives, has held hearings on the subject of biotechnology and has expressed the view that additional legislation may be needed. For example, he has stated:

> "Everyone wants the system to be simple and efficient, and to permit growth of a new industry and the marketing of new products.
>
> However, I am concerned about early errors that may cloud the future.
>
> 1. Making the regulatory climate the central issue overemphasizes the hazards. The public is immediately reminded of toxic wastes, radiation, and carcinogens. And not reassured!!

2. There is a perverse determination to insist that the existing law will work. The Administration asserts that no new legislation is needed even though most of our public health and environmental laws predated biotechnology. Obviously there is a desire to avoid uncertainty but, in the absence of new law, industry faces substantial uncertainty. The application of existing laws is far from clear and will be tested in the courts.

3. Finally, the assertion that genetic manipulation is entirely safe seems premature. In the past it has taken many years to learn the health effects of radiation, of industrial chemicals, of pesticides, and of toxic wastes. There were many premature assertions of safety. Most often there was no evidence of a hazard only because no one had looked" (30).

It is not clear, however, whether Congressman Dingell's views will lead to legislative changes.

CONCLUSIONS

To regulate food substances that are produced by modern biotechnology, FDA intends to apply existing requirements and procedures that are applicable to food additives and GRAS substances. The agency does not intend to establish new requirements or procedures specifically for food substances that are developed with new techniques of biotechnology.

Agency statements indicate that FDA may treat food substances that are produced by biotechnology as food additives, requiring agency approval of a food additive petition and issuance of a food additive regulation prior to use. However, legal precedents suggest that at least some of these substances may qualify as GRAS, or come within the scope of existing food additive regulations, and not require additional agency approval prior to marketing.

The agency states that it intends to apply the food additive and GRAS provisions of the FDC Act to the new products of biotechnology on a case-by-case basis. There has been some uncertainty and concern within the regulated industry about how the food additive and GRAS provisions will be applied in the future in various types of situations. There is a more general concern, also, that FDA regulation not become an unreasonable impediment to the development of useful new products. At least, it is reassuring to hear from senior levels of the Federal Government expressions of support on behalf of the need to foster the development of biotechnology.

REFERENCES

1. Executive Office of the President of the United States, Office of Science and Technology Policy, "Proposal for a Coordinated Framework for Regulation of Biotechnology," 49 Fed. Reg. 50856 (Dec. 31, 1984)

2. Title 21, United States Code (U.S.C.), sections 301-392
3. 21 U.S.C. 321(s) [Emphasis added.]
4. 21 C.F.R. 182.1(a)
5. 21 C.F.R. 182-186 [in general]
6. 21 C.F.R. 170.30(d) [Emphasis added.] To same effect, 21 C.F.R. 182.1(a)
7. 21 U.S.C. 331-334
8. 21 U.S.C. 348(a)(2)
9. 21 U.S.C. 342(a)(2)(C)
10. 21 U.S.C. 348(a)(1), 348(i)
11. 21 C.F.R. 171.1(c)
12. 21 C.F.R. 171.1(i)
13. 21 U.S.C. 348(c)(1), (2); 21 C.F.R. 171.100(a)
14. 49 Fed. Reg. 50859 (Dec. 31, 1984)
15. 49 Fed. Reg. 50856 (Dec. 31, 1984)
16. 49 Fed. Reg. 50856-57 (Dec. 31, 1984)
17. 49 Fed. Reg. 50878-80 (Dec. 31, 1984)
18. 49 Fed. Reg. 50878 (Dec. 31, 1984)
19. 49 Fed. Reg. 50879 (Dec. 31, 1984)
20. Frank E. Young, M.D., Ph.D., Official text of presentation to "Panel Discussion on New Regulations," Biotech 85 Europe, Geneva, Switzerland, May 22, 1985
21. Remarks at Center for Energy and Environmental Management Conference on Biotechnology, Arlington, Virginia, April 23, 1985
22. Remarks before Council of Scientific Society Presidents, Washington, D.C., May 14, 1985
23. Presentation at "Panel Discussion on Regulatory Standards," Biotech 85 Europe, Geneva, Switzerland, May 23, 1985
24. Remarks before Instituto Superiore di Sanita, Rome, Italy, May 24, 1985
25. Remarks at National Institutes of Health, Director's Advisory Committee (location not identified in official text of presentation), June 24, 1985
26. 21 C.F.R. 170.30(f)(1), (2) [Emphasis added.]
27. 49 Fed. Reg. 50880 (Dec. 31, 1984)
28. "Regulatory Program of the United States Government, April 1, 1985-March 31, 1986", Executive Office of the President of the United States, Office of Management and Budget, pages 556-7 (issued Aug. 8, 1985)
29. Food Chemical News. pg 2. Oct. 14, 1985
30. Statement of the Honorable John D. Dingell, "On Biotechnology," April 22, 1985

Chapter 3

Federal Regulation of Food Biotechnology

Richard Ronk

INTRODUCTION

Perhaps the major task occupying FDA's time at present is the implementation of our action plan. When Dr. Frank Young became Commissioner of Food and Drugs, Secretary of Health Education and Welfare, Margaret Heckler charged him to prepare the Agency for the 21st century. In developing the plan, the Agency extended considerable effort to involve not only its internal membership but its outside constituencies as well. Ideas, suggestions and criticisms were sought and the fruit of these labors is a genuinely first-rate plan. The Agency's major areas of emphasis as listed in the plan are as follows:

1) New Drug Review Process
2) Medical Device Program
3) Postmarketing Surveillance
4) Health Fraud
5) Internal Management
6) Risk Assessment
7) Risk Management
8) Food Safety
9) Nutrition
10) New Technologies

As Dr. Young has said many times, planning is only 10%; implementation is 90%. Dr. Young and FDA Deputy Commissioner, John Norris, have met with each organization responsible for a piece of the Action Plan, and a comprehensive implementation effort is underway. This chapter will focus on some the areas that are most relevant especially the topic of biotechnology and food processing.

NUTRITION ISSUES

The nutrition activities of FDA will be devoted to significantly broadening the program to advance the science of nutrition, to increase consumer awareness, and to provide continued assurance of the nutritional value of the nation's food supply. Here are some examples of ways that this initiative will be pursued: 1) implement a plan to study the effects of food technology; 2) conduct a Food Label and Packaging Surveillance Survey to rank trends in nutrition labeling generally, and sodium, fat, and cholesterol labeling specifically; and, 3) conduct a weight loss practices survey.

FOOD SAFETY

In food safety, work will continue towards the resolution of food safety issues by administrative or legislative means, as appropriate. This goal will involve close coordination with other federal agencies to determine the present interest in developing food safety legislation. FDA also plans to complete its review and evaluation of the Food Safety Working Group recommendations.

BIOTECHNOLOGY

FDA has plans to build its internal capabilities in new technologies most likely to lead to new scientific and medical developments, placing emphasis on biotechnology and microelectronics. The area of biotechnology is of special personal and professional interest to Dr. Young, as well as anyone who will be confronting this new technology, both in the laboratory and in the regulatory areas. The promises found in laboratories must be translated into practical applications in a hungry world. The contrast between the potential of new technology to increase and improve the world's food supply and the reality of malnutrition and starvation is our challenge. Even though biotechnology has been used for centuries in food production, it holds tremendous new promise for both the future quality and quantity of our food supply.

THE REGULATORY ISSUE

Let us look ahead to the year 2000, a millennium just 15 years away. Where will we find ourselves? Will farmlands, fisheries, and food factories be displaced by biotechnological laboratories?

A basic general principle of food safety and nutrition is that "food should be safe, sound, wholesome and fit for human consumption." Biotechnology will certainly advance mankind towards this goal—how far we are able to go in a decade and one-half remains to be seen. Yet the popular publications now beginning to herald the "exciting involvements of biotechnology in providing, securing, and improving the world's food supply" too often ignore a caution which should be apparent. Alteration of one food attribute in a positive, beneficial manner could be risky if it alters another attribute negatively. For instance, increasing the amount of an essential nutrient in a particular food may be beneficial in theory, but not in practice if it predisposes the food to growth of either pathogenic or spoilage bacteria, or if it interferes with another nutrient.

We have developed codes of hygienic practice for foods that have long histories of use and economic importance. Now, the new biotechnology may force us to reconsider past practices. Instead of a traditional crop or meat, the new food may be the result of industrial synthesis, never exposed to such common sources of contaminants as soil, water or wind.

The year 2000 will see new foods and new ways of food processing. Already, processes for genetically engineered and chemically synthesized foods have been patented, with more developing daily. These advances will improve the production and availability of food, and provide a greater variety of foods from which to choose. However, each presents the FDA with new regulatory challenges and increased workloads. From a regulatory point of view, there is the problem of determining at what point a traditional food stops being traditional, or in other words, with apologies to Gertrude Stein, "When is a carrot not a carrot not a carrot".

Certain new innovative methods may need to be developed to assess any novel health and safety concerns which arise. The Center for Food Safety and Applied Nutrition has found detailed chemical analysis helpful for evaluating irradiated foods, and it is exploring development of analytical patterns and computer recognition techniques for traditional foods, which can be compared with those of the "new" ones. Thus, in the future they may be able to define the nutritional quality of new products, while identifying any potentially toxic properties. However when a working group of the International Council of Scientific Unions concedes that the scientific community is in general, (quote), "profoundly ignorant of risks associated with traditional foods," we clearly have far to go before addressing the anticipated ones.

In 1983, Dr. Jarvis of the United Kingdom published a Delphi forecast on biotechnology and the food industry. From his prediction of the specific years these events seem most likely to occur, it is clear that the food industry must already be applying new biotechnology to food production and is planning future applications as well. Microbes, specifically tailored for certain tasks, will be principal sources of new, undegradable thickening agents, vitamins, antibiotics, colors, flavors, fertilizers, pesticides and herbicides, peptides, and enzymes. Enhanced microbial enzyme production and new enzymes will be applied increasingly in food processes such as continuous fixed-bed fermentations and effluent treatments. These advances will result in new types of fermented foods and wider availability of reliable starter cultures with specific attributes.

THE PROMISE OF BIOTECHNOLOGY

Microbes will not be the only targets of the new biotechnologies. As knowledge concerning genetic manipulation increases, totally new species of animals, plants and perhaps fusions of the two may emerge, permanently altering the ecology of our agriculture and mariculture. Plants may be reared in a smaller, more productive form such that hydroponic gardening could be profitable in climates presently unsuitable for growing. Possible new shapes and textures of fruits and vegetables could lend themselves to more efficient processing and packaging, yet retain more of their natural flavor, consistency and nutritive value.

Biotechnology may alter the chemical composition of foods for specific health purposes. Foods with even lower cholesterol or lower natural caloric content than our present diet products may become available. As the nutritional sciences advance, as we gain a greater understanding of chronic disease processes and the influence of diet on such processes, it may be possible to produce foods capable of altering the disease process itself. For example, if the benefits of the omega-3 polyunsaturated fatty acid, eicosapentanoic acid, on clinical manifestations of rheumatoid arthritis are at some point proven scientifically, foods with increased levels of such fatty acids may be engineered for arthritis sufferers. Such accomplishments will need careful scrutiny, as beneficial changes to one organ system need not always be beneficial to the whole organism. Advances of this nature will also rekindle the debate within regulatory agencies as to when a food becomes a drug. It is imperative that the regulators be prepared for these technological advances so the right questions can be asked.

Along these lines, the basic structure of the proteins of foods may be altered specifically to decrease their allergenicity. For instance, it may be possible to reduce the allergenicity of soybean protein by modifying the amino acid composition of the three major allergenic proteins without substantially reducing their nutritive value. Custom-tailored enzymes may be capable of altering the allergenicity of cow's milk to the benefit of infants afflicted with intolerance to the responsible protein in the milk. If biotechnology makes a wider variety of safer foods available, will it also help us to solve the new problems that will undoubtedly arise?

GENE PROBES

One of the products of the new biotechnology that will surely facilitate microbiological analysis is the gene probe. These probes permit one to detect pathogenic microorganisms in foods because they bind specifically to the microorganism's DNA. The genus and species name of a microorganism is assuming less and less importance in judging pathogenic potential, while the ability to produce a toxin or to be invasive is becoming the real issue. The development of genetic probes to detect these abilities has already begun.

To date, the FDA's experience with such gene probes has been successful. The basic methodology has been tested collaboratively and accepted by the Association of Official Analytical Chemists, and the field personnel have been trained in these techniques. The FDA pathogen surveillance programs for several foods are already applying gene probes to differentiate non-hazardous *Escherichia coli* in foods from those capable of causing human disease. FDA's capabilities are expanding in this area to include: *Yersinia*, *Campylobacter*, *Listeria* and other pathogens of concern. It is noteworthy that FDA's Center for Devices and Radiological Health has recently approved for clinical use the first DNA probe, used for detecting *Legionella*. It is hoped that this technology will also enable the FDA to be more effective in those areas where current methods do not work well, namely, in finding human viruses in foods.

RISK ASSESSMENT

In all this, the question of risk must be considered. The very idea of altering the genetic makeup of things considered "natural", coupled with the complex jargon of the recombinant DNA technologist, engendered concern in many people's minds during the mid 1970's. The fears that super-germs would be created have been allayed by ten years of research experience with rDNA. Yet employing the technology for manipulating foods, is only now becoming a reality. The technology available now is generally capable of defining the exact changes that will occur. For example, duplicate additional genes for a desired enzyme can be placed in bacteria or yeasts with no excess genetic material being incorporated into the recipient. In the agency's limited experience thus far, genetic engineering companies have gone to extreme lengths to assure the FDA of the safety of their procedures and products.

It must be remembered that the overall probability of risk is never zero, much as one might prefer a risk-free environment. Potential risks must always be considered along with potential benefits. The risks originally conjectured for recombinant DNA technology have, by and large, failed to appear. Concern now has shifted justifiably more towards issues regarding production technologies. Although demonstrable harm due to applications of new biotechnologies has not yet surfaced, everyone must be vigilant to ensure that it will not.

FDA'S POSITION

FDA has developed a philosophy of regulating food products that have been produced by genetically modified microorganisms. Although advances in biotechnology demand that FDA build its science base, FDA does not believe that the use of the new biotechnological techniques requires any additional laws or regulations. Although the laws under which FDA operates have no statutory provisions or regulations that address biotechnology directly, Congress has provided FDA the authority to regulate products regardless of how they are manufactured. Under the laws that FDA enforces — the Federal Food, Drug, and Cosmetic Act and the Public Health Service Act — FDA's regulatory review of those products using biotechnology is based on the intended use of each product on a case-by-case basis. The agency possesses extensive experience with the regulatory regimens as applied to the products of biotechnological processes, new and old.

Within the framework of FDA's statutes and regulations, strategies have been developed for the evaluation of various kinds of "biotechnological" or "genetically engineered" products, as well as for other products. These strategies are product-specific rather than technology-specific. For example, review of the production of human viral vaccines routinely involves a number of considerations including the purity of the media and the serum used to grow the cell substrate, the nature of the cell substrate, and the characterization of the virus.

Because approval of a product application also involves the review and, in many cases, the approval of the sponsor's processing techniques, FDA must determine whether the quality assurance within the manufacturing process is adequate to detect deviations that might occur. For example, mutations during fermentation of a microorganism to produce a polypeptide product could, in theory, give rise to a subpopulation of molecules with an anomalous primary structure and altered activity. This is a potential problem inherent in the production of polypeptides in any fermentation process. FDA has a variety of mechanisms to deal with these situations such as the application of release specifications to products.

THE FUTURE

The principal focus of this discussion is in respect to where are we going, and what impact will this have on the future? We are now in one of the most exciting periods in the history of scientific and technological progress. For the first time, there is an opportunity to make a major impact on the health and future of all peoples in such a way as to provide for not only an increased life span, but for better quality of life. Safe food of good nutritional quality is not only the foundation of public health but also a basic human right. To assure that right, the problems have to be approached rationally, without promising what cannot be delivered and, at the same time, without excessive caution because of unfamiliarity with the potential hazards involved. In such an undertaking, Machiavelli in 1513 said, "It must be remembered that there is nothing more difficult to plan, more doubtful of success and more dangerous to manage than the creation of a new system. For the initiator has the enmity of all who would profit by the preservation of the old institutions and merely lukewarm defenders in those who would gain by the new ones." Nevertheless, Oliver Wendell Holmes said: "Life is action and passion. It is required of a man that he should share the action and passion of his time at the risk of being judged not to have lived."

Chapter 4

Enzymology and Food Processing

Saul Neidleman

PURPOSE

The food processing industry is steeped in enzymology. It has been for thousands of years, is at present, and will be in the future. The purpose of this review is to present selected details of today and to anticipate and speculate as to some promises of the future. Special attention will be paid to new directions being taken in biocatalysis since, it is certain, these will imprint their novelty on the traditional body of food processing enzymology.

BACKGROUND

Enzymes are protein catalysts produced by living systems. It has been estimated that 25,000 different enzymes exist (1). Of these, some 2,100 have been officially recognized by the International Union of Biochemistry (2). Simply stated, then, most of the native enzymes have not been discovered or definitively characterized.

The recognized enzymes have been categorized into six classes (2) as shown in Table 4-1. Almost all the enzymes known to be significant in the food processing industry are in the first three categories, with the hydrolases in the overwhelming majority. Examples of these enzymes and their applications will be given below.

In addition to these native enzymes, another group of protein catalysts must be mentioned: the xenozymes. These are foreign enzymes, not occurring in nature, created in the laboratory by techniques of chemical modification, random mutation, and site-specific genetic or protein engineering (3). Native enzymes are evolved and synthesized to solve process problems in nature, not industry. Food processing, as practiced by microorganisms, plants, and animals in the wild, differs dramatically from that in our commercial facilities. Consumer acceptance in the marketplace and home likewise diverges from nature's way. In many cases, therefore, a native biocatalyst must be altered to a xenozyme to optimize its commercial function and the properties of the final product.

Table 4-1. Enzyme Classes

Class	Function	# Recognized
hydrolases	hydrolysis of C-C, C-N, C-O, etc. bonds	490
oxido-reductases	oxidation-reduction involving oxygenation or hydrogen atom addition	537
isomerases	isomerizations, racemizations	98
transferases	transfer of acyl, sugar, phosphoryl, methyl, aldehydic, ketonic or sulfur-containing groups from one molecule to another	559
lyases	addition to or formation of C=C, C=O, C=N bonds	231
ligases	formation of C-C, C-O, C-S, C-N bonds with ATP or other nucleoside triphosphate cleavage	83

Furthermore, novel reaction environments for enzymatic activity are becoming more prevalent. The xenozymes and unusual reaction conditions will be discussed below to illustrate that the traditional utilization of native enzymes will, for certain applications, be supplanted by inventive and more productive process design.

ENZYMES IN FOOD PROCESSING: INTRODUCTION

There are a number of general principles that have to be realized in order to enjoy the full complexity of biocatalysis in food processing. These are concerned with the origin of the enzymes, the actions of the enzymes, and the judgment as to the merits of what the enzyme hath wrought.

In the context of this discussion, enzyme origin refers to whether it is: 1) endogenous in the foodstuff; or, 2) exogenous to the foodstuff through either a conscious addition of an enzyme or enzyme producing organism or through accidental contamination, particularly by microorganisms (4,5).

Regardless of origin, an enzyme may synthesize, degrade, or alter a component that is to be part of the composition of the foodstuff. Finally, it is obvious that this spectrum of events may be good news or bad, a desirable or an undesirable effect. To a greater or lesser degree, these principles will form the basis for the chapter.

It is beyond the scope of this chapter to discuss, except in a few cases, the use of whole microorganisms such as yeast, fungi, or bacteria in baking, beer-making, amino acid synthesis, cheese making, production of flavors and flavor enhancers (6). The focus of this chapter will be on enzymes isolated from the producing species. The use of enzymes in food analyses (6,7), waste disposal (2), and the cloning of enzymes with subsequent expression in suitable, commercially useful hosts (8,9) will not be covered.

EXOGENOUS ENZYMES IN FOOD PROCESSING

Enzymes are added during food processing to invoke a wide range of effects. Among these are the control of texture, appearance and nutritive value, as well as the generation of desirable flavors and aromas, or their precursors. Further processing such as baking or cooking causes synthesis of additional flavor and aroma elements through nonenzymatic effects such as browning reactions. The successful production of an acceptable final product is often as dependent on experience and art, as it is on science. Selected examples of such enzymes and their functions are illustrated in Table 4-2 (2,6,7,10-27).

A few points should be emphasized. The application of these enzymes to generate a desirable end product is often a balancing act in which the degree of enzymatic modification of the foodstuff must be carefully controlled. For example, the use of proteases in hydrolyzing proteins, such as that from soybeans, can cause production of bitter peptides if hydrolysis proceeds too far (28). One way of eliminating such bitter peptides is by treatment with α-chymotrypsin to yield a protein-like material, plastein. Research is continuing for most of these enzymes related to discovering improved enzymes in nature or xenozymes in the laboratory and inventing improved reaction conditions. Most of the enzymes of interest to the food processing industry are hydrolases that degrade polymers and other large molecules to their component parts. In many cases, these reactions are reversible and can yield novel products. This will be discussed below. Finally as enzyme-catalyzed reactions go, many of the processes superficially indicated in Table 4-2 and subsequent tables are bewildering in their complexity. Not only are there unanswered questions as to the identity of the enzyme's substrates and products, there is, in addition, the network of secondary enzymatic and nonenzymatic reactions that follow. Foodstuffs are hardly a purist's paradise, carbohydrate products react with peptide and amino acid products, lipid products seek other reactive species, and environmental conditions such as temperature, moisture content, salt concentration, and pH vary during process stages. Frustration and challenge are constant companions.

ENDOGENOUS ENZYMES IN FOOD PROCESSING

In dealing with exogenous enzymes, reasonable control of the myriad of effects is approachable if not attainable. Enzyme concentrations and reaction conditions can be optimized to favor the desired reaction—not always, however, to the complete exclusion of the undesirable.

Table 4-2. Applications of Selected Exogenous Enzymes in Food Processing

Enzyme	Class*	Source**	Substrate	Function
aminoacylase	H	B	D,L-amino acids	L-amino acid production
α-amylase	H	B,F,P	starch	liquefaction to dextrins, brewing, proper volume in baked goods, confectionery
β-amylase	H	P	starch	maltose production, brewing, proper volume of baked goods
anthocyanase	H	F	anthocyanine glycoside	decolorization of juice/wine
catalase	OR	F,M	hydrogen	milk sterilization, cheese making
cellobiase	H	F	cellobiose	ethanol production, juice clarification
cellulase	H	F	cellulose	ethanol production, juice clarification, extraction processes
cysteine desulfhydrase	H	B	β-chloro-L-alanine + sodium sulfide	L-cysteine synthesis
glucoamylase	H	F	dextrins	degradation to glucose
D-glucose isomerase	I	B	D-glucose	high fructose corn syrup
D-glucose oxidase (±catalase)	OR	F	D-glucose, oxygen	flavor and color preservation in eggs, juices
hemicellulase	H	B,F	hemicellulose	clarification of plant extracts
hesperidinase	H	F	hesperidin glycoside	juice clarification
hydantoinase	H	B	hydantoins of D-amino acids in D,L mixtures	D-amino acid production
invertase	H	Y	sucrose	production of invert sugar, chocolate manufacture

(continued)

Enzyme	Class*	Source**	Substrate	Function
lactase	H	F,Y	lactose	glucose production from cheese whey, improve milk digestibility
lipase	H	B,F,M	lipid	cheese ripening, chocolate manufacture, modify milk fat for sausage curing
lipoxidase	OR	P	carotene	bleaching agent in baking
melibiase	H	F	raffinose	improve sucrose production from sugar beets
naringinase	H	F	naringin glycoside	juice debittering
pectinase	H	F	pectin	wine/fruit juice clarification, viscosity reduction in fruit processing, coffee and tea processing
protease	H	F,M,P	protein	meat tenderizer
		F	protein	condensed fish solids
		B,F,M	casein	cheese making
		F	gluten	dough conditioner
		B	protein	sausage curing
		P	protein	beer haze removal
		M	protein	peptone manufacture
		B	protein	soy sauce manufacture
pullulanase	H	B	amylopectin	beer production, improve glucose and maltose production
L-tryptophanase	H	B	indole + serine, or indole + pyruvic acid + NH_4+	L-tryptophan production
β-tyrosinase	H	B	phenol + pyruvic acid + NH_4+	L-tyrosine production

* H = hydrolase; OR = oxidoreductase; I = isomerase
** B = bacteria; F = fungi; M = mammals; P = plants; Y = yeast

In the case of endogenous enzymes, they may also cause desirable or deleterious effects in texture, aroma, flavor, and appearance, just as their exogenous relatives. For example, natural food flavors such as terpenes, hydrocarbons, alcohols, aldehydes, ketones, esters, lactones, amines, and sulfur-containing compounds are enzymatically produced in fruits and vegetables (29). The major difference is that these enzymes are already in the foodstuff and so control may be more difficult. Some examples of desirable reactions, other than those of the "more common" enzymes (the proteases, amylases, and lipases) are listed in Table 4-3.

Table 4-3. Desirable Activities of Endogenous Enzymes in Foods

Enzyme	Source	Effect	Ref.
alliinase	garlic, onion	characteristic odor	30,31
collagenase	beef	tenderizer	32
myrosinase	mustard, cabbage, cress	pungent taste	30
lipoxygenase + aldehyde lyase + alcohol dehydrogenase + aldehyde oxidase + esterases	fruits, vegetables	volatile flavor compounds: acids, alcohols, aldehydes, ketones, esters	30
polyphenol oxidase	cocoa, coffee, tea	desirable color and aroma	7

From the negative side, Table 4-4 shows examples of undesirable endogenous enzyme reactions. It should to be emphasized that some enzymes can be both useful or deleterious, depending on the situation. Proteases, lipases, polyphenol oxidase, and lipoxygenase are examples of

Table 4-4. Undesirable Activities of Endogenous Enzymes in Foods

Enzyme	Source	Effect	Ref.
alliinase	onion	bitter flavor	24
carbon-sulfur lyase	yeast	excess hydrogen sulfide in beer and wine	24
lipase	milk	rancidity	7
lipoxygenase	legumes, cereals	off-color, off-flavor	7
polyphenol oxidase	fruits, vegetables	off-color, off-flavor	7
proteases, lipases	fish tissue	autolysis	5
trimethyamine demethylase	fish tissue	toughens tissue	7
trimethyl-N-oxide reductase	fish tissue	overly fishy	24
xanthine oxidase	milk	oxidative rancidity	7

such. Enzymes make no judgments as to good or evil; they just catalyze. The food technologist makes judgments as to the applicability.

There have been many proposals to use exogenous enzymes to counter the deleterious effects of endogenous enzymes, including those listed in Table 4-5. Following the principle that there is always room for improvement, Table 4-6 lists some suggestions from the literature for desirable enzyme enhancement using gene technology.

Table 4-5. Use of Exogenous Enzymes to Eliminate Undesirable Enzymatic Products in Food

Enzyme	Function	Treated Food	Ref.
aldehyde oxidase	remove off-flavors	soybean	24
caffeinase	decaffeination	coffee	33
diacetyl reductase	off-flavor reduction	beer	30
limoninase	bitterness elimination	citrus juice	34,35
protease	remove bitterness	high protein	24,30
ribonuclease	reduce odor	fish	24
sulfhydryl oxidase	off-flavor reduction	UHT milk	36
urease	remove bitterness	shark meat	24

Table 4-6. Suggestions for Improved Enzymatic Activity through Genetic Technology

Enzyme	Application	Useful Improvement	Ref.
α-amylase	starch liquefaction, saccharification	acid-tolerant and thermostable	37
amyloglucosidase	high fructose corn syrup	immobilized with higher productivity	38
esterases, lipases, proteases, etc.	flavor development	more specificity	39
glucose isomerase	high fructose corn syrup	increased thermo-stability, lower pH optimum	40
limoninase	fruit juice debittering	more complete limonin degradation	41
protease	beer chill proofing	more specific	42
pullulanase	high fructose corn syrup	thermostable	38

NONTRADITIONAL BIOCATALYSIS: POTENTIAL FOR THE FOOD PROCESSING INDUSTRY

As noted earlier, the majority of enzymes added to foodstuffs and all of the endogenous enzymes present in foodstuffs are native enzymes. They are the biocatalysts evolved in nature, for nature. These enzymes and the traditional conditions for their use do not necessarily represent an optimized situation. Certainly, the range of native enzymes has not been fully explored. As noted earlier, the majority of naturally occurring enzymes have not yet been characterized. Organisms and their enzymes from exotic ecosystems or rare organisms from mundane ecosystems, in an optimistic view, must contain enzyme variants with commercial usefulness. Environments of extreme temperature, pressure, pH, salt, and solvent concentration certainly harbor novel biocatalysts. As more of these enzymes are isolated and characterized, a more complete understanding of the relationship between structure and function will develop. This information will advance xenozyme synthesis into a more rational sphere than the largely random approach that presently dominates. As more is learned about structure and function, the more directed the synthesis of xenozymes with desirable properties will become.

Among the techniques presently available for xenozyme synthesis are: 1) chemical modification of native enzymes; and, 2) genetic or protein engineering or mutation. Not all the examples to be cited are for catalysts relevant to food processing, but they serve to illustrate the possibilities.

XENOZYMES FROM CHEMICAL MODIFICATION

Chemical modification of enzymes can be roughly divided into three major types: 1) reactions with ions or small molecules; 2) reactions with water soluble polymers; and, 3) reactions with water insoluble polymers and matrices (enzyme immobilization). Only examples where improved or new catalytic activity results from the treatment will be mentioned.

An instance of enzyme modification involving ions is the effect of metal ion exchange on the activity of bovine carboxypeptidase A. The enzyme occurs naturally as a zinc protein and has both esterase and peptidase activity. When zinc is replaced by manganese or cadmium, the esterase activity is increased, and the peptidase activity is decreased. Replacement of zinc by cobalt improves both esterase and peptidase activity, the latter more than the former (43). Additionally, it has been shown that replacing the native zinc with cobalt in the aminoacylase of *Aspergillus oryzae* alters the substrate specificity of the enzyme (44), and that sequential replacement of native zinc in *Aeromonas* amino peptidase by copper or nickel and then zinc results in an enzyme containing two metals with an increased activity of up to 100-fold (45).

Some examples wherein chemical modification with a small molecule has a positive effect on enzyme activity are shown in Table 4-7. Examples such as these indicate that the activities associated with native enzymes can be increased and useful properties such as thermostability can be altered.

Since the total laboratory synthesis of protein catalysts from component parts is still in the future, the creation of new catalytically active sites on existent proteins has become a shorter term goal in enzyme engineering. Among the terms used to describe such work are "chemical mutations"

(51,52), "Bio-syn-cat" (biological synthetic catalyst) (53,54), and "semisynthetic bioinorganic enzyme" (55). These diverse approaches have the common objective of xenozyme synthesis.

Table 4-7. Effects of Chemical Modification on Enzyme Activity

Enzyme	Chemical Modification	Effect	Ref.
α-amylase (*Bacillus subtilis*)	acetylation with p-nitrophenyl acetate	increased thermostability above 70°C, reduced thermostability below 67°C	46
carboxypeptidase A (mammals)	acetylation or iodination of active site tyrosine	increased esterase and eliminated peptidase activity	47
rennet (*Mucor pusillus*)	acylation with anhydrides	up to 2-fold increase in milk coagulating activity	48
rennet (*Mucor* species)	methionine oxidation as with H_2O_2	decreased thermostability for easier inactivation during pasteurization in cheese making	49
thermolysin (*Bacillus thermoproteolyticus*)	acylation with amino acid N-hydroxysuccinimide esters	increase in activity up to 70-fold	50

A number of different approaches to enzyme semisynthesis have been reported. Flavins covalently attached to the protease, papain, yielded proteins with oxidoreductase activity (51,52). Reactions of α-chymotrypsin in the presence of pyridoxal resulted in the new ability to hydrolyze D-aromatic amino acid esters (56). Hemoglobin in the presence of ascorbic acid or dihydroxyfumaric acid carried out aromatic hydroxylations (57). In "Bio-syn-cat" (53,54), the natural conformation of a host protein is altered in the presence of a competitive inhibitor of the desired reaction. The new conformation is stabilized by chemical cross-linking and the inhibitor is then removed. In the presence of indoles, esterase activity has been induced in bovine serum albumin and bovine pancreatic ribonuclease; while with trypsin, chymotrypsin activity resulted. With cellobiose, amylase was induced to show β-glucosidase activity. Whether any of these xenozymes attain commercial interest is an open question; nevertheless, such techniques contribute to the overall understanding of biocatalysis.

Similar modification studies have also been carried out using water-soluble polymers. In native enzymes such as yeast internal invertase and fungal cellulases from *Humicola insolens*, it has been shown that partial removal of enzyme-associated carbohydrate resulted in significant decreases in pH stability and thermostability (58,59). The protease subtilisin had increased thermostability after modification with dextrins and dextrans.

It has been demonstrated that a number of enzymes including peroxidase, chymotrypsin, catalase, and lipase, when modified with the amphipathic polymer polyethylene glycol, become soluble and active in solvents such as benzene. One fascinating observation was that, with the modified lipoprotein lipase from *Pseudomonas fluorescens*, the temperature optimum for the synthesis of methyl laurate from methanol and lauric acid was at $-3°C$ (61). Further, it has been shown that the substrate specificity of β-galactosidase was altered by modification with polyethylene glycol (62).

Another aspect of the chemical modification of enzymes is their immobilization on water-insoluble matrices (21,63,64). Purified enzymes or enzymes present in living or dead cells may be so treated. Five general methods for immobilization are available: covalent attachment, adsorption, encapsulation, entrapment, and cross-linking. Each method may have desirable advantages over other procedures, depending upon whether an enzyme or a cell is being immobilized, and on the cost-benefit picture for the process under consideration.

The major advantages that may result from immobilization are: 1) the biocatalytic system may be readily separated from the products and reused; and, 2) immobilization often affords improved enzyme stability against temperature, pH, and other environmental denaturants. These effects allow for an increase in the useful lifetime of enzymes. While many present applications of immobilization may be classified as traditional, future developments in immobilization will allow for a more interactive role between enzyme and the immobilization matrix. The chemical and physical nature of the solid matrix will be modified to afford additional stabilization effects, for example, hydrogen peroxide degradation to protect sensitive oxidase, and microenvironments to favor enzyme reactions in solvents. Heterogeneous matrices, allowing a mosaic of hydrophilic and hydrophobic environments will allow for the simultaneous or sequential immobilization of multienzyme systems to produce a variety of chemicals (2). Sophisticated immobilization methods will serve as an adjunct in optimizing the use of the enzyme variants discussed throughout this paper.

XENOZYMES FROM GENETIC (PROTEIN) ENGINEERING

Eventually, commercial enzymes will be machine-made with specific properties engineered into molecules by programs based on a detailed understanding of structure-activity relationships. At present, we know too little, but our information base is growing. Still, despite these limitations, there have been positive successes in which a desirable change has been achieved in catalytic activity or function (64). Two methods that have shown promise are: random mutation of the enzyme gene *in situ* in the producing organism (RM) and site-specific mutation of the isolated genetic material (SSM). A third approach that may yield positive results is random mutation of the isolated enzyme gene.

Some positive alterations in enzyme properties are shown in Table 4-8.

Table 4-8. Positive Effects of Genetic Engineering on Enzyme Activities

Enzyme	Method	Modification	New Property	Ref.
subtilisin	SSM	methionine^{222}→alanine	greater bleach stability	65
	SSM	glycine^{166}→aspartic, glutamic acids	altered substrate specificity	66
T4 lysozyme	SSM	isoleucine^{3}→cysteine, then chemical cross-linking	increased thermo-stability	67
trypsin	SSM	glycine^{226}→alanine	altered substrate specificity	68
tyrosyl-tRNA synthetase	SSM	cysteine^{35}→serine	K_m for ATP lowered, increased enzyme activity	69
amidase	RM	serine→phenylalanine and others	change in substrate range	70
xanthine dehydro-genase or purine hydroxylase I	RM	alteration in relative positions of catalytic and orienting sites	change in substrate range	71

EFFECT OF REACTION VARIABLES

Nontraditional biocatalysis may also be mediated through parameters other than the nature of the enzyme such as: 1) use of organic solvents; 2) increased pressure; 3) temperature variations; and, 4) unexpected substrates.

In contrast to traditional dogma is the fact that a variety of enzymatic reactions are more effectively, and in some cases only, carried out in the presence of organic solvents. Actually, this should not be surprising since in terms of enzyme stability and kinetic arguments it is to be expected. Many enzyme reactions in nature are associated with hydrophobic, lipid-rich cellular locations such as membranes. Therefore, enzymes thrive in "organic solvents" as part of their reaction milieu. Other enzymes may be relatively indifferent to the denaturing effects of organic solvents, even though their apparent natural functions may not indicate such a need (72).

The kinetic truth is that it is possible to shift the direction of a normally hydrolytic or degradative reaction toward a synthetic function by reducing the water concentration or water activity in the reaction phase.

Another effect that might favor a reaction in the presence of an organic solvent is the increase in solubility and availability of a substrate only poorly soluble in water. Dependent on the nature of an enzyme's stability in organic solvents, either a biphasic system, utilizing water and a water-immiscible solvent, or a monophasic system, involving water and a water-miscible solvent, may be preferred. The appropriate system will often be that in which the balance of substrate solubility and enzyme stability is most favorable (73). Table 4-9 indicates the types of reactions that have shown novel or improved performances in organic solvents.

Table 4-9. Enzyme Reactions in Organic Solvents

Biocatalyst	Reaction	Ref.
20 β-hydroxysteroid dehydrogenase	cortisone reduction	73
Nocardia rhodocrous cells	various steroid dehydrogenations	74
Rhodotorula minuta cells	DL-menthol resolution	74
proteinases	peptide synthesis	75
lipase	ester synthesis	2,76
lipase	triglyceride synthesis	2,77
lipase	lipid inter- or transesterification	2,78,79,80
lipase	lipid hydrolysis	81
tannase	gallic acid esterification	82

Increasing the concentration of the organic solvent dimethyl-sulfoxide increases the esterase and decreases the amidase activity of bovine α-thrombin. These effects are presumed to be mediated via a change in enzyme conformation (83).

Tradition would suggest that enzymes in the presence of organic solvents are less stable than in aqueous solutions. In fact, it has been shown that porcine pancreatic lipase and the lipoprotein lipase from *Candida cylindracea* were remarkably more thermostable and active at 100°C in heptanol/tributyrin, as compared to in water. Furthermore, it was demonstrated that as little as 0.02% water sufficed to maintain enzyme activity (84,85).

These nontraditional types of reactions need further study. As mentioned previously, Takahashi et al. (61) found that the same lipoprotein lipase from *Pseudomonas fluorescens* when solubilized in benzene with polyethylene glycol was found to be active. Whether enzymes need to be solubilized or may be used in an insoluble form for commercial reactions in solvents remains to be seen and may have to be assessed on a case-by-case basis. In some measure, this situation resembles a water-based reaction for which a decision must be made between using an enzyme in a soluble or insoluble (immobilized) form. The enzyme may be active in either condition and factors of economics decide the issue, along with convenience, safety, available capital equipment, etc.

Enzyme reactions in solvents have great promise for future commercial applications, but the safety and toxicity aspects of such reactions and their products will require clarification. A reaction product for human use that resulted from a reaction in benzene, for example, would have to be free of toxic effects due to the solvent. Pressure is another reaction variable that may positively affect the course of an enzyme reaction. It has been demonstrated that the half-life of immobilized amyloglucosidase from *Aspergillus niger* at 60°C was increased from 5 days at atmospheric pressure to 22 days at 3,000 psig, a greater than 400% increase (86).

The temperature at which a microbial reaction is performed may dramatically affect the chemistry of the product. For example, the level of unsaturation in the wax or long chain esters produced from ethanol or N-eicosane (C-20 alkane) by the bacterium, *Acinetobacter* sp. H01-N was significantly increased by decreasing temperature (73,87) as seen in Table 4-10.

Table 4-10. Influence of Temperature on Unsaturation of Esters from Ethanol or n-Eicosane by *Acinetobacter* sp. (73,87)

	Wax Esters from Ethanol		
Temp.(°C)	Di-ene products (32:2, 34:2, 36:2) Total %	Mono-ene products (32:1, 34:1, 36:1) Total %	Saturated Products (32:0, 34:0, 36:0) Total %
17	72	18	10
24	29	40	31
30	9	25	66
	Wax Esters from n-Eicosane		
Temp.(°C)	**Di-ene products (32:2, 34:2, 36:2) Total %**	**Mono-ene products (32:1, 34:1, 36:1) Total %**	**Saturated Products (32:0, 34:0, 36:0) Total %**
17	71	24	5
24	41	37	22
30	35	40	25

This phenomenon of increasing unsaturation in fatty acid composition with decreasing temperature is a general effect noted throughout living systems. The major purpose for this biochemical response is to maintain membrane fluidity and function (87). Another example is the fatty acid distribution in soybean seeds as a function of growth temperatures as shown in Table 4-11 (88).

Table 4-11. Effect of Day/Night Temperature Cycles on Fatty Acid Unsaturation in Soy Beans (88)

Fatty Acid	Day/Night Temperature °C 18/13	24/19	27/22	30/25	38/28
	Fatty Acid (%) of Total				
Palmitic (16:0)	12	12	12	12	12
Stearic (18:0)	4	4	4	4	4
Oleic (18:1)	13	20	22	29	38
Linoleic (18:2)	56	54	54	48	40
Linolenic (18:3)	17	10	8	6	5

A number of examples of unexpected substrates for a variety of enzyme systems have been reported recently. The wax ester synthesizing system of *Acinetobacter* sp. H01-N as discussed above was able to produce a broad spectrum of end products that depended upon the chemistry of the substrate. It was found that the bacterium can simulate other biological sources of wax esters, and that the range of substrates affording similar wax esters was surprising (73,87). Some of these results are shown in Table 4-12.

Table 4-12. Comparison of the Chain Length of the Major Wax Esters Produced by *Acinetobacter* sp H01-N on Different Substrates to Those From Other Sources (73,87)

Source	Number of C Atoms in Major Wax Ester
orange roughy	38
sperm whale	32
jojoba bean	42
Acinetobacter sp. H01-N	
n-hexadecane	32
n-octadecane	36
n-eicosane	38
n-docosane	42
acetic acid	34
ethanol	34
propionic acid	34
propanol	34

Table 4-13 lists some examples of enzymes and their traditional and nontraditional substrates. The first three reactions in Table 4-13 may be of value in the selective separation of enantiomers, cis-trans isomers, and positional isomers of specialty chemicals. The significant lesson is that many enzymes may have broader catalytic activity than traditionally credited to them.

Table 4-13. Traditional and Nontraditional Substrates of Certain Enzymes

Enzyme	Traditional Substrate	Nontraditional Substrate	Ref.
cholinesterase	O-acylcholine	O-acyl-D-carnitine	89
D-galactose oxidase	D-galactose	aliphatic and aromatic alcohols	84,90
xanthine oxidase	xanthine, hypoxanthine	aliphatic and aromatic aldehydes	84,90
D-glucose-1-oxidase	D-glucose	D-glucosone	91
chloroperoxidase	alkenes	alkynes, cyclopropanes	92
chloroperoxidase	ethyl alcohol	allyl alcohol	93

CONCLUSIONS

Enzymes are in food processing to stay. They have proven their worth, but they have much more to prove. They will. There are growth limiting factors including: 1) the cost of the basic research to develop new enzymes and reaction conditions; 2) the resistance to change and "taking chances": you don't alter a good thing; 3) the capital expenses involved in developing and using new enzyme processes by the enzyme producers and users; 4) the time and expense involved in satisfying governmental clearance for enzyme additives, especially xenozymes; and, 5) the absence of scientific definition of what enzymes are doing in food processing—ignorance of the facts makes improving them difficult (15).

On the other hand, the sweet smell, taste, and appearance of profit will crumble these deterrents. It is clear that new enzyme processes will be better in terms of product qualities and commercial success. This is irresistible. A recent poll carried out by Bioprocessing Technology supports this optimistic view, although with the conservatism typical of "predict the future" questionnaires (94). The results showed that 53% of the respondents believed that one or more new commercially viable enzyme-based processes (not pharmaceutical) would exist by the year 2000; 29% believed several such processes would exist. Secondly, 72% asserted that one or more new immobilized enzymes would be available for commercial use while 44% thought that several would be available. Almost 63% agreed that one or more microbially produced flavor/fragrance would be on the market; while only 21% said there would be several.

There is a strong likelihood that these opinions are grossly underestimating the influx of new biocatalytic processes into the food processing industry alone. The answer will be known by the year 2000.

REFERENCES

1. Kindel, S. Enzymes, the bio-industrial revolution. Technology 1:62-74 (1981)
2. Neidleman, S.L. Applications of biocatalysis to biotechnology. Biotechnol. Genetic. Eng. Rev. 1:1-38 (1984)
3. Neidleman, S.L. Nontraditional biocatalysis. The World Biotech. Report USA 2:189-199 (1985)
4. Ruttloff, H. Biotechnology and aroma production. Nahrung 26: 575-589 (1982)
5. Shahani, K.M. Enzymatic aspects of nutritional, flavor, and sanitary qualities of fish products. In: "Microbial Safety of Fishing Products" (Chichester, C.O., and Graham, H.D., eds). Academic Press, New York, pp. 137-150 (1973)
6. Scott, D. Enzymes, industrial. In: "Kirk-Othmer Encyclopedia of Chemical Technology". John Wiley and Sons, New York, 9:173-224 (1980)
7. Schwimmer, S. Biochemistry in the food industry: molecular pabulistics and thanatobolism. Trends Biochem. Sci. 8:306-310 (1983)
8. Moir, D., Duncan, M., Kohno, T., Mao, J., and Smith, R. Production of calf chymosin by the yeast *S. cerevisiae*. The World Biotech. Report USA 2:190-197 (1984)
9. White, T.J., Meade, J.H., Shoemaker, S.P., Koths, K.E., and Innis, M.A. Enzyme cloning for the food fermentation industry. Food Technol., p. 90, 92-5, 98 (Feb. 1984)
10. Anonymous. Why growth is slow for industrial enzymes. Chem. Week, pp. 30-34 (Nov. 30, 1983)
11. Aunstrup, K. Enzymes of industrial interest: traditional products. In: "Annual Reports on Fermentation Processes" (Perlman, D., and Tsao, G.T. eds.). Academic Press, New York, pp. 125-154 (1978)
12. Dubois, D. What is fermentation? It's essential to bread quality. Baker's Digest, pp. 11-14 (Jan. 10, 1984)
13. Godfrey, T., and Reichelt, J. "Industrial Enzymology". The Nature Press, New York (1983)
14. Home, S., Maunula, H. and Linko, M. Cellulases - a novel solution to some malting and brewing problems. Proc. 19th European Brewery Convention (London). IRL Press Limited, pp. 385-392 (1983)
15. Kilara, A. Enzyme-modified lipid food ingredients. Process Biochem. 20:35-45 (Apr. 1985)
16. Leuchtenberger, W., Karrenbauer, M., and Plocker, U. Scale-up of a enzyme membrane reactor process for the manufacture of L-enantiomeric compounds. Ann. N.Y. Acad. Sci. 434:78-86 (1984)
17. Leunser, S.J. Microbial enzymes for industrial sweetener production. Dev. Ind. Microbiol. 24:79-95 (1983)
18. McLellan, M.R., Kime, R.W., and Lind, L.R. Apple juice clarification with the use of honey and pectinase. J. Food Sci., 50:206-208 (1985)
19. Pilnik, W. Enzymes in the beverage industry. Util. Enzymes Technol. Aliment. Symp. Int. pp. 425-450 (1982)
20. Posorske, L.H. Industrial-scale application of enzymes to the fat and oil industry. J. Amer. Oil. Chem. Soc. 61:1758-1760 (1984)
21. Poulsen, P.B. Current application of immobilized enzymes for manufacturing purposes. Biotechnol. Eng. Rev. 1:121-140 (1984)
22. Rattray, J.B.M. Biotechnology and the fats and oils industry - an overview. J. Amer. Oil. Chem. Soc. 61:1701-1712 (1984)
23. Reimerdes, E.H. Process for producing aroma-containing food products. U.S. Patent #4,432,997 (1984)
24. Schwimmer, S. "Source Book of Food Enzymology". AVI, Westport Conn. (1981)

25. Ter Haseborg, E. Enzymes in flour and baking applications, especially waffle batters. Proc. Biochem. 16:16-19 (Aug./Sept. 1981)
26. Yamada, H. Enzymatic processes for the synthesis of optically active amino acids. Enzyme Eng. 6:97-106 (1982)
27. Yamamoto, Y. Industrial applications of enzymes. Hakkokogaku Kaishi 56:656-661 (1978)
28. Ward, O.P. Proteinases. In: "Microbial Enzymes and Biotechnology" (Fogarty, W.M., ed.). Applied Science, New York, pp. 251-317 (1983)
29. Mabrouk, A.M. Flavor of browning reaction products. Food Taste Chem. 115:205-245 (1979)
30. Chase, T., Jr. Flavor enzymes. Adv. Chem. Ser. 136:241-266 (1974)
31. Block, E. The chemistry of garlic and onions. Sci. American. 252: 114-119 (1985)
32. Whitaker, J.R. Some present and future uses of enzymes in the food industry. In: "Enzymes - The Interface Between Technology and Economics" (Daneby, J.P., and Wolnak, B., eds.). Marcel Dekker, Inc., New York, pp. 53-73 (1980)
33. Weetall, H.H., and Zelko, J.T. Applications of microbial enzymes for production of food-related products. Dev. Ind. Microbiol. 24:71-77 (1983)
34. Anonymous. New methods remove citrus bitterness. Chem. Eng. News. pp. 58-59 (May 27, 1985)
35. Herman, Z., Hasegawa, S., and Ou, P. Nomilin acetyl-lyase, a bacterial enzyme for nomilin debittering of citrus juices. J. Food Sci. 50:118-120 (1985)
36. Swaisgood, H.E. Sulphydryl oxidase: properties and applications. Enzyme Microb. Technol. 2:265-272 (1980)
37. Fogarty, W. Some recent developments in starch-degrading enzymes. Curr. Dev. Malting, Brew. Distill. (Proc. Aviemore Conf., Priest, F.G., and Campbell, I., eds.). Inst. Brew., London, pp. 83-110 (1983)
38. Reichelt, J. Starch. In: "Industrial Enzymology" (Godfrey, T., and Reichelt, J., eds.). The Nature Press, New York, pp. 375-396 (1983)
39. Godfrey, T. Flavouring and colouring. In: "Industrial Enzymology" (Godfrey, T. and Reichelt, J., eds.). The Nature Press, New York, pp. 375-396 (1983)
40. Hollo, J., Laslo, E., and Hoschke, A. Enzyme engineering in starch industry. Bie Stärke, 35:169-175 (1983)
41. Janda, W. Fruit juice. In: "Industrial Enzymology" (Godfrey, T., and Reichelt, J., eds.). The Nature Press, New York, pp. 315-320 (1983)
42. Godfrey, T. Brewing. In: "Industrial Enzymology" (Godfrey, T. and Reichelt, J., eds.). The Nature Press, New York, pp. 315-320 (1983)
43. Vallee, B.L. Zinc and other active site metals as probes of local conformation and function of enzymes. Carlsberg Research Communications 45, 423-441 (1980)
44. Gilles, I., Loffler, H.G., and Schneider, F. Co^{+2}-substituted acylamino acid amidohydrolase from *Aspergillus oryzae*. Zeitschrift fur Naturforschung 36C:751-754 (1981)
45. Prescott, J.M., Wagner, F.W., Holmquist, B., and Vallee, B.L. One hundred fold increased activity of *Aeromonas* aminopeptidase by sequential substitutions with Ni (II) or Cu (II) followed by zinc. Biochem. Biophys. Res. Commun. 114:646-652 (1983)
46. Urabe, I., Nanjo, H., and Okada, H. Effect of acetylation of *Bacillus subtilis* α-amylase on the kinetics of heat inactivation. Biochem. Biophys. Acta 302:73-79 (1973)
47. Riorden, J.F., and Muszynska, G. Differences between the conformations of nitrotyrosyl-248 carboxypeptidase A in the crystalline state and in solution. Biochem. Biophys. Res. Commun. 57:447-451 (1974)

48. Cornelius, D.A., Asmus, C.V., and Sternberg, M. Acylation of *Mucor pusillus* microbial rennet enzyme. US Patent #4,362,818 (1982)
49. Cornelius, D.A. Decreasing the thermal stability of microbial rennet. US Patent #4,348,482 (1982)
50. Blumberg, S., and Vallee, B.L. Superactivation of thermolysin by acetylation with amino acid N-hydroxysuccinimide esters. Biochemistry 14:2410-2419 (1975)
51. Kaiser, E.T., and Lawrence, D.S. Chemical mutation of enzyme active sites. Science 226:505-511 (1984)
52. Slama, J.T., Rodziejewski, C., Oruganti, S.R., and Kaiser, E.T. Semisynthetic enzymes: characterization of isomeric flavopapains with widely different catalytic efficiencies. J. Am. Chem. Soc. 106: 6778-6785 (1984)
53. Maugh, T.H. II. Semisynthetic enzymes are new catalysts. Science. 223:154-156 (1984)
54. Saraswathi, S., and Keyes, M.H. A systematic approach to induce new enzyme activities in naturally occurring proteins. Polym. Mater. Sci. Eng. 51:198-203 (1984)
55. Lee, V. Creating new enzymes with chemistry. Biotechnology 2: 549-551 (1984)
56. Kraicsovits, F., Previero, A., and Otvos, L. The pyridoxal effect on the specificity of α-chymotrypsin. Bulgarian Acad. Sci., 32: 386-390 (1981)
57. Guillochon, D., Esclade, L., Cambou, B., and Thomas, D. Hydroxylation by hemoglobin-containing systems: activities and regioselectivities. Annals N.Y. Acad. Sci. 434..214-218 (1984)
58. Chu, F.K., Trimble, R.B., and Maley, F. The effect of carbohydrate depletion on the properties of yeast internal invertase. J. Biol. Chem. 253:8691-8693 (1978)
59. Hayashida, S.W., and Yoshioka, H. The role of carbohydrate moiety on thermostability of cellulases from *Humicola insolens* YH-8. Agric. Biol. Chem. 44:481-487 (1980)
60. Barker, S.A., and Gray, C.J. The role of carbohydrates in enzymes. Biochem. Soc. Trans. 11:16-17 (1983)
61. Takahashi, K., Yoshimoto, T., Tamaura, Y., Saito, Y., and Inada, Y. Ester synthesis at extraordinary low temperature of -3°C by modified lipase in benzene. Biochem. International 10:627-631 (1985)
62. Naoi, M., Kiuchi, K., Sato, T., Morita, M., Tosa, T., Chibata, I., and Yagi, K. Alteration of the substrate specificity of *Aspergillus oryzae* β-galactosidase by modification with polyethylene glycol. J. Appl. Biochem. 6:91-102 (1984)
63. Klibanov, A.M. Immobilized enzymes and cells as practical catalysts. Science 219:722-727 (1983)
64. Hutlin, H.O. Current and potential uses of immobilized enzymes. Food Technol. 37:66-82, 176 (Oct. 1983)
65. Bott, R.R., Ferrari, E., Wells, J.A., Estell, D.A., and Henner, D.J. Procaryotic carbonyl hydrolases, methods, DNA vectors and transformed hosts for producing them and detergent compositions containing them. Eur. Pat. Appl., Publication #0130756 (1985)
66. Estell, D.A., Miller, J.V., Graycar, T.P., Powers, D.B., and Wells, J.A. Site-directed mutagenesis of the active site of subtilisin BPN. The World Biotech. Report USA 2:181-187 (1984)
67. Perry, L.J., and Wetzel, R. Disulfide bond engineered into T4 lysozyme: stabilization of the protein toward thermal inactivation. Science 226:555-557 (1984)
68. Craik, C.S., Largman, C., Fletcher, T., Roczniak, S., Barr, P.J., Fletterick, R., and Rutter, W.J. Redesigning trypsin: alteration of substrate specificity. Science 228:291-297 (1985)

69. Fersht, A.R., Shi, J.-P., Wilkinson, A.J., Blow, D.M., Carter, P., Waye, M.M.Y., and Winter, G. Analysis of enzyme structure and activity by protein engineering. Angew. Chem. Int. Ed. Engl. 23..467-473 (1984)
70. Paterson, A., and Clarke, P.H. Molecular basis of altered enzyme specificities in a family of mutant amidases from *Pseudomonas aeruginosa*. J. Gen. Microbiol. 114:75-85 (1979)
71. Scazzocchio, C., and Sealy-Lewis, H.M. A mutation in the xanthine dehydrogenase (purine hydroxylase I) of *Aspergillus nidulans* resulting in altered specificity. Eur. J. Biochem. 91:99-109 (1978)
72. Zaks, A., and Klibanov, A.M. Enzyme-catalyzed processes as in organic solvents. Proc. Natl. Acad. Sci. USA 82:3192-3196 (1985)
73. Neidleman, S.L., and Geigert, J. Reaction conditions: determinants in biocatalytic expression. J. Dev. Ind. Microbiol. 25:19-30 (1984)
74. Soda, K. Bioconversion of lipophilic compounds by immobilized biocatalysts in organic solvents. Trends in Biochem. Sci. 8:428 (1983)
75. Oyama, K., and Kihara, K. A new horizon for enzyme technology. Chemtech pp. 100-105 (Feb. 1984)
76. Neidleman, S.L., and Geigert, J. Biotechnology and oleochemicals: changing patterns. J. Am. Oil. Chem. Soc. 61:290-297 (1984)
77. Patterson, J.D.E., Blain, J.A., Shaw, C.E.L., Todd, R., and Bell, G. Synthesis of glycerides and esters by fungal cell-bound enzymes in continuous reactor systems. Biotechnol. Letters 1:211-216 (1979)
78. Cesti, P., Zaks, A and Klibanov, A.M. Preparative regioselective acylation of glycols by enzymatic transesterification in organic solvents. Appl. Biochem. Biotechnol. 11(5) (1985)
79. Nielsen, T. Industrial application possibilities for lipase. Fette Seifen. Anstrichm. 87:15-19 (1985)
80. Matsuo, T., Sawamura, N., Hashimoto, Y., and Hashida W. Method for enzymatic transesterification of lipid and enzyme used therein. US Patent #4,472,503 (1984)
81. Kim, K.H., Kwon, D.Y., and Rhee, J.S. Effects of organic solvents on lipase for fat splitting. Lipids 19:975-977 (1984)
82. Weetall, H. Enzymatic gallic acid esterification. Biotechnol. Bioeng. 27:124-127 (1985)
83. Pal, P.K., and Gertler, M.M. The catalytic activity and physical properties of bovine thrombin in the presence of dimethyl sulfoxide. Thrombosis Research 29:175-185 (1983)
84. Cambou, B., and Klibanov, A.M. Unusual catalytic properties of usual enzymes. Annals N.Y. Acad. Sci. 434:219-223 (1984)
85. Zaks, A., and Klibanov, A.M. Enzymatic catalysis in organic media at 100°C. Science 224:1249-1251 (1984)
86. Rohrbach, R.P., and Maliarik, M.J. Increasing the stability of amyloglucosidase. US Patent #4,415,656 (1983)
87. Ervin, J.L., Geigert, J., Neidleman, S.L., and Wadsworth J. Substrate-dependent and growth temperature-dependent changes in wax ester compositions produced by *Acinetobacter* sp. H01-N. In: "Biotechnology for the Oil and Fats Industry" (Ratledge, C., Rattray, J.B.M., and Dawson, P.S.S., eds.), American Oil Chemists' Society, Chapter 19, pp. 217-22 (1984)
88. Wolf, R.B., Cavins, J.F., Kleiman, R., and Black, L.T. Effect of temperature on soybean seed constituents: oil, protein, moisture, fatty acids, amino acids and sugars. J. Amer. Oil Chem. Soc. 59: 230-232 (1982)
89. Dropsy, E.P., and Klibanov, A.M. Cholinesterase-catalyzed resolution of D,L-carnitine. Biotechnol. Bioeng. 26:911-915 (1984)

90. Klibanov, A.M. Unconventional catalytic properties of conventional enzymes: applications in organic chemistry. Basic Life Science 8:497-518 (1983)
91. Geigert, J., Neidleman, S.L., Hirano, D.S., Wolf, B., and Panschar, B.M. Enzymatic oxidation of D-arabino-hexos-2-ulose (D-glucosone) to D-arabino-2-hexulosonic acid ('2-keto-D-gluconic acid'). Carbohydrate Res. 113:163-165 (1983)
92. Neidleman, S.L., and Geigert, J. Biological halogenation and epoxidation. Biochem. Soc. Symposia. 48:39-52 (1983)
93. Geigert, J., Dalietos, D.J., Neidleman, S.L., Lee, T.D., and Wadsworth, J. Peroxide oxidation of primary alcohols to aldehydes by chloroperoxidase catalysis. Biochem. Biophys. Res. Communs. 114:1104-1108 (1983)
94. Anonymous. Survey results reflect reader's diverse opinions on bioproduct/process outlook. Bioprocessing Technol. pp. 4-5 (July, 1985)

Chapter 5

Protein Engineering: Potential Applications in Food Processing

Ronald Wetzel

INTRODUCTION

Proteins are important in food processing in two fundamental ways. First, proteins are a component of foods, and as such contribute both to their nutritional value and their physical properties. In some cases it may be desirable to improve the nutritional balance of the protein components of a foodstuff. In others, the physical behavior of a protein preparation may need to be improved, either for aesthetic reasons or because the protein is unusable as a food ingredient without some change. In the past, efforts to improve the physical properties of food proteins have centered on attempts to change their character through physical or chemical alteration of their structures during processing steps. However, chemical treatments may have the undesirable side effect of reducing nutritional value. The only way to increase nutritional value in the past has been to supplement the food preparation or the diet.

The second way in which proteins are important is as enzymes in food processing operations. Improvements in such enzymes in the past might have been sought by looking for replacement enzymes in nature, or for chemical or enzymatic derivatives of the traditional enzyme.

In the past 10 years recombinant DNA techniques have revolutionized many fields of fundamental and applied research involving proteins. The revolution has come more or less quickly in different fields depending on technical barriers in particular applications and on research prioritization which, in the commercial sphere, might in turn be influenced by market factors and regulatory issues. This review deals with a set of techniques, growing out of the recombinant DNA revolution, known as protein engineering (1-4). The techniques and methodological approaches embodied in this term will be summarized and illustrated with examples from the recent literature. In addition, the technical barriers to exploiting these new methods in food science, and prospects for overcoming these barriers, will be discussed.

PROTEIN ENGINEERING

In the past, biochemists interested in structure/function relationships of proteins have been limited to relatively inflexible and/or imprecise methods of generating protein analogs to test hypotheses. *De novo* chemical synthesis is possible for polypeptides of up to 40 to 50 amino acids in length. In some cases it has been possible to enzymatically prepare useful analogs by semi-synthetic routes utilizing large fragments derived from natural polypeptides (5). Chemical modification (6) of natural molecules has also provided useful derivatives, but the general lack of specificity in the reagents, and the requirement for difficult and tedious purification and characterization to insure homogeneity, severely limits the power of the method when it is rigorously applied. As mentioned above, chemical methods in food processing suffer from their tendency to reduce nutritive value by destroying particular amino acids (7).

In the late 1970s, molecular geneticists demonstrated that there were no formal constraints on the ability of a cell to produce unnatural and heterologous polypeptides (8-10). This suggested the possibility of producing previously scarce proteins in large amounts, ending the strict dependence of medicinal and food biochemists on the availability in nature of interesting proteins. It also presaged a time when organisms such as food producing plants might be precisely genetically engineered to improve performance, by adding or deleting natural genes.

More recently, molecular biology techniques founded in this breakthrough of heterologous expression have made possible the precise alteration and subsequent expression of cloned genes to yield protein variants of defined sequence. These new methods have provided a kind of nucleation site for the growth of the field of protein engineering. Recent advances in our ability to efficiently collect and decipher X-ray diffraction data also contributed to this rapid development. The rise of sophisticated, powerful computer graphics systems, which facilitate X-ray structure refinement as well as analysis of existing structures, also played a major role. Finally, modern cloning techniques have given new power and applications to the traditional field of microbial genetics, which can now also play a more important role in the fundamental and applied study of protein structure/function relationships.

The following sections outline the two major approaches to protein engineering which rely on combinations of the above methods.

Structure-Mechanism Approach

The approach which has received the most attention in the past few years relies on a knowledge of some level of structure in the protein of interest, preferably the three-dimensional structure as well as primary sequence. It is in this approach that both computer graphics analysis and site-directed mutagenesis are especially valuable. This structure/mechanism approach is attractive because it both relies on, and tests, a hypothesized structure/function relationship. One way of schematically illustrating the logical progression and technical interrelationships of this approach is shown in Figure 5-1.

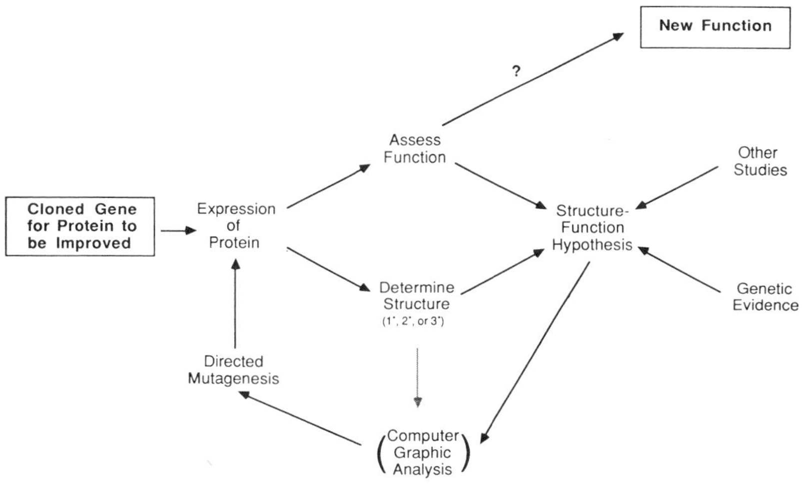

Figure 5-1. Scheme for Protein Engineering *via* a Directed Mutagenesis Approach Based on Protein Structure and Mechanism

The central feature of this pathway is the requirement for a working model for the relation between structure and function. For an enzyme, this may be a chemical mechanism based on a knowledge of the structure of the active site, on kinetic studies of the enzymatic reaction, on studies on model chemical systems, etc. For a protein folding problem, it may involve a folding model based on packing densities in the native structure, or folding studies on fragments or chemically modified protein. Data from classical genetic studies can also provide mechanistic clues. The point is that efficient methods for systematically changing and characterizing structures are powerless unless the ideas behind the design of mutants are sound. Lacking a well developed structure-function model, it may be advisable to forego directed mutagenesis in favor of other studies which may lead more quickly to a testable hypothesis.

Clearly, the first step in the structure/mechanism-based approach is to obtain structural information if it is not already available. This does not always have to be a three dimensional structure. For example, many of the structural components of proteins susceptible to chemical reactions have been characterized (7,11). Knowing which amino acids in a sequence are responsible for sensitivity to oxidation or hydrolysis makes it possible to immediately design variant sequences which may be of improved stability. In the same way, nutritional balance might be designed into a protein at the primary sequence level. Of course, such alterations in sequence may have unpredictable, negative consequences on tertiary structure and consequently on other properties, and a knowledge of the native structure might facilitate choice of replacements which will minimally perturb the structure.

Given a working model for the molecular determinants of function, computer graphic analysis of the tertiary structure can be used to design amino acid replacements predicted to alter function with only local changes in structure. These replacements, as well as more drastic changes in structure such as deletions or insertions of sequence, can be carried out by altering the DNA sequence of the cloned gene. Most commonly this is accomplished by directed mutagenesis methods based on the *E. coli* phage M13 (12). Expression of these mutated genes generates protein variants whose behavior in an assay for the activity of interest will, at best, satisfy the requirements for the desired new function, or, at worst, allow one to refine the structure/function hypothesis which in turn can serve as the basis for further mutagenesis experiments.

Random Mutagenesis/Selection-Screening

In cases in which it is possible to devise a host expression system and genetic selection related to the desired functional improvement in a protein, the most direct approach to an improved protein may be that of random mutagenesis. This may be true even if there is considerable structural information. Figure 5-2 summarizes how various classical and modern techniques can be spliced together into a powerful protein engineering approach complementary to the molecular mechanistic approach described above. The core of the method is a genetic selection, which demands improved function in the protein of interest for survival of the producing organism under certain conditions. Alternatively one might use a screening method efficient enough to allow sifting through hundreds or thousands of randomly generated mutants for those exhibiting improved properties. The concepts behind these methods are central to classical microbial genetics, but they gain new power with the ability to relatively rapidly construct required genetic backgrounds by recombinant DNA techniques, and the existence of improved methods for random mutagenesis and mutation enrichment (13,14).

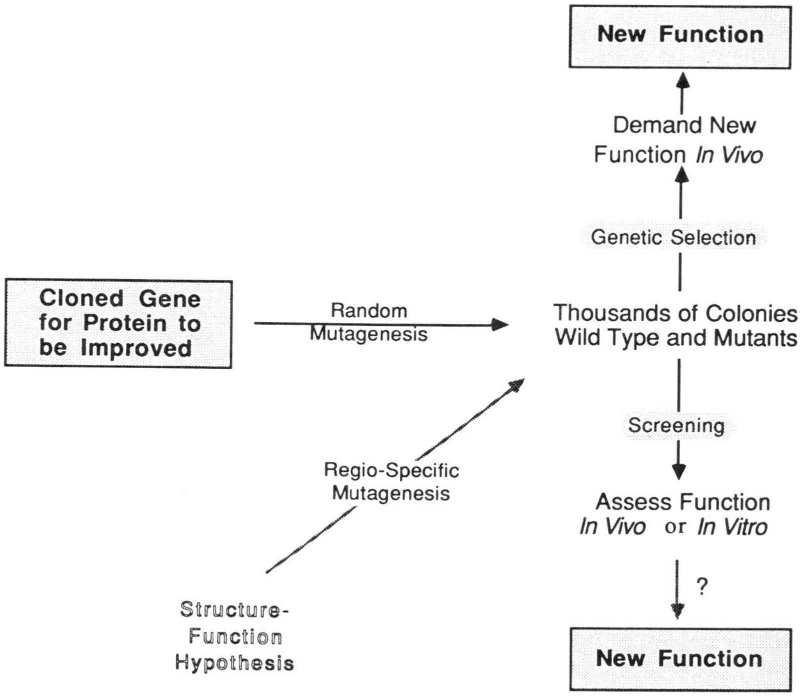

Figure 5-2. Scheme for Protein Engineering *via* Random Mutagenesis with Genetic Selection or Screening for Altered Protein Function

As Figure 5-2 shows, there is at least one amalgam of the two approaches. A structure/hypothesis may implicate a particular residue or short amino acid sequence in a particular function or property, but not be capable of predicting how certain replacements might alter that property. In a case like this, the nucleotide sequence for the region of interest can be set up in several ways for "regio-specific" mutagenesis (random mutagenesis of a short stretch of oligonucleotides) or "site-saturation" (replacement of an amino acid with all 19 possible alternatives). The products of these reactions, predicted to be particularly good candidates for functional improvement, are now subjected to a selection or screen for identification.

The two figures suggest that there are other ways in which information and materials generated by one approach may be utilized or further refined by the other. For example, a product improved in one property by a structure/mechanism-based approach may serve as a starting point for random mutagenesis for further improvement of the original or of a second property. Products of the random mutagenesis approach might serve as starting points for the directed approach. Finally, random mutagenesis can generate vast quantities of genetic data which can help form the basis of the structure/function hypothesis required for efficient progress in the mechanistic approach.

EXAMPLES OF PROTEIN ENGINEERING

Subtilisin

The first description of the use of the above techniques on an enzyme, and subsequently the most detailed mechanistic studies, have been those of Fersht and Winter on the enzyme tyr-tRNA synthetase (15-17). A large body of data is also being generated by several groups on the enzyme subtilisin. Owing to the commercial potential of this enzyme, some of the subtilisin results will be reviewed here. Subtilisin is an alkaline, serine proteinase secreted by many species of *Bacillus*. Its tertiary structure indicates a spacial homology in its active center to the catalytic ensemble of mammalian serine proteinases such as trypsin. Because subtilisin is a component of some household detergents, there is more than fundamental interest in understanding and altering some of its properties.

One limitation to some uses of subtilisin as a component of laundry detergents is its sensitivity to oxidation by bleach. Peptide mapping studies have demonstrated that the oxidation of a single residue, Met222, is largely responsible for the inactivation of subtilisin by hydrogen peroxide. Based on this knowledge, Estell et al. generated a series of subtilisin analogs in which Met222 was replaced with all 19 possible alternative amino acids (18). Several of these 222-variants exhibited specific activities from 25% of wild type to higher than wild type. Most of those tested were dramatically more stable to peroxide oxidation than wild type; the results suggest that several 222-variants should remain active significantly longer than wild type in a detergent containing bleach. The importance of residue 222 can be rationalized to some extent based on its position in the tertiary structure; nonetheless, in this application primary structural information was sufficient to design a successful protein engineering experiment.

Because wild type subtilisin has relatively broad specificity, which is presumably employed to advantage in its commercial use, improvements toward one class of substrates might reduce its activity toward others in such a way as to reduce its overall performance. In this complex situation, how can one identify structural changes which will, in balance, improve the enzyme toward the commercially targeted substrates? If it is acknowledged that practical improvement will most likely result from changes in specificity, but that the most effective specificity change cannot be predicted, then the best course is to generate as many specificity mutants as possible and screen them for efficacy. Examination of the binding surface of the enzyme for substrate indicates which enzyme residues contribute directly to substrate binding. These are the residues whose alteration is most likely to affect enzyme specificity. "Saturation mutagenesis" of these residues allows replacement with all possible alternative amino acids. The products can be tested directly in the intended application and/or in a model assay . This approach is indicated in Figure 5-2 as the mechanism-based "regio- specific" mutagenesis route. One such series is that of the 166 variants of subtilisin (19). In the wild type, Gly166 is located in a position where an added side chain might affect binding of the substrate. . .specifically, the substrate amino acid which contributes its carbonyl group to the peptide bond to be cleaved (the P-1 site). The results obtained from saturation mutagenesis at position 166 show that this position is important for specificity. For some P-1 substrates, the catalytic efficiencies of the enzyme variants vary by up to three orders of magnitude. Several enzymes have been generated which are individually either more or less active than wild type against particular substrates, supporting the notion that the practical value of a variant will be derived from a favorable balance of such rate enhancements and suppressions against the peptide bonds of the commercially important substrate(s).

T4 Lysozyme and Thermal Stability

Stability against exposure to elevated temperatures can be a desirable feature of an enzyme used in industrial applications like food processing. It may also be desirable to have an enzyme that is enzymatically active at elevated temperatures. Sensitivity and response to heat is an important feature of proteinaceous components of foods as well. The example of boiling an egg, a process mediated by the irreversible thermal denaturation of some of its protein components, is in fact often used to introduce the subjects of protein stability and denaturation. Changes in the sensitivity of particular proteinaceous components of a food may affect how the character of the material changes after denaturative processing steps such as heating or shifting the pH. From this point of view, it may at times be desirable to decrease the stability of such proteins to heat or pH, in order to achieve equivalent effects with less expenditure of time and energy, and less exposure of the rest of the system to harsh conditions (20).

In one effort to engineer thermal stability into a protein, we have introduced a disulfide bond into T4 lysozyme. This protein normally contains no disulfides and exhibits a limited stability toward heat at neutral pH. Computer graphic analysis of the X-ray crystal structure facilitated the choice of a position in the structure where the presence of two cysteine residues should make possible the formation of a disulfide bridge. Exposure of such a T4 lysozyme variant to a thiol/disulfide exchange buffer promoted formation of a disulfide between residues 3 and 97 (21). After mutagenic removal of an unpaired cysteine in the same molecule (22), a crosslinked lysozyme was generated which was substantially more stable than wild type to irreversible thermal inactivation. In fact, this T4 lysozyme variant was shown to be essentially as stable as hen egg white lysozyme to heat at neutral pH (23).

This demonstrates at least a partial success of the structural/ mechanistic approach in engineering stability. . . "partial" because, while graphic analysis clearly facilitates introduction of a disulfide, the rules guiding the stabilizing effects of disulfides are yet to be fully elucidated. The 3-97 bond is located at an attractive site: away from the enzyme's active center and linking the N- and C-terminal lobes of the molecule. In the future it will be important to assess the stabilizing effects of other disulfides introduced elsewhere in the molecule.

The 3-97 disulfide seems to work by eliminating a particularly aggregation-prone unfolding intermediate from the thermal unfolding pathway of the enzyme; while the wild type decays by a mechanism related to the thermal unfolding of the molecule, the crosslinked variant is resistant to this mechanism (23). Many other proteins probably undergo irreversible thermal inactivation in the 40-80°C range by mechanisms similar to that of wild type T4 lysozyme. Some of these may also be stabilized by the introduction of a well placed disulfide bond. A more general method of providing more modest stabilizing effects for such proteins should be to stabilize the proteins toward initial unfolding, which occurs for each protein at a characteristic melting temperature (Tm). For this reason, as well as to address fundamental questions concerning how an amino acid sequence determines a unique folding pattern and stability, there is considerable interest in improving our understanding of the effects of amino acid replacements on conformational stability.

Other work on T4 lysozyme illustrates the state of progress in this area, and also serves as an example of the use of both random mutagenesis and a structural approach to protein engineering. Several methods are available which allow the characterization of lysozyme activity *in situ* in cells producing lysozyme. Using these methods it is possible to isolate variants (generated by random mutagenesis) with altered activities as well as with altered sensitivities to exposure to heat. Such screening methods have generated both temperature-sensitive (24) and temperature-stable mutants (25). Characterization of such mutants should facilitate development of a structure/function hypothesis (Fig. 5-1), which in turn can promote a graphics design/directed mutagenesis approach to the problem.

Unfortunately, it has proved difficult to rationalize some of the effects on stability of structural changes in lysozyme and other proteins. Comparison of the x-ray structure of one ts mutant of T4 lysozyme with that of the wild type enzyme showed only minor, local effects (26). It is not yet possible to reliably calculate the consequences of small structural perturbations on packing forces. To improve the data base, Matthews and coworkers are using the saturation mutagenesis approach, at sites indicated by genetic studies (above) to be sensitive to replacements (27). A series of variants differing in side chain structure at one position, and their stabilities and structures, should help to elucidate general rules guiding the determination of the conformational stability of proteins.

Until such rules are elucidated, random mutagenesis approaches may be the best general methods for the isolation of proteins of modified thermal stabilities. Besides screens such as those developed for T4 lysozyme (see above) and staphylococcal nuclease (28), genetic selection may sometimes be possible. A recent example is the generation of thermally stable mutants in kanamycin resistance by selecting mutants at high temperatures in transformed strains of the thermophile *B. stearothermophilus* (29).

POTENTIAL OF PROTEIN ENGINEERING IN FOOD PROCESSING

Enzymes

When an enzyme is well characterized and its action in food processing is well understood, such as the use of chymosin in cheesemaking, the prospects for improving the enzyme by a structure-based protein engineering approach are relatively good. Chymosin cleaves a specific peptide bond in kappa-casein to initiate coagulation and curd formation; further hydrolysis, which is especially significant in microbial versions of this enzyme, tends to generate peptides which impart off-flavors to the cheese product. It might be possible to engineer a specificity change into a microbial version of the enzyme so that it retains its primary, desired activity against casein but loses secondary activities. It might also be possible to introduce, by protein engineering methods, a lower stability into these microbial enzymes so that the activity is lost during storage before it becomes a problem (30). Another approach, that of expressing bovine chymosin in microorganisms in attempts to produce it for lower cost, has been taken by several biotechnology firms (see, for example, ref. 31). Even when the expression of an active enzyme has already been demonstrated, protein engineering to increase the efficiency with which the enzyme is produced in the recombinant microbe, or to alter or introduce other properties, may prove a productive approach.

Despite the exciting and very encouraging results of the past few years in the engineering of many relatively simple, well characterized proteins, the prospect of an attempt to introduce specific new properties into a particular protein by these methods looms a formidable task. Still, in comparison to the technical barriers to engineering food proteins (see below), modification of enzymes is relatively straightforward. Many enzymes are well behaved, soluble, regular structures amenable to structural characterization. The chemistry of their action in food processing is also accessible for study in many cases. Some are microbial in origin; others can be produced in microbes by recombinant DNA techniques.

Thus, it should generally be possible to express engineered genes for these proteins and observe their altered properties. In microbial expression systems, they are also more likely to be amenable to classical genetic selection/screening methods.

In surveying the field for candidates for engineering, one should be considering not only enzymes currently in use, but also enzymes which have been tried in the past and found wanting. Whatever limited their applicability before (availability; stability to heat, pH, or shear forces; too broad or too narrow substrate specificity; inadequate kinetic parameters, etc.) it is now possible to consider improving these properties through protein engineering.

Protein Components of Foods

There are tremendous opportunities for the engineering of improved properties into the proteinaceous constituents of food sources. While most cultures have developed processes which take good advantage of many of the special properties of traditional food sources, one can still hope to adjust these features to make further improvements. In addition, where both traditional and modern methods have failed to exploit some nutritive sources because of limitations in the molecular architecture of their component proteins, one can now attempt to remove these constraints by protein engineering.

There are, however, significant impediments to the rapid implementation of these methods to alter the composition and properties of food proteins. First, the development of techniques for the productive introduction of new genes into animals (32) and higher plants (33), while great progress is being made, still lags significantly behind the state of the art for the introduction and expression of genes in bacteria and yeast. While the latter systems might, and probably should, serve as intermediate expression systems for the iterative exploration of structure/function relationships outlined in Figures 5-1 and 5-2, at some point it will usually be necessary to transfer the gene for the engineered protein into the genome of the plant or animal which serves as a food source; the new gene will need to be in a genetic context which supports expression comparable to the wild type protein, and the wild type gene may have to be removed or otherwise inactivated at the same time.

The other major barrier to immediate exploitation of protein engineering methods in food proteins is the lack of well developed structure/function ideas in this field, and the technical barriers responsible for this deficiency. While significant progress has been made in the elucidation of the role of particular proteins in the behavior of milk in cheese making, or wheat flour in bread making, the complex interactions of proteins, other macromolecules, and small molecules including water, in these systems is far from understood. Ideally, one would like to see: 1) models for these and other systems in which the flour or milk could be reconstituted from purified, well characterized components; 2) detailed structural information of these isolated components, and of the intermediates and products of their interactions during processing; and, 3) chemical and physical tests for the measurement of properties directly associated with the desired macroscopic properties of the processed material. Only with such systems in place could one efficiently explore functional improvements by a structure-mechanistic approach.

For these reasons a genetic selection/screening approach may have some promise. If one has a clear idea of the kind of functional improvement one wishes to confer on a protein component (which itself implies a highly developed structure/function hypothesis for the protein as part of the whole system), it may be possible to conduct selection or screening of mutants in a foreign host system such as *E. coli.* This would have the advantages of speed and simplicity. On the other hand, the properties which a cloned plant protein imparts to *E. coli* cells or extracts may not relate to the properties it confers in the natural milieu. In these cases it is probably better to do the genetics on the natural system, linked to an efficient screening procedure. While this may sound remarkably similar to traditional genetic methods (plant hybridization, for example), modern molecular genetics does contribute significant improvements. One of these is the ability to control the nature of the genetic change and to confine it to the gene of interest. Another is the ability to construct a host organism that will maximize screening efficiency.

One important goal of some workers in food chemistry is the enhancement of nutritional value in the protein components of major food sources; for example, increasing the lysine content of wheat proteins. Where should the amino acid replacements or additions be made to improve nutritive balance while retaining key functional features of the protein? Lacking a three dimensional structure, one might start from the knowledge that charged residues tend to occur in surface-exposed loops in globular proteins, and that therefore the least disruptive change might be the substitution of lysine for glutamate or aspartate in the primary sequence. If the tertiary structure is known, one could make more informed guesses as to which replacements or additions might produce the least disruption in functional properties. There are still many mysteries, however, in how substitutions affect the properties of proteins of known structure. This suggests that some lysine-enriched proteins may express unpredictable, undesirable attributes. It is at this point that a genetic approach might supplement a directed approach. The gene for a poorly-behaved lysine-enriched protein might be exposed to random mutagenesis and screening or selection in search of second-site revertants that suppress the poor functional feature caused by the original lysine substitution.

How much structural information is necessary to make the directed mutagenesis approach effective? Some attributes of proteins, notably sensitivity toward chemical degradation (which certainly can be a consequence of food processing steps), are largely a function of the amino acid residues in the primary sequence. There are several examples of the successful engineering of stability into proteins by simple modification of the primary sequence (18,34,35). Disulfide bonds play important roles in the primary and tertiary structure of proteins, and thus comprise an intermediate area in which, at least in some cases, it is sufficient to understand primary structure, including disulfide arrangements, to do effective protein engineering. Examples of the removal of a compromising cysteine residue exist in the area of medically important proteins (36,37) as well as in the case of the T4 lysozyme model system discussed above. Since thiol-disulfide chemistry is very important in the functional properties of proteins in wheat flour and in milk (38) it is possible that effective protein engineering experiments involving altered thiol content can be designed and monitored in the absence of tertiary structural information.

CONCLUSIONS

While peptide chemists have made and studied analogs of natural polypeptides for decades, the incredible power of molecular genetics methods has only been applied to protein structure/function studies for the past several years. Considering this short history, the harvest of important information and exciting innovation is quite impressive. The rapidity with which these methods have been adopted and creatively exploited in so short a time promises increasingly important fundamental, and applied, breakthroughs in the future. We might look to the area of industrial (including food processing) enzymes for early successes in applied research. In medicine (4), where proteins must run a gauntlet of complex biochemical and regulatory pressures, creation and acceptance of effective derivatives take more time. A later frontier will likely be so-called protein "functionality" in food processing, where fundamental knowledge as well as technical breakthroughs seem to be required before the existing methodologies of directed and random gene and protein alteration can be effectively utilized. At the same time, if there is one truism in molecular genetics, it is that major surprises may lie just around the corner. It is less than 10 years since "back-of-the-envelope" calculations demonstrated that microbially produced insulin could never be practical; by 1986 human insulin produced by *E. coli* will have been on the market for three years. While the time frame is open to conjecture, it seems clear that we can look forward to major contributions from protein engineering and related biotechnology in the availability and quality of food sources and processes.

ACKNOWLEDGEMENTS

I wish to thank Tom Richardson for helpful discussions, and Rey Gomez, Jonathan MacQuitty, and Tony Kossiakoff for reviewing the manuscript. I gratefully acknowledge Wayne Anstine for help with the illustrations and manuscript.

REFERENCES

1. Rastetter, W.H., Enzyme engineering: applications and promise. Trends Biotech. 1:80, pp. 1-5 (1983)
2. Inouye, M. and Sarma, R. (eds.) "Protein Engineering". Academic Press, New York, (1986)
3. Oxender, D. (ed.) "Protein Structure, Folding and Design". Alan R. Liss, Inc., New York, (1986)
4. Wetzel, R. Medical applications of protein engineering. In: "Protein Tailoring and Reagents for Food and Medical Uses" (R.E. Feeney and J.R. Whitaker, eds.). Marcel Dekker, New York (1986)
5. Fruton, J.S. Proteinase-catalysed synthesis of peptide bonds. Advs. Enzymol. 53:239-306 (1982)
6. Means, G.E. and Feeney, R.E. "Chemical Modification of Proteins" Holden-Day, San Francisco, (1971)
7. Feeney, R.E. Chemical modification of food proteins. In: "Food Proteins: Improvement through Chemical and Enzymatic Modification" (R.E. Feeney and J.R. Whitaker, eds). American Chemical Society, Washington, D.C. pp. 1-91 (1977)
8. Struhl, K., Cameron, J.R., and Davis, R.W. Functional genetic expression of eukaryotic DNA in *Escherichia coli.* Proc. Natl. Acad. Sci. USA 73:1471-1475 (1976)
9. Itakura, K., Hirose, T., Crea, R., Riggs, A.D., Heyneker, H.L., Bolivar, F. and Boyer, H.W. Expression in *Escherichia coli* of a chemically synthesized gene for the human hormone somatostatin. Science 198:1056-1063 (1977)
10. Wetzel, R. and Goeddel, D.V. Synthesis of polypeptides by recombinant DNA methods. In: "The Peptides: Analysis, Synthesis, Biology" (E. Gross and J. Meienhofer, eds.) Academic Press, New York, pp. 1-64 (1983)
11. Ahern, T.J. and Klibanov, A.M. The mechanism of irreversible enzyme inactivation at 100 degs. C. Science 228:1280-1284 (1985)
12. Zoller, M.J. and Smith, M. Oligonucleotide-directed mutagenesis of DNA fragments cloned into M13 vectors. Methods in Enzymol. 100:468-500 (1983)
13. Myers, R.M., Lerman, L.S., and Maniatis, T. A general method for saturation mutagenesis of cloned DNA fragments. Science 229:242-247 (1985)
14. Botstein, D. and Shortle, D. Strategies and applications of *in vitro* mutagenesis. Science 229:1193-1201 (1985)
15. Fersht, A.R., Shi, J.-P., Wilkinson, A.J., Blow, D.M., Carter, P., Waye, M.M.Y., and Winter, G. Analysis of enzyme structure and activity by protein engineering. Angew. Chem. (Eng.) 23:467-473 (1984)
16. Fersht, A.R., Shi, J.-P., Knill-Jones, J., Lowe, D.M., Wilkinson, A.J., Blow, D.M. Brick, P. Carter, P., Waye, M.M.Y. and Winter, G. Hydrogen bonding and biological specificity analysed by protein engineering. Nature 314:235-238 (1985)
17. Wells, T.N.C. and Fersht, A.R., Hydrogen bonding in enzymatic catalysis analyzed by protein engineering. Nature 316:356 (1985)

18. Estell, D.A., Graycar, T.P. and Wells, J.A. Engineering an enzyme by site-directed mutagenesis to be resistant to chemical oxidation. J. Biol. Chem. 260:6518-6521 (1985)
19. Wells, J.A., Powers, D.B., Bott, R.R., Katz, B.A., Ultsch, M.H., Kossiakoff, A.A., Power, S.D., Adams, R.M., Heyneker, H.H., Cunningham, B.C., Miller, J.V., Graycar, T.P., and Estell, D.A. Protein engineering of subtilisin. In: "Protein Modification and Design" (D. Oxender, ed.) Alan R. Liss, New York (1986)
20. Kinsella, J.E. Relationships between structure and functional properties of food proteins. In: "Food Proteins" (P.F. Fox and J.J. Condon, eds.) Applied Science Publ., London p. 55 (1982)
21. Perry, L.J. and Wetzel, R. Disulfide bond engineered into T4 lysozyme: Stabilization of the protein toward thermal inactivation, Science 226:555-557 (1984)
22. Perry, L.J. and Wetzel, R. Unpaired Cys54 interferes with the ability of an engineered disulfide to stabilize T4 lysozyme, Biochemistry (In press)
23. Wetzel, R. Investigation of the structural roles of disulfides by protein engineering; a study with T4 lysozyme. In: "Protein Engineering" (M. Inouye and R. Sarma, eds). Academic Press, New York (1986)
24. Hawkes, R., Gruetter, M.G. and Schellman, J. Thermodynamic stability and point mutations of bacteriophage T4 lysozyme. J. Mol. Biol. 175:195-211 (1984)
25. Alber, T. and Wozniak, J.A. A genetic screen for mutations that increase the thermal stability of phage T4 lysozyme. Proc. Natl. Acad. Sci. (USA) 82:747-750 (1985)
26. Gruetter, M.G., Hawkes, R.B. and Matthews, B.W. Molecular basis of thermostability in the lysozyme from bacteriophage T4. Nature 277:667-669 (1979)
27. Alber, T., Gruetter, M.G., Gray, T.M. Wozniak, J.A., Weaver, L.H., Chen, B.-L., Baker, E.N. and Matthews, B.W. Structure and stability of mutant lysozymes from bacteriophage T4. In: "Protein Structure, Folding and Design". (D. Oxender, ed.) Alan R. Liss, New York (1986)
28. Shortle, D. and Lin, B. Genetic analysis of Staphylococcal nuclease: Identification of three intragenic "global" suppressors of nuclease-minus mutations. Genetics 110:539-555 (1985)
29. Liao, H., McKenzie, T. and Hageman, R. Isolation of a thermostable enzyme variant by cloning and selection in a thermophile. Proc. Natl. Acad. Sci. USA (1985)
30. U.S. Patent 4,357,357. Thermal destabilization of microbial rennet. Assignee Novo Industri A/S (Nov. 2, 1982)
31. Marston, F.A.O., Lowe, P.A., Doel, M.T., Schoemaker, J.M., White, S. and Angel, S. Purification of calf prochymosin (prorennin) synthesized in *Escherichia coli.* Biotechnology 2:800-804 (1984)
32. Kucherlpati, R. and Skoultchi, A.I., Introduction of purified genes into mammalian cells. CRC Critical Reviews in Biochemistry, Vol. 16, #4, p.349-379 (1985)
33. Caplan, A., Herrera-Estrella, L., Inze, D., Van Haute, E., Van Montagu, M., Schell, J. and Zambryski, P. Introduction of genetic material into plant cells. Science 222:815-821 (1983)
34. Courtney, M., Jallat, S., Tessier, L.-H., Benavente, A., Crystal, R.G., and Lecocq, J.-P. Synthesis in *E. coli* of alpha-one antitrypsin variants of therapeutic potential for emphysema and thrombosis. Nature 313:149-151 (1985)
35. Rosenberg, S., Barr, P.J., Najarian, R.C. and Hallewell, R.A. Synthesis in yeast of a functional oxidation-resistant mutant of human alpha-one antitrypsin. Nature 312:77-80 (1984)
36. Wang, A., Lu, S.-D. and Mark, D.F. Site-specific mutagenesis of the human interleukin-2 gene: Structure-function analysis of the cysteine residues. Science 224:1431-1433 (1984)

37. Mark, D.F., Lu, S.D., Creasey, A.A., Yamamoto, R., and Lin, L.S. Site-specific mutagenesis of the human fibroblast interferon gene. Proc. Natl. Acad. Sci. (USA) 81:5662-5666 (1984)
38. Kinsella, J.E. Relationships between structure and functional properties of food proteins. In: "Food Proteins" (P.F. Fox and J.J. Condon, eds.) Applied Science Pubs., London, p. 71, p. 91 (1982)

Chapter 6

Biopolymers and Modified Polysaccharides

Anthony Sinskey, Spiros Jamas, David Easson Jr. and ChoKyun Rha

INTRODUCTION

Polysaccharides are biopolymers of major importance with applications in the food, cosmetics, chemical, medical, waste treatment and oil industries (Table 6-1). The roles they play in biological systems and in nature make it evident that polysaccharides possess unique chemical and physical properties which provide a wide variety of functions. The specific functional properties of polysaccharides are the result of their physical characteristics which are primarily controlled by the molecular structure, i.e., monomeric components and the linkages joining them.

The ability of polysaccharides to dissolve or disperse in water, thus changing the properties of their aqueous environment coupled to their low toxicity is responsible for most of their uses as hydrocolloids in the food and health care industries. These hydrocolloids are often used as thickeners, viscosifiers, or drag reducing agents. Since these functional properties, which are in fact rheological effects, are primarily rendered by the molecular structure of the polysaccharides, a systematic study of structure-property-function relationships would allow us to establish the basic principles for predicting the performance and application of polysaccharides. This will then lead to the engineering of molecules with a specific desired function. Applications based on the matrix-forming properties of polysaccharides will follow.

Often, the function of a biopolymer is to provide a microstructure or matrix of desired diffusional and mechanical properties. Utilization of these properties has not been practiced to date even though the applications implied by microstructures of defined functional properties scan both the food industries in controlled flavor release, and the high technology separations and biomedical industries.

Table 6-1. Some Commercial and Industrial Uses of Hydrocolloids

Functions	Applications
Adhesive or binding	Paper, pulp and textile products, construction uses
Bulking or carrier	Drugs, germicides, fungicides, insecticides, fertilizers
Chelating	Metal recoveries, reaction controls, pollution controls, water purifications
Coagulant	Protein precipitations or recoveries, industrial and food processing, waste treatment, clarification of beverages
Dispersant, stabilizer or surface active agents	Cosmetic and personal care products, drugs, photographic products, paints, pigments, inks, food products
Finishing	Textile, paper, and pulp products, surgical adjuncts
Gellant	Food products, oil-field applications, cleaning agents, personal care products, cosmetics, drugs, lubricants
Matrix	Separation or chromatographics, structure matrix for food or sustained release, immobilization of enzymes or cells, biomedical or surgical materials
Membrane or films	Separations or filtrations, edible films or packages

The availability of recombinant and classical DNA techniques to manipulate the genetics of an organism allows the development of strategies to alter the polymer structure, *in vivo*, and hence its function. The application of these technologies to the production of microbial polysaccharides could lead to structural manipulation at the genetic level and the possibility of an unlimited potential for the development of new biopolymers.

A clear requirement therefore exists to establish guidelines and the fundamental principles in the field of biopolymer engineering. The objectives of biopolymer engineering can be separated into two general areas, however, it is vital that these areas are not approached independently, as a continuous flow of information between them is necessary to achieve the final goal. These areas are the systematic study of structure-property-function relationships and the application of modern biotechnology for the synthesis and production of molecular conformations optimal for a desired use.

REVIEW OF MICROBIAL BIOPOLYMERS IN THE FOOD INDUSTRY

Biologically synthesized polymers, especially polysaccharides, have many commercial applications as was shown in Table 6-1. These products are derived from plants, bacteria, fungi and higher eukaryotes. Unique viscosifying, emulsion stabilizing and surface tension modifying properties make polysaccharides attractive for a wide variety of uses. Other advantages of biopolysaccharides include the fact that they have low toxicity, are environmentally safe, are naturally derived and are viable alternatives to petroleum based polymers. Bacterial polysaccharides represent a small fraction of the current biopolymer market but have one of the largest potentials for development of novel and improved products with the successful application of new genetic engineering technologies. Microbial polysaccharides with current or potential applications in the food industry are summarized in Table 6-2 with specific research areas which will require development. Xanthan gum is an example of a microbial polysaccharide that already has commercial significance in the food industry.

Table 6-2. Microbial Food-Related Biopolymers

Biopolymer	Organism	Composition and Linkage	Molecular Weight	Areas for Immediate Development
Cellulose	*Acetobacter* *Agrobacterium* *Alcaligenes*	D-glucose β(1-4), linear	$2x10^6$	Fermentation Genetic systems
Dextran	*Lactobacillus* *Leuconostoc* *Streptococcus*	D-glucose α(1-6) α(1-3),branching	10^4-10^8	Genetic systems Biosynthetic and hydrolytic enzyme biochemistry
Alginate	*Azotobacter* *Pseudomonas*	D-mannuronate L-guluronate β(1-4), linear	$1.5x10^5$	Fermentation Genetic systems
Curdlan	*Alcaligenes* *Agrobacterium*	D-glucose α(1-3), linear	—	Producing strains Product recovery
Pullulan	*Aureobasidium*	D-glucose α(1-6),α(1-4) linear	$2.5x10^5$- $1x10^7$	Fermentation Functional properties
Xanthan	*Xanthomonas*	D-glucose β(1-4) D-glucose- D-mannose- D-glucuronate β(1-4), β(1-2) α(1-3) branching	$2x10^6$- $1.5x10^7$	Genetic systems Structure-function properties
Chitosan	*Mucorale*	D-glucosamine	$1.7x10^4$- $1.3x10^5$	Fermentation Biosynthesis Genetic Systems

Xanthan Gum

Xanthan gum is produced commercially by *Xanthomonas campestris*, and is a polyanionic exopolysaccharide composed of D-glucose, D-mannose and D-glucuronic acid in a molar ratio of 2:2:1, respectively (1). The repeating unit is a pentasaccharide (Figure 6-1) consisting of a β(1-4)-linked glucose backbone with branches occurring every other unit consisting of a mannose-β-1,4-glucuronic acid-β-1,2-mannose-α-1,3 linked trisaccharide. Pyruvic acid is ketally linked to the external mannose and the internal mannose is 0-acetylated (2). These two moieties occur at varying proportions. Reported titers of xanthan in batch culture generally exceed 25 g/liter with yields on carbon source ranging from 56% to 70% (3,4,5).

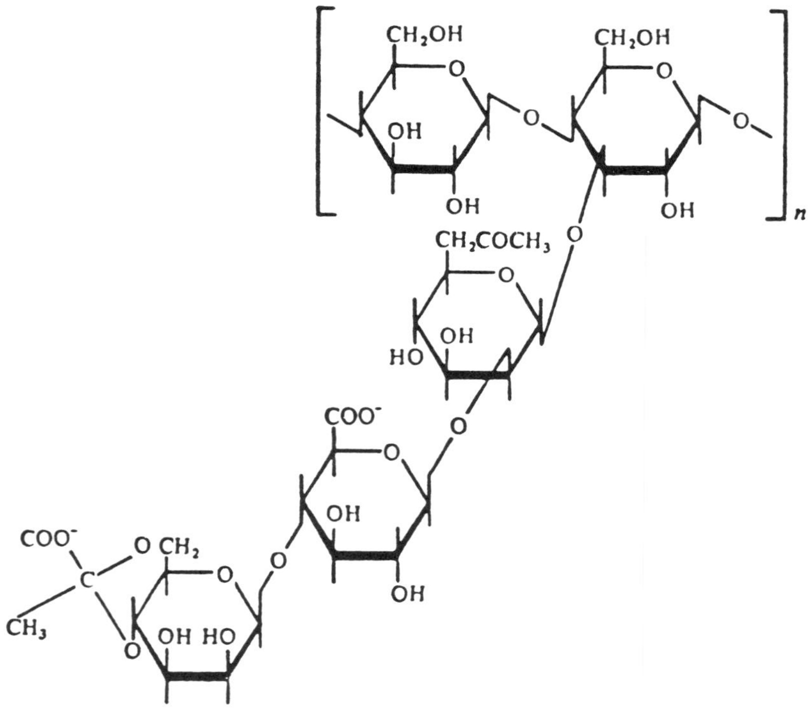

Figure 6-1. Structure of Xanthan Gum (2)

Xanthan gum has unique functional properties that have led to many commercial applications in the food, cosmetics, pharmaceuticals and textile industries and in oil recovery. In the food industry it is used as a viscosifier, suspending and bodying agent, emulsion stabilizer and settling agent. Specific food applications include uses in bakery fillings, canned foods, dry mixes, frozen foods, pourable dressings, sauces, gravies, processed cheeses, and juice drinks. Xanthan gum forms highly viscous, non-thixotropic and pseudoplastic solutions that are stable to wide ranges of temperature, pH and salt concentration. The gum also reacts synergistically with galactomannans such as locust bean or guar gum. When xanthan and guar gums are mixed, a viscosity increase occurs higher than would be expected for the individual gums.

Limited fundamental investigations are available that elucidate the genetic and biochemical basis for synthesis and over-production of xanthan. With *Xanthomonas campestris* pv. *campestris*, recent studies on cloning of genes involved in plant pathogenicity demonstrate the feasibility of recombinant DNA technology for the molecular cloning of xanthan biosynthetic genes (6).

Alginate

Alginates are heteropolysaccharides containing D-mannuronic acid and L-guluronic acid that are β(1-4) linked (Figure 6-2). They are produced by brown seaweeds and several bacteria including *Azotobacter vinelandii* and *Pseudomonas aeruginosa*. The bacterial alginate differs from the algal alginate in that a fraction of the D-mannuronic acid residues are 0-acetylated (8). The proposed pathway for the synthesis of alginate in *Azotobacter vinelandii* is shown in Figure 6-3. In batch culture of *A. vinelandii*, alginate production has been reported to be 5 to 7 g per g dry cell weight corresponding to a yield on carbon of approximately 25% (10).

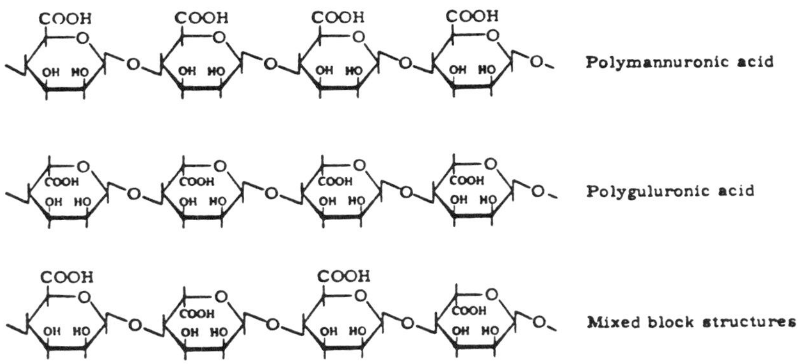

Figure 6-2. Structure of Polymer Segments Contained in Bacterial Alginate. A portion of the D-mannuronosyl Residues Are 0-acetylated (7)

Commercially, alginates are used extensively for their water holding, gelling, stabilizing and emulsifying properties (11). In the food industry

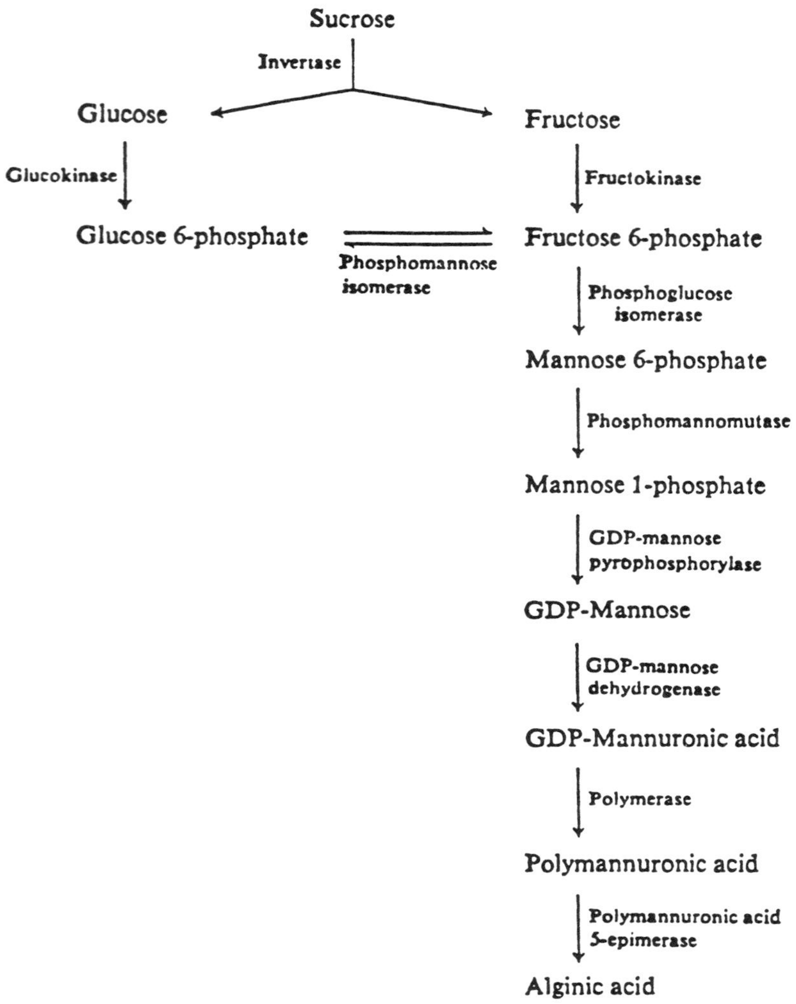

Figure 6-3. Biosynthesis of Alginate in *A. vinelandii* (9)

alginates are used as stabilizers in ice cream, bodying agents in bakery fillings and icings, and gelling agents in puddings and dessert gels.

Currently, large scale production is more economical for the algal alginate, however, production of microbial alginate would have the advantages of constant composition, constant yields and less pollution (2), and could become feasible if the process is optimized and the above advantages result in a more desirable product. The potential for applications of genetic approaches for investigating alginate biosynthesis is high. Several investigators have initiated a genetic approach with *P. aeruginosa* that has led to the cloning of several genes involved in alginate biosynthesis (12,13,14).

Dextrans

Dextrans are glucose homopolysaccharides that have an α(1-6) linked main chain with α(1-3) linked branches (Figure 6-4). Dextrans are synthesized by *Lactobacillus, Streptococcus, Leuconostoc* and other bacterial species. Its biosynthesis is unique in that it does not require nucleotide bound precursors. Glucose for biosynthesis is provided by sucrose, which is extracellularly converted to dextran by the enzyme dextransucrase.

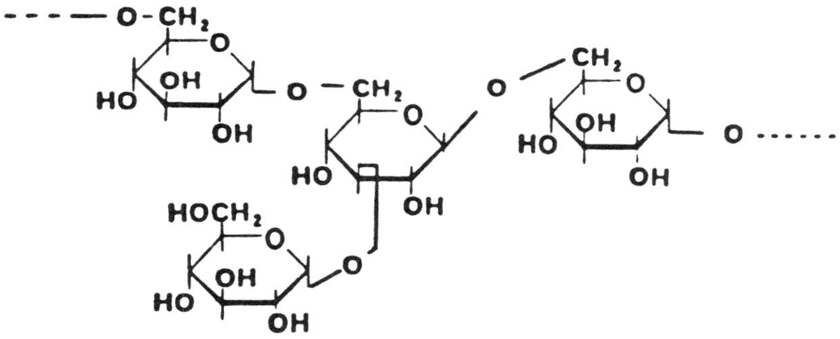

Figure 6-4. A Portion of the Structure of Dextran (7)

Dextran and its derivatives are used in the pharmaceutical and chemical industries as plasma extenders, anticoagulants (sulfated dextrans) and adsorbents (DEAE dextran). Since dextrans produced by oral microorganisms play an important role in the etiology of dental diseases, extensive studies have been conducted recently on the development of genetic systems for such organisms (15,16). Fundamental studies on the genetic and biochemical basis for biopolymer formation by lactic bacteria in food fermentation processes is an underdeveloped area of science and technology.

Cellulose

Cellulose is a linear, β(1-4) linked glucose homopolysaccharide. Multiple chains associate by intermolecular hydrogen bonding resulting in a water-insoluble crystalline microfibril. Cellulose is found predominantly in plants and trees but also occurs as a microbial product. Cellulose for industrial use is isolated mainly from trees, cotton and flax, and is always associated with a variety of other polysaccharides such as starch, pectin, lignin and hemicellulose.

Industrial applications include the use of cellulose in encapsulation of drugs for delivery systems or immobilized cells for use as catalysts. Cellulose fibers are used in paper manufacturing. Derivatives of cellulose are used extensively and include DEAE-cellulose for enzyme purification (17), carboxymethylcellulose, methylcellulose, hydroxyethyl cellulose and microcrystalline cellulose (Avicel[R]) for use in foods and feeds, cosmetics, paper, textiles, petroleum products, paint, pharmaceuticals, detergents, adhesives and sealants.

The production of cellulose by bacteria offers an interesting alternative to its traditional source. The bacterium *Acetobacter xylinum* produces extracellular cellulose microfibrils that are devoid of any other contaminating polysaccharides. The crystallinity of the cellulose can be controlled reversibly (18) by treatment with the dye Cellufluor (Polysciences Chemicals) which does not interfere with the polymerization or synthesis of the chains. Other strategies for altering and controlling cellulose synthesis can be envisioned utilizing genetic engineering technologies. Thus, the application of biotechnology principles (many of which have to be established) to the microbial production of cellulose could result in new and improved structures that are produced in a relatively pure form from an economically attractive process.

Curdlan

Curdlan is a linear β(1-3)-D-glucan produced by *Alcaligenes faecalis*. Yields of 50% on glucose have been reported (19). The commercial significance of curdlan is its ability to form a nonreversible gel upon heating of aqueous solutions. Proposed food applications include use as a gelling agent, thickener or stabilizer (7).

Other Bacterial Exopolysaccharides

Many other bacterial species produce extracellular polysaccharides that are diverse in composition, structure and physical properties. Relatively few have been studied extensively, so there is still a great deal of potential that exists for the development of new commercial polysaccharides. Important questions have to be addressed, however, for investigating new biopolymers for food systems. For example:

1) Is the biochemical pathway understood?
2) Can recombinant DNA technology be applied to the organism?
3) Is the biopolymer produced by a food grade organism?
4) What are the unique structure/function relationships of the biopolymer?
5) What innovation is required to reduce production costs?

Examples of polysaccharides that can be derived from microbial sources that require further study include microbially derived chitosan and dextran, yeast glucans, and exopolysaccharides from gram-negative organisms such as *Zoogloea ramigera*.

BIOSYNTHESIS OF BACTERIAL EXOPOLYSACCHARIDES

The synthesis of bacterial exopolysaccharides, with the exception of dextrans, levans and mutan, involves activated sugars in the form of nucleotide diphosphate intermediates. A possible biosynthetic pathway for a hypothetical glucose and galactose polysaccharide from glucose is shown in Figure 6-5. Glucose is transported into the cell as glucose-6-

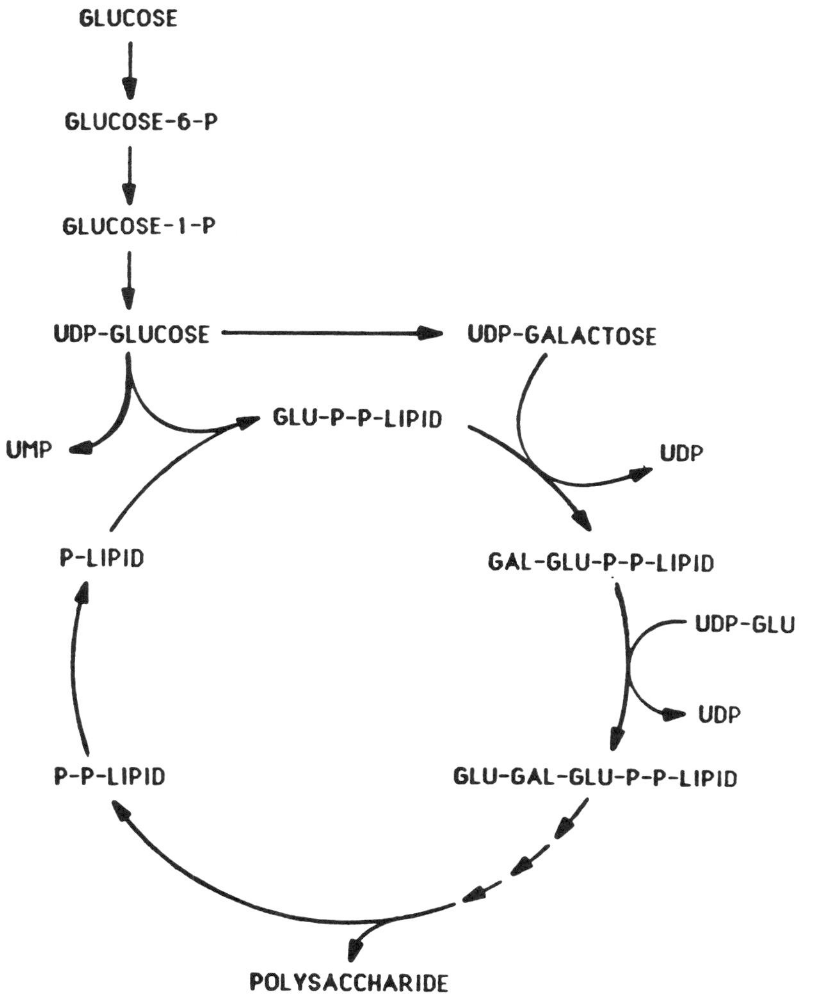

Figure 6-5. Proposed Biosynthetic Pathway for a Glucose and Galactose Exopoly-asccharide in Gram-negative Bacteria

phosphate, which is then converted to glucose-1-phosphate by phosphoglucomutase (20). A sugar nucleotide diphosphate is formed by a nucleotidyl transfer to the sugar phosphate, forming, in this case, UDP-glucose. At this stage, the activated sugar can be interconverted into a variety of sugar nucleotides. Reactions that accomplish these interconversions include epimerization, dehydration and decarboxylation. Almost without exception, the presence of a particular monosaccharide in a polysaccharide requires the existence of that particular monosaccharide nucleotide diphosphate as a precursor. In this example, UDP-glucose is epimerized to UDP-galactose. The reactions leading to polymerization of the sugar nucleotide are varied and not completely understood, however, there is strong evidence supporting the presence of lipid intermediates in some systems (7). Troy et al. (21) proposed the transfer of sugar-1-phosphate to isoprenoid lipid phosphate, followed by the addition of sugars from sugar nucleotides to form the lipid bound oligosaccharide repeat units. Multiples of the repeat unit are then formed by transfer to the reducing end of the growing polysaccharide. Subsequently, the exopolysaccharide must be released so that the isoprenoid lipid can be returned to the pool, and then it must pass through the hydrophobic outer membrane. In gram-negative bacteria, the probable mechanism involves adhesion sites known as Bayer sites in which the outer and inner membranes are associated with each other (7). It is proposed that exopolysaccharides are exported through these channels (22) in a way that is analogous to the import of some substrates through porins (23).

It must be emphasized that the scheme below is a compilation of proposed pathways for several exopolysaccharide producing organisms much of which remains to be proven. One variation on the scheme has been documented in the production of alginate by *Azotobacter vinelandii* in which there is no evidence supporting the involvement of lipid intermediates. In this pathway, GDP-mannuronic acid is polymerized to poly-mannuronic acid by alginic acid polymerase followed by epimerization to alginic acid (9).

NEW GENERATION OF ENGINEERED BIOPOLYMERS

Polysaccharide Structure-Function Relationships

Introduction: The nature of the glycosidic linkage in a polysaccharide chain will determine the relative orientation of adjacent pyranose rings and in turn the rheological properties of the molecule.

Conformational energy maps (24,25) have shown that 90-95% of the possible conformations that two adjacent sugar units can take are forbidden by steric limitations. The degree of conformational freedom allowed for the adjacent sugar units is determined by the glycosidic bond linking them. A polysaccharide contains a large number of these linkages, therefore, the overall shape of the molecule is the result of these individual conformational states. The greater the freedom of movement about a glycosidic oxygen at each linkage, the greater the number of probable conformations an individual chain can assume since the entropic drive will force a chain to adopt a disordered conformation as encountered with random coiling macromolecules (26,27). Most types of glycosidic linkages, however, impose such severe steric limitations that chains tend to adopt ordered shapes which are thermodynamically stabilized by the presence of non-covalent interactions (i.e., hydrogen bonding and ionic interaction). This "cooperative interaction" of non-covalent bonds along extended sequences of the polysaccharide thus outweighs the conformational entropy.

The ability of polysaccharides to form a wide variety of ordered structures in an aqueous environment allows specific structure-function properties which are of paramount importance for their biological function as structural, energy-reserve and lubricating agents. The inherent flexibility of configurations in polysaccharide segments provides the most powerful tool in the engineering of biopolymers with pre-determined properties for specialized industrial applications.

The relationships between structures and properties of polysaccharides are presently only poorly understood. The rational development of biotechnology in polysaccharide engineering requires a detailed knowledge of the chemical structure (composition and sequence of sugar moieties, glycosidic linkages, crosslinking or branching). Also, standardized rheological and chemical techniques need to be implemented in order to establish the criteria for predicting the complex solution properties of polysaccharides. In the past, structural determinations of polysaccharides by methylation analysis and Smith degradation produce variable and unreliable results due to the complex, tertiary configurations of polysaccharide chains preventing quantitative results. It is important to understand the relationship of primary chemical structure to tertiary inter and intramolecular configurations. New techniques such as high field proton and ^{13}C NMR spectroscopy and circular dichroism in conjunction with rheological studies could provide the detailed information on structure-configuration relationships which is essential in the prediction of properties.

The following sections outline specific rheological parameters which describe polymer solution behavior as related to configuration. Rheological studies will provide essential information for the evaluation and prediction of polysaccharide functional properties.

Conformation of Polysaccharide Chains in Solution: The most common property of polysaccharides is their ability to impart high viscosities in solution at low concentrations, which is generally explained by the overall extended conformation of the individual chains. The dimensions of a polymer chain can be defined by the root-mean-square end-to-end (r.m.s.) distance $(\langle r^2 \rangle)^{1/2}$ derived by the random flight theory (27):

$$\langle r^2 \rangle = nL^2 \qquad (1)$$

assuming: 1) Linear not branched chains
2) Succession of n bonds each of length L
3) Bond i+L can make any random direction with respect to the ith bond (limitless flexibility)
4) Bonds have zero thickness

i.e., the probable random flight of the chain can be defined using Gaussian mechanics as applied to the motion of a gas molecule. Now, assumptions 3 and 4 can be revised to account for valence and rotation angle restrictions (steric limitations), and for the fact that inter-residue linkages occupy a defined volume thus imposing further spatial limitations. Therefore, the r.m.s. end-to-end distance is now given by:

$$\langle r^2 \rangle = C_{\infty}\, nL^2 \qquad (2)$$

where, C_{∞} = characteristic ratio and gives a measure of the degree of extension and randomness in the chain conformation.

Therefore, chains that form extended ribbon structures have characteristic ratios in the range 10-100 whereas helical chains are less than 10 (28) as indicated in Table 6-3.

Table 6-3. Characteristic Ratio, C_{∞} for Typical Polysaccharides

Polysaccharide	Linkage	C_{∞}
Pullulan	α(1-6); α(1-4)	5.0
Amylose	α(1-4)	2.5-4.8
Dextran	α(1-6)	<10
Cellulose	β(1-4)	>0
Galactomannan	β(1-4) (α(1-6) branching)	12.6

The measurement of C_{∞} for polysaccharides is hindered by the ability to determine the unperturbed coil dimensions. To do this, the "theta" conditions must be found under which the polymer conformation approaches a random coil behavior. This condition is naturally found by changing temperature in order to disrupt H-bonding or by using poor solvents in which monomer-solvent and monomer-monomer interactions are balanced (27,29,30). These conditions, however, are not experimentally possible. It is possible to characterize C_{∞} by studying a range of samples with different average chain lengths and extrapolating results to zero molecular weight, since interactions between distant residues in the chain become negligible at small chain lengths.

Polysaccharide molecular dimensions in solution can also be characterized by the intrinsic viscosity, $[\eta]$, which represents the hydrodynamic volume of an individual molecule in solution (see Table 6-4). Intrinsic viscosity can be used to determine molecular weight and provides a good index of conformation. Intrinsic viscosity can also be correlated to coil dimensions as in equation (3) or to molecular weight (M) according to equation (4) below:

$$[\eta] = \phi[<r^2>]^{3/2}n^{-1} \qquad (3)$$

$$[\eta] = KM^{\alpha} \qquad (4)$$

The constants ϕ and K are derived empirically and depend on polymer-solvent interactions. The Mark-Houwink exponent, α , provides a reliable index for the conformation of the macromolecule in the solvent of interest.

The Mark-Houwink exponent, α, is between 0.5 and 0.8 for random coil polymers and the overall shape of the hydrodynamic volume is spherical. Mark-Houwink exponents greater than 0.8 reflect an expanded conformation which may be rod-like in shape. Low values of the exponent, significantly less than 0.5, reflect a more globular conformation. An arrangement of the chain segments which results in a globular conformation of polysaccharides can be caused by local ordering which is due either to attractive hydrogen bonding forces or to electrostatic repulsive forces.

Table 6-5 shows the Mark-Houwink exponent (α) for some polysaccharides in solution. Polyelectrolyte polysaccharides such as chitosan in dilute acid in the absence of salts are more compact than other polysaccharides. This is due to local ordering perhaps due to hydrogen bonding or Van der Waal's forces resulting in a "quasi-globular" conformation. However, at high ionic strength and in the presence of concentrated urea, both electrostatic and hydrogen bonding forces are disrupted so that the chitosan conformation resembles that of the more typical random coil. The ability of electrostatically charged polysaccharides to adopt the more compact conformation results from the flexibility of the backbone chain. Alginate and hyaluronate have Mark-Houwink exponents approaching 1.0 reflecting a high degree of chain extension in solution.

For polyelectrolyte polysaccharides, the Mark-Houwink exponents are more dependent on pH and ionic strength because of the covalent backbone structure and, in conjunction with the charge state of the polyion, ultimately determine the hydrodynamic volume. Therefore, the numerous possible modifications of monomer units, for instance monomer units

Table 6-4. Intrinsic Viscosity [η] of Polysaccharides

Polysaccharide	[η] ml/g	Molecular Weight	Solution Conditions
Alginate	1190	150,000	NaCl
Amylopectin	127	90,000,000	
Amylose	81	488,000	0.33 M KCl
	154	1,750,000	
Carboxymethyl amylose	174	1,000,000	0.42 M NaCl
Carboxymethyl-	22400	970,000	H_2O
cellulose	1040	970,000	NaCl
Cellulose	1230	346,000	
Chitosan	800	130,000	pH 2.5, 0.1 M NaCl
	770	130,000	pH 2.5, 0.2 M NaCl
	640	170,000	1% acetic acid, 2.8% NaCl
	470	17,000	1% acetic acid, 2.8% NaCl
	700	170,000	Trifluoroacetic acid
	360	17,000	Trifluoroacetic acid
Guar gum	230	268,000	
	675	850,000	
Gum arabic	12.5	320,000	
	25.4	1,185,000	
Hyaluronic acid	1340	500,000	pH 4.5, 0.1 M NaCl
	850	500,000	pH 4.5, 1.0 M NaCl
Locust bean gum	1000	1,200,000	
Sodium alginate	225	112,000	
	3100	1,550,000	
Sodium carboxymethyl	151	45,000	
Xanthan gum	19000	2,000,000	H_2O
	1900	2,000,000	NaCl

Table 6-5. Mark-Houwink Exponent of Polysaccharides

Polysaccharide	Solvent	α
Amylose	Water	0.5
Chitosan	Trifluoroacetic acid	0.296
Chitosan	Acetic acid 1%, NaCl 2.8%	0.147
Chitosan	Acetic acid 1%, LiCl 2%	0.186
Chitosan	Acetic acid, 0.2 M, NaCl 0.1 M, urea 4 M	0.71
Hyaluronic acid	pH 6, 0.2 M NaCl	0.82
Hyaluronic acid	pH 6, 0.5 M NaCl	0.78
Pectin (low D.E.)	Water	0.8
Sodium alginate	Water	1.00

of the electrostatic charge state, provide an unlimited potential for the engineering of polysaccharides with desired hydrodynamic properties.

Although secondary forces may play a significant role, it is primarily the relative orientation of the sugar residues around the glycosidic linkage which allows rearrangements in the presence of secondary forces (31). As mentioned previously, the bulky nature of the sugar rings constrains the rotation about the glycosidic linkage with the result that polysaccharides tend to be relatively stiff in comparison to synthetic polymers in solution.

The conformation of simple periodic chain sequences as indicated by solid state X-ray diffraction are found to be regular, periodic and helical (26,32,33,34). Distinct ranges of height (h) and the number of monomer units (n) in each turn of the helix can be predicted for different homopolymers as schematically presented in Figure 6-6.

Type A	Type B	Type C	Type D
Extended ribbon	Hollow helix	Crumpled ribbon	Flexible coil
e.g.,	e.g.,	e.g.,	all
β(1→4) glucan	α(1→4) glucan	β(1→2) glucan	α(1,6) and
α(1→3) glucan	β(1→3) glucan	α(1→2) glucan	β(1→6) polymers

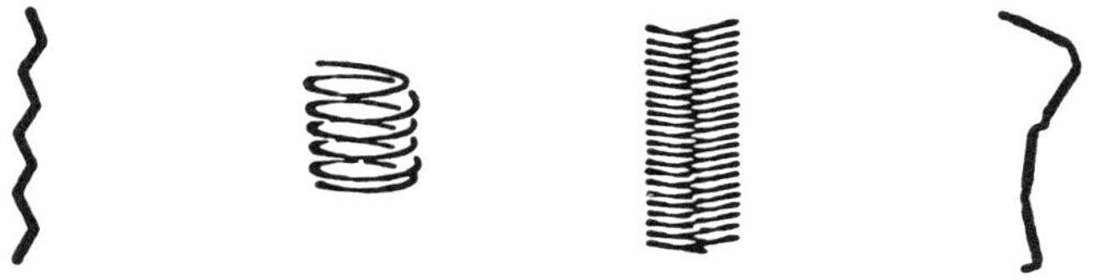

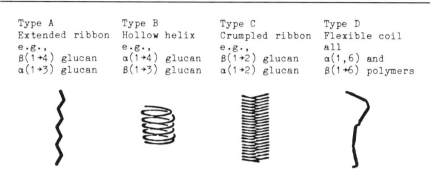

Figure 6-6. Schematic Representation of the Regular Conformation Predicted for Homopolysaccharides by Conformational Analysis (34)

The type A, extended ribbons, have 2-4 n and each sugar is nearly parallel to the helical axis. Cellulose, an β(1-4) glucan has a 'bent-chain' conformation with a 2-fold axis of symmetry, allowing adjacent residues to be H-bonded between 0-3 and 0-5. This chain shape polysaccharide results in dense packaging into insoluble arrays with efficient H-bonding within and between organized layers.

The type B hollow-helix resembles coiled springs in various states of extension to the helix axes. Amylose, an α(1-4) glucan, is of this type and is known to exist in many forms (32,33). Hollow helices normally adopt a random coil conformation in solution to eliminate an empty void along their helical axis (26). The void is filled by the formation of an inclusion complex with a small molecule or alternatively, they combine with another chain to form a double helix. The type C crumpled ribbons are not normally found in nature as repeated sequences of the required linkage type (for example, an α(1-2) glucan) would cause extension clashes between monomer units. Ordered conformations are thus difficult for this chain type. The type D flexible coil polysaccharides in α(1-6) linkages have an extra bond between residues and thus, more freedom of rotation.

Conformational predictions for the highly ordered form in the solid state, as discussed above, cannot be assumed to hold for fully hydrated chains in aqueous solutions. That is because there is an increase in entropic drive to disorder due to the fluctuations of polysaccharides about the polymer backbone. On the other hand, noncovalent interactions such as hydrogen bonding and ionic interactions may counteract to fix the polysaccharides cooperatively into an ordered conformation. Because of these opposing effects, conformational predictions from the solid state may be considered a rational base for the first approximation, but it is extremely difficult to predict the fundamental solution behavior of polysaccharides in solution. It is even more difficult to predict solution behavior when the polysaccharides are not homopolymers of a single linkage type but also contain branching. The frequency and length of the branches further complicate the structure of the molecules. Therefore, experimental data are required for the understanding of the properties of polysaccharides.

The nature of the glycosidic bond between monomeric units is the major determinant of the rheological behavior of polysaccharides. In general, α(1-4) type polymers are less extended than those of an equivalent molecular weight β(1-4) linked polymer. This is exemplified by the lower value of Mark-Houwink exponent for amylose ($\alpha = 0.5$) compared with that for chitosan ($\alpha = 0.71$) when ionic and hydrogen bond effects are suppressed (Table 6-5) The difference in the conformation of β(1-4) and α(1-4) linked polymers can also be represented by the value of the characteristic ratio of amylosic materials ($C_\infty = 5$) and cellulose derivatives ($C_\infty = 35$) (35).

Chain Stiffness: An estimation of chain stiffness can be made using the chain stiffness parameter, B, (36) where equation (5) shows:

$$S = B([\eta]_{I=0.1})^{1.3} \tag{5}$$

S is the slope of log $[\eta]$ versus log $I^{-1/2}$ where I is the ionic strength of the solution. The constant, B, is independent of the charge state and molecular weight of the polysaccharide and thus provides a comparison of the backbone stiffness. The value of B ranges from 0.005 for rigid rods to greater than 0.24 for polyelectrolytes with simple carbon-carbon backbones. Table 6-6 shows values of the chain stiffness parameter for some polysaccharides.

The wide range of chain flexibility makes polysaccharides an especially versatile material because it imparts an enhanced sensitivity to the external solution conditions. In the absence of backbone constraints, it is possible to obtain a wide range of rheological responses, at high amplification, by controlling pH and ionic strength. On the other hand, the rigid backbone would allow mechanical strength or an enhanced rheological effect in solutions.

The values of stiffness parameter "B" indicate that β(1-4) linkages are more rigid than α(1-4) linkages (40,41,42) as is evident from the comparison of, for instance, carboxymethyl cellulose to carboxymethyl amylose

Table 6-6. Chain Stiffness Parameter of Polysaccharides

Polysaccharide	I = 0.1	S	B	Reference
Alginate	5.9	0.42	0.04	36
Carboxymethyl amylose			0.2	36
Carboxymethyl cellulose (DS = 0.73)	8.7	0.85	0.045	36
Chitosan	8.4	1.5	0.1	37
Dextran sulfate	1.27	0.3	0.23	36
Hyaluronic acid	8.1	0.9	0.065	38
Hyaluronic acid	13.4	2.2	0.07	39

(see Table 6-6). These differences in molecular backbone strongly affect the rheological behavior of aqueous solutions of these polymers (43). Thus, based solely on this fact, it can be estimated that a β(1-4) linked polymer will produce a thicker solution than those from an α(1-4) linked polymer consisting of the same monomer of similar molecular weight and concentration.

Modification of Polysaccharide Structure

The functional properties of any polymer are dependent upon its composition and structure. The ability to manipulate these characteristics will facilitate the engineering of biopolymers to meet specific functional criterion. This requires elucidation of the role that is played at the genetic level in determining biopolymer production, composition and structure. Manipulation of the genes and mechanisms controlling polymer synthesis will permit the development of stable microorganisms producing unique biopolymers.

Detailed knowledge of the biosynthetic pathway and its regulation are crucial to the successful manipulation of an exopolysaccharide. The application of classical mutagenesis to change the specific activities or functions of enzymes involved or alter their control mechanisms could lead to the development of new polysaccharides with unique properties. Unfortunately, no methods exist for selection of mutants with altered polysaccharide composition or structure, and one must rely on screening colonies for altered morphology or dye binding properties (e.g., Cellufluor, Polysciences Chemicals).

A more eloquent strategy for manipulating biopolysaccharides involves the isolation of the polysaccharide genes on a plasmid after which study and manipulation is greatly simplified.

In the following sections two research projects in our laboratory are summarized that are currently being applied to the manipulation of microbial polysaccharides. One involves an extracellular bacterial polysaccharide and the other an insoluble glucan matrix derived from yeast. Some of the data presented is in press (44) as well as in preparation for publication (45,46).

Strategies to Control Exopolysaccharide Biosynthesis

The ability to understand and control the structure and function of biopolymers will make it possible to develop unique, well-defined polymers for specific applications. This section summarizes research studies on the polysaccharide of *Zoogloea ramigera*, its production and the genes coding for its production as a model system for exocellular polysaccharide production in gram-negative bacteria. The primary objective of this investigation is the development of a system in which polysaccharide structure and function can be controlled through genetic manipulation of the producing organism.

Introduction: *Z. ramigera* can grow on a wide variety of carbon and nitrogen sources producing an exocellular polysaccharide which causes flocculation of the cells and the ability to accumulate heavy-metal ions. *Z. ramigera* is one of the major contributors to sludge flocculation and is able to substantially decrease the BOD in waste water.

Z. ramigera is a gram-negative, rod-shaped, floc-forming, single polar flagellated, obligate aerobe found in aerobic waste treatment facilities and natural aquatic habitats. They are distinguished from other gram-negative pseudomonads, by the presence of an exocellular polymer which causes flocculation and occurs, in some strains, as a zoogloeal or capsule-like matrix. In nature, the polymer is thought to function by concentrating nutrients around the cell flocs enabling them to grow in nutrient deficient environments. Heavy metal ions are adsorbed by this matrix, and it was found that matrix forming strains have a higher affinity for heavy metal ions compared to nonmatrix forming strains (47).

Z. ramigera isolate 115 is a zoogloeal matrix forming strain, which when grown in a nitrogen limiting medium, converts 60% (w/w) of the available glucose substrate into exopolymer (48) where yields of more than 15 g/L have been reported in batch culture (48,49). The polymer producing cells are able to adsorb Cu^{++} up to 34% of their weight and 25% of their weight of Co^{++}. They also have a high affinity for Fe^{++}, Zn^{++}, and Ni^{++} (47).

Recent studies report that the biopolymer produced by *Z. ramigera* 115 is a highly branched heteropolysaccharide composed of glucose, galactose and pyruvate in a molar ratio of 11:3:1.5 with a molecular weight of approximately 10^5 (Figure 6-7) (50). The main chain of the polymer is proposed to be a glucose and galactose polysaccharide that is predominantly β(1-4) linked, with branching at the galactose units which consists of short β(1-4) linked glucose chains, some of which contain pyruvate moieties. The negatively charged carboxyl groups of the pyruvate are thought to be primarily responsible for the biopolymer's high affinity for heavy metal ions (50). A more recent investigation contradicts this data on several points. Franzen and Norberg (51) found that the polysaccharide contains glucose and galactose in a ratio of approximately 2:1 and a pyruvic acid acetal accounting for 3.6% of the total polysaccharide. Although a complete structure was not proposed, several linkages were determined on the basis of methylation analysis. Some of these linkages are not in agreement with the data of Ikeda et al. (50) and at least one α linkage was proposed.

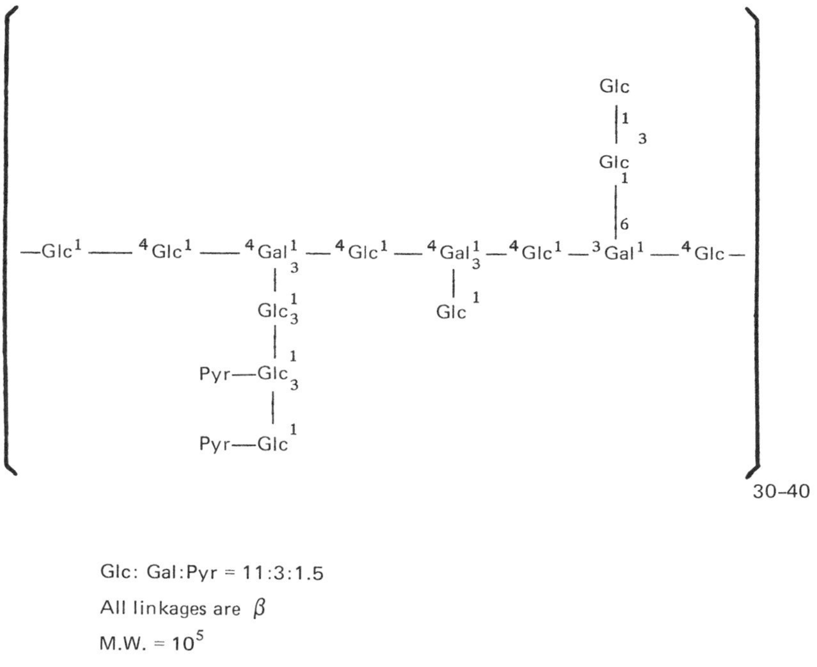

Figure 6-7. Proposed Structure of the Exopolysaccharide of *Z. ramigera* 115 (50)

To date no studies on the genetic manipulations of *Z. ramigera* have been reported. However, genetic manipulations in *Pseudomonas* and *Rhizobium* which are both taxonomically similar to *Zoogloea* and also produce extracellular polysaccharides have been reported utilizing broad host range cloning vectors. Goldberg and Ohman (12) have reported the cloning and expression in *P. aeruginosa* of a gene involved in alginate production. Similarly, Darzins and Chakrabarty (13) have reported the cloning of genes controlling alginate biosynthesis in *P. aeruginosa*. Friedman et al. (52) have reported the construction of a broad host range cloning vector and construction of a *Rhizobium meliloti* gene bank in this vector from which they isolated recombinants able to complement *R. meliloti* auxotrophs.

Presently, attempts are being made to clone and express the *Z. ramigera* genes involved in polysaccharide synthesis in *E. coli* and in *Z. ramigera* non-producing strains, in conjuction with chemical and rheological characterization of the polysaccharide. Some preliminary results are presented below.

Polysaccharide Characterization: Isolation of purified polysaccharide from strain 115 has been achieved at a yield of 1 g/L, the purity defined at 97%, and the monosaccharide composition determined. The results indicate that glucose and galactose are present in a ratio of approximately 2:1 confirming the data of Franzen and Norberg (51) but somewhat different from the 11:3 glucose to galactose ratio reported by Ikeda et al. (50). Franzen and Norberg (51) conditions used or a difference in the *Z. ramigera* strain. Infrared analysis indicated that the polymer is a polysaccharide that contains ionized carboxyl groups. The monosaccharide composition data and the IR "fingerprint" scan of the polysaccharide will be compared in the future to the composition and IR scans of polysaccharides from mutant or genetically manipulated strains to detect changes in composition and structure.

Rheology: The exocellular polysaccharide of *Z. ramigera* 115 exhibits some interesting rheological properties. Figure 6-8 shows the apparent viscosity of the polysaccharide as a function of concentration. At low concentration (0.1 to 0.5%) the polymer solutions show pseudoplastic behavior (Figure 6-9). The polymer is shear thinning when increasing stress is applied and its viscosity returns when the stress is removed. This property is advantageous when pumping solutions is required. At concentrations greater than 1.0%, strain 115 polysaccharide solutions show viscoelastic behavior. These solutions demonstrate the Weissenberg effect and possess a shear stress yield value. Other researchers have shown this polymer to be heat stable, pH stable, salt compatible and able to lower the surface tension of water (53).

Cloning of the Polysaccharide Genes in *E. coli*: A *Z. ramigera* I-16-M, cosmid gene library was transduced into *E. coli* HB101 and screened directly for exopolysaccharide production on the basis of increased fluorescence in the presence of the dye, cellufluor. Several fluorescent candidates were isolated which upon sub-culturing lost the ability to fluoresce. Although the possibility of the *Z. ramigera* polysaccharide genes being expressed in E. *coli* with proper excretion of the polysaccharide seemed remote, the simplicity of the procedure combined with the potential benefits warranted carrying out the experiments. It appears that perhaps some polysaccharide may have been produced at one stage in the growth of the original fluorescent candidates, which later lost this phenotype possibly due to deletions in the cosmids. The following evidence supports this theory but no proof is offered at this stage: 1) the I-16-M gene library transduction was repeated 3 times each time

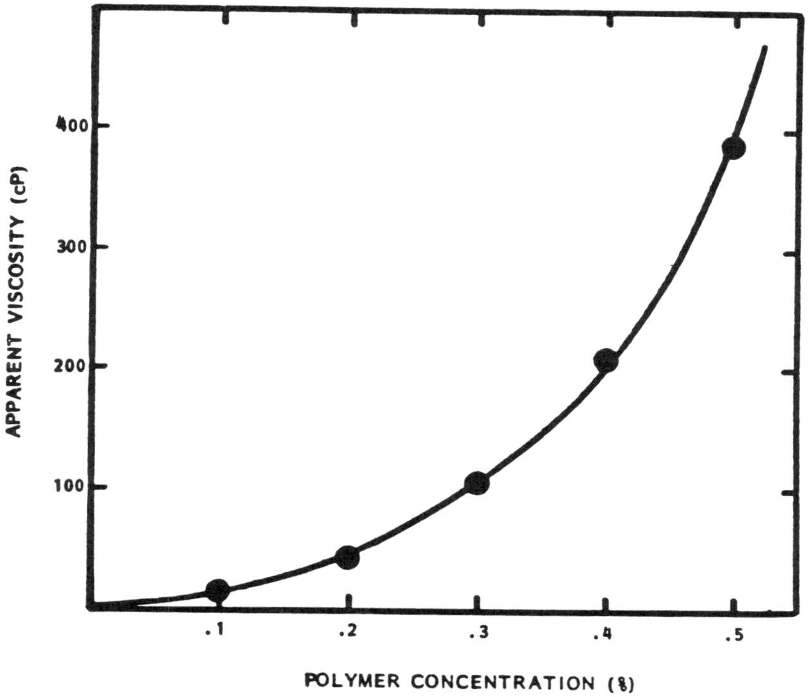

Figure 6-8. Apparent Viscosity as a Function of *Z. ramigera* 115 Exopolysaccharide Concentration. Measured Using a Brookfield LVT Viscometer with UL Adapter at 1.5 rpm (46)

resulting in the same frequency of fluorescent candidates; 2) hybridization experiments show that all of the cosmid inserts of the fluorescent candidates contain large regions of homologous *Z. ramigera* DNA; 3) restriction enzyme fragments of the fluorescent cosmids add up to only 30-35 kb which are too small to be packaged efficiently and could indicate the occurrence of deletions; and, 4) hybridization of the fluorescent cosmids to identically digested *Z. ramigera* I-16-M chromosomal DNA and cosmid DNA show more bands than would be expected in the chromosome that are not present in the cosmids, which again could indicate deletions. Further work is being carried out to prove or disprove this theory and to determine the significance of the fluorescent *E. coli* candidates.

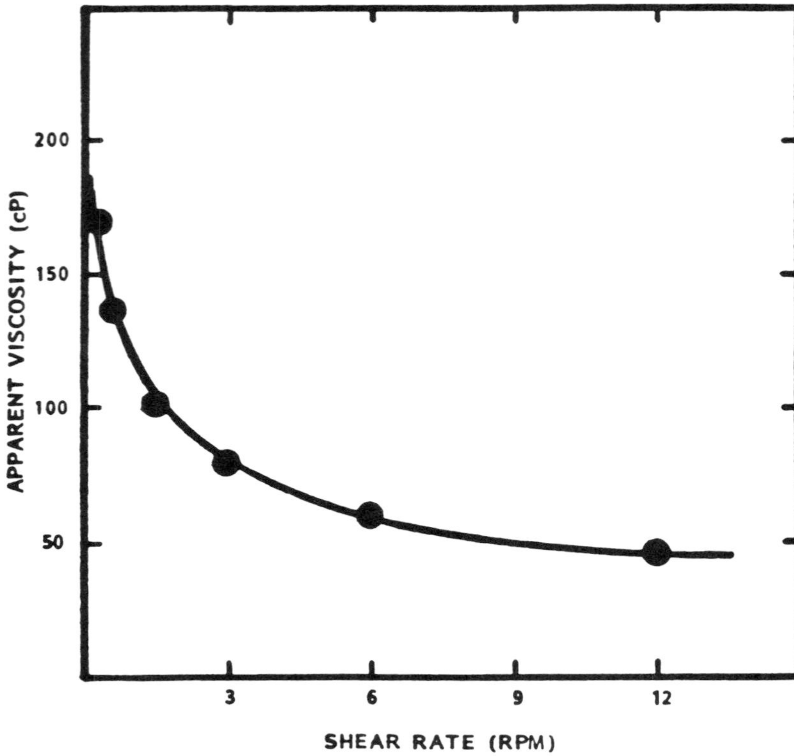

Figure 6-9. Apparent Viscosity of a 0.3% *Z. ramigera* 115 Polysaccharide Solution as a Function of Shear Rate. Measured Using a Brookfield LVT Viscometer with UL Adapter (46)

Cloning of Polysaccharide Genes in *Z. ramigera*: A strategy has been developed for the cloning of polysaccharide genes in *Z. ramigera* or a related organism (e.g., *Pseudomonas*) and is shown in Figure 6-10. To accomplish this successfully, the genes for polysaccharide synthesis must be ligated onto a plasmid which can then be introduced into a *Z. ramigera* nonproducing strain (or related organism), be expressed and be identified by some screening technique. The crucial part of this scheme is the introduction of plasmid DNA into *Z. ramigera*, which has previously never been reported for *Z. ramigera* but has been reported for taxonomically similar organisms. The strategy for cloning in *Z. ramigera* involves the conjugal transfer of a broad host range cloning vector (pCP13) from *E. coli* to *Z. ramigera*; this has been succussfully demonstrated. A *Z. ramigera* gene library is currently being constructed in pCP13, for subsequent transfer into a polysaccharide non-producing strain and direct

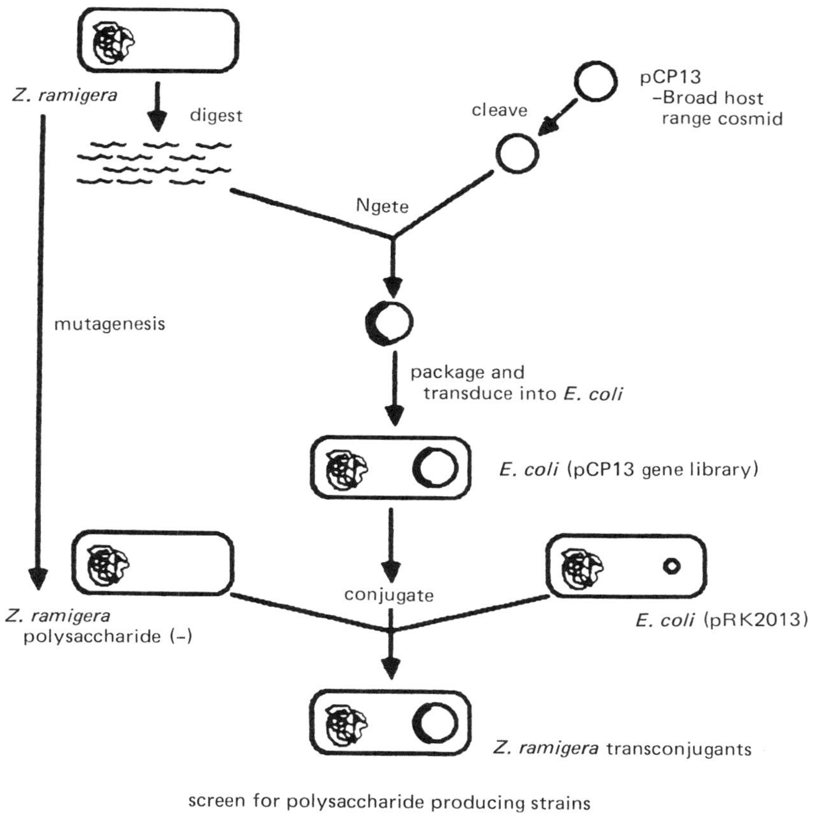

Figure 6-10. Strategy for Isolating the *Z. ramigera* Genes Coding for Exopolysaccharide Production (45)

screening of the transconjugants for polysaccharide production on cellufluor plates. Candidates that have regained polysaccharide production presumably contain the gene or genes involved in its production on the plasmid, which can then be studied and manipulated in a variety of ways.

An alternate strategy that is also being pursued involves the use of transposon mutagenesis. Transposons are self-mobilizing DNA fragments that carry an identifiable gene, e.g., antibiotic resistance. The insertion of a transposon into a gene results in a mutational event. The gene into which the transposon has been inserted is thus tagged with an antibiotic resistance gene. If a gene bank is made from the DNA isolated from a transposon mutagenized cell, the tagged gene can be identified by the antibiotic resistance.

A transposon introduced into *Z. ramigera* on a suicidal conjugative plasmid would be forced to hop into the chromosome causing a mutation. Polysaccharide mutants can be detected by screening for altered colony morphology or cellufluor binding properties. The gene can then be identified in a clone bank made from the DNA of the polysaccharide mutant by screening for the antibiotic resistance gene carried by the transposon. The sequences flanking the transposon is then used as a probe for identifying the intact gene from a bank constructed from the parent strain.

Conclusions: The isolation and characterization of exopolysaccharide biosynthetic genes can open up new opportunities in strain improvement and development. Manipulation of these genes to control polysaccharide structure and function could ultimately lead to the ability to design and produce unique polysaccharide structures for specific functional applications. Such a system could provide a virtually limitless number of novel biopolymers with diverse properties and many potential applications.

Control of the Structure-Function Properties of a Yeast Glucan Matrix

Introduction: The glucose polymer, β-glucan is the most abundant polysaccharide occurring in the cell walls of yeast and comprises approximately 12-14% of the dry cell weight. Glucans provide the structural rigidity of yeast cell walls, hence maintaining the specific morphology and integrity of the cells. The cell wall formed by glucans is insoluble and encapsulates the cell contents. Yeast cell wall glucan is a homopolymer of glucose linked through either β(1-3)-D- or β(1-6)-D-glycosidic bonds.

Glucans are grouped into fractions on the basis of their solubility in alkali (54,55). The alkali soluble fractions are a minor component (15-20% w/w of total glucan) which is of little structural importance to the cell wall. The major glucan component is insoluble in alkali. This fraction is responsible for the structure and integrity of cell walls. They are organized to form membrane networks encapsulating the whole cell. The alkali insoluble glucan fraction consists of a β(1-3) linked backbone of high molecular weight (240,000) containing 3% β(1-6) interchain linkages (56). This fraction maintains the intact morphology of the cells after the extraction. This fact has some important implications. For example, the extracted whole cell wall particles can be a model for the morphology of the viable wild type and mutant cells, as well as the cell wall matrix structure.

As a model system for the engineering of polysaccharides to serve as ultrastructures, the directed biosynthesis of yeast glucans with known structure-function properties are being studied. Yeast glucans have great potential for use in the chemical and food industries. Yeast must historically carry the most prominent role as a food grade and industrial organism, therefore, by-products from the growth of many yeasts will probably be free from government health-related constraints which govern many such new products.

The system of interest is the whole cell wall particulate material derived from yeast, which may have novel functional applications. The whole cell wall particulate material is comparable in shape and size to whole yeast cells and composed almost entirely of the polysaccharide β-glucan. Through chemical or enzymatic treatment, or by genetic modification of the biosynthetic pathway of this material, one can control its hydrodynamic properties. Thus, certain desirable functional properties can be built into these particles. This may allow additional utility of the currently viable processes involving large-scale yeast fermentations. The applications implied in such a system of microsize particles with defined functional properties are, for example, in controlled or sustained release, encapsulation of small molecules such as flavors or oxygen for increased mass transfer in fermentations, non-caloric food thickening, and in support systems for the separation technology (57).

Two major bioengineering considerations strengthen the choice of this system. First, processes involving large scale yeast fermentations (brewing, bread, SCP for animal feed) are being perfected, and manufacturers are becoming interested in additional commercial gains that can be exploited using the biomass which is produced in a fermentation. The second consideration falls under the problem of product recovery, probably the most serious problem affecting the commercialization of most biological products. Current efforts to produce polysaccharides such as xanthan and alginate are focused on the biosynthesis by fermentation and the extraction of the exocellular product from a complex fermentation broth. Hence, the recovery steps initially involve the separation of the biomass from the complex growth medium, and then costly methods need to be applied to recover exocellular polysaccharides from dilute solution. In addition, care must be taken in the choice of solvents and conditions so as to avoid product degradation. The production of glucans differs from that of a typical exocellular product as glucans constitute the yeast cell wall. Thus, once the biomass has been harvested from the complex fermentation broth, it is subjected to a vigorous alkali extraction in which all the intracellular components such as proteins and nucleic acids, and the minor constituents of the cell wall (mannoproteins, chitin) are degraded and solubilized yielding an easily separated pure glucan residue. In other words, this system employs the recovery of the product from a relatively concentrated starting material and in a fraction (approximately 1/500th) of the total fermentation volume.

Yeasts are the best characterized eukaryotic organisms in genetic terms, and procedures for the introduction of specific genetic changes using recombinant DNA technology have been established. The ability to modify the structure of yeast glucans, and therefore, their physical properties would greatly enhance potential commercial applications. This work therefore sets out to define the basic principles in polymer engineering as applied to yeast glucans. The fulfilment of the proposed objectives will require close interrelation of physical property characterization through rheology, genetic characterization through classical and recombinant DNA techniques and chemical characterization. The extent to which each of the above areas is exploited must always be kept in the perspective defined by the objectives - the design of biopolymers for specific functions.

Hydrodynamic and Functional Properties of Glucans: Glucans are polymers of glucose linked by β(1-3) and β(1-6) linkages. Hollow helix and flexible coil conformations are predicted for β(1-3) and β(1-6) homopolyers, respectively, in the solid state as discussed in earlier sections.

The helical conformation of β(1-3) glucan segments persists to varying degrees in solution (depending on the frequency of β(1-6) branching), and this allows unique hydrodynamic properties resulting in viscosity tolerance to a broad range of conditions such as temperature, pH and salt. This high performance resulting from the unique intercomplexing of glucan chains has never been achieved with synthetic macromolecules.

We can predict that glucans with β(1-3) and β(1-6) linkages between glucose units may have unique rheological properties and, hence, effective functional properties. With the main backbone of β(1-3) type, the conformation in solution may well be organized in a spring like coil at one extreme or a relatively rigid rod similar to carboxymethyl cellulose β(1-4) type and hyaluronic acid β(1-3) when the order is disrupted by solvation and secondary interactive forces. This wide range of possible conformations may be further complicated by the β(1-6) lateral side chains. The higher frequency and longer length of the side chains would disrupt the helical conformation and thus make it more likely random coils or free chains. Xanthan is a good example of a polysaccharide with regular side chains which have a stiffening effect on backbone with a high intrinsic viscosity value (Table 6-3) and a high Mark-Houwink constant probably close to 1.8, more like a rigid rod than any other polysaccharides known. However, to what extent the side chains have back-bone stiffening or randomizing effects, and what types of side chains are predominantly responsible for these effects needs to be determined experimentally.

The nature of the side chains will determine the solution behavior of the glucan and will play an important role in the structure stabilization processes. This is one of the main features to be investigated in the proposed study. The effect of non-ionic interactions between neighboring segments of the polymer or the solution behavior will also be examined.

A glucan which contains only β(1-3) linkages, curdlan, is insoluble in cold water, but forms gels in aqueous suspensions upon heating (58). Homopolymeric sequences of β(1-3)-D-glucosyl units will give rise to chains of the hollow helix type because of the "u turn" relationship between the bridging oxygen atoms on each sugur unit. The gelling capability of curdlan is attributed to an ordered structure of a triple helix (59). Curdlan is soluble in dilute alkalies but forms a gel when dialyzed against water (60). The transition from an ordered form to a random coil occurs with an increase in alkali concentration at 0.2 M NaOH (60). The ordered form does not seem to occur for degree of polymerization lower than 25 (61).

With only structural information for β(1-3) and β(1-6)-D-glycosidic linkages, rough predictions for glucan properties may be made as follows:

1) The glucan may be an efficient thickener (viscosifier). The intrinsic viscosity may be higher than 1000 ml/g in aqueous solution for M.W. 250,000.
2) The glucan may have a flexible coil like conformation. The Mark-Houwink exponent (α) would probably be higher than 0.7 and the stiffness parameter (B) higher than 0.03.
3) Glucan may have a relatively high yield value as a result of both intermolecular association and branching.
4) Glucan is likely to be highly shear thinning due to orientation of the macromolecules under shear.
5) The non-ionic nature of glucan may endow mechanical strength and stability at a wide range of pH and/or salt concentrations.

Rheological Analysis - Establishing Structure-Function Properties: The first stage in this work involves identification and description of a glucan matrix in yeast cell walls via rheological characterization of whole cell wall glucan particles derived from the wild type strain *S. cerevisiae* A364A, and the cell division cycle (*cdc*) mutants 374 and 377. The capillary viscosity data as a function of concentration for glucan particle suspensions were evaluated using equation (6):

$$\frac{1}{C} = \frac{Jv}{\ln(\eta_r)} + v/\phi_m = \frac{k_1}{\ln(\eta_r)} + k_2 \qquad (6)$$

where:
C = concentration of suspended particles
η_r = relative viscosity
J = shape factor
v = hydrodynamic volume
ϕ_m = maximum packing fraction
k_1, k_2 = constants

This equation allows the quantitative determination of the hydrodynamic volume and maximum packing fraction of the whole cell wall glucan particles in an aqueous suspension, thus explaining the gross structural differences in glucans from different strains and upon chemical and enzymatic modifications.

The extraction of whole cell wall glucan particles with 0.5 M acetic acid solubilizes the highly branched β(1-6) glucan component. Balances indicated that the solubilized component comprises 3-4% w/w of the initial whole glucan. After this extraction, the thickening properties of glucans from *cdc* mutants 374 and 377 were greatly enhanced as seen in Figure 6-11. The effect of this extraction on the hydrodynamic parameters of the glucan particles was used to determine the role of β(1-6) crosslinking on the rheological properties as seen in Table 6-7.

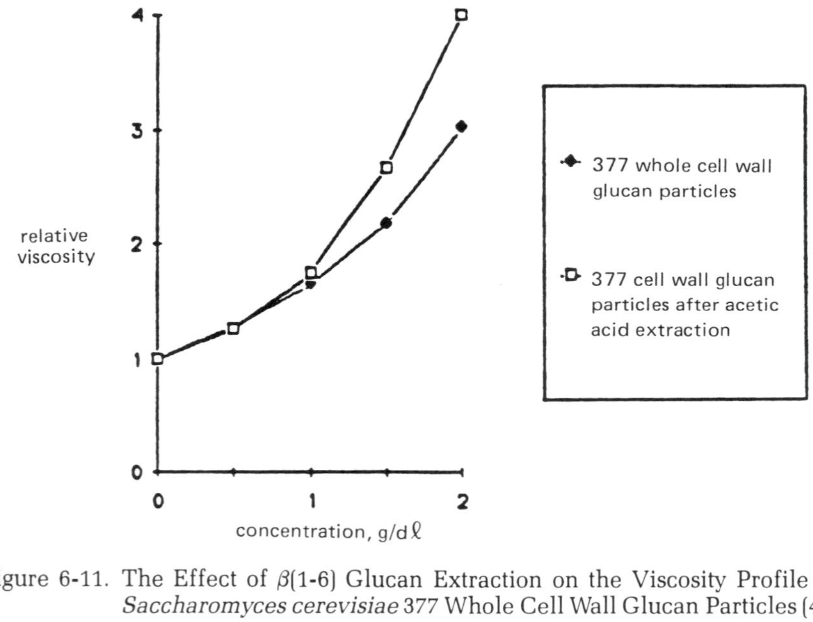

Figure 6-11. The Effect of β(1-6) Glucan Extraction on the Viscosity Profile of *Saccharomyces cerevisiae* 377 Whole Cell Wall Glucan Particles (44)

Figure 6-12 illustrates the effect of a 4 hour laminarinase digest on whole glucan particle suspensions. The profile for A364A whole cell wall glucan particles is compared with the profile of the *cdc* mutant, 377. Laminarinase has endo-β(1-3) glucanase activity, therefore, breakdown of the glucan particles is expected. The viscosity profiles indicate that for both strains, the thickening properties of the glucan suspensions had decreased after digestion. It is clear from these results that laminarinase has a more prono ghost cells. In other words, the β(1-3) glucan backbones in sample are more susceptible to lysis by the enzyme.

Analysis of these results using the linear model indicates that the constant k_2 (= v/ϕ_m) remains unaffected after digestion and the constant k_1

Table 6-7. Hydrodynamic Properties of Whole Cell Wall Glucan Matrix Particles After Extraction (44)

Whole cell wall matrix particles of *S. cerevisiae* strain	Shape Factor J	Hydrodynamic Volume v(dl/g)	Max. Packing Fraction ϕ_m	%Δv
A364A NaOH extracted	2.5	0.092	0.63	−2
A364A NaOH and acetic acid extracted	2.5	0.090	0.61	
374 NaOH extracted	4.1	0.088	0.60	+17
374 NaOH and acetic acid extracted	4.1	0.103	0.44	
377 NaOH extracted	4.1	0.091	0.46	+13
377 NaOH and acetic acid extracted	4.1	0.103	0.45	

(= Jv) is directly related to digestion time. Since the shape factor, J, as determined by microscopy is 2.5, which is the minimum value, the reduction in the thickening properties could only be due to a decrease in the hydrodynamic volume.

Resistance of the glucan matrix to degradation by laminarinase must have a proportional relationship to the degree of β(1-6) glucosidic cross-linking as seen in Table 6-8. This has been confirmed by end-group analysis and ^{13}C-NMR. The results from the viscometry study support this hypothesis. Analysis of the effect of the acetic acid extraction using the rheological model attributed it to an increase in the hydrodynamic volume of the glucan particles (see Table 6-7). Accompanied by infra-red spectra, this indicated that the removal of the β(1-6) component resulted in the occlusion of more water molecules within the glucan matrix, the β(1-3) chains of which could now be partially solvated or expanded.

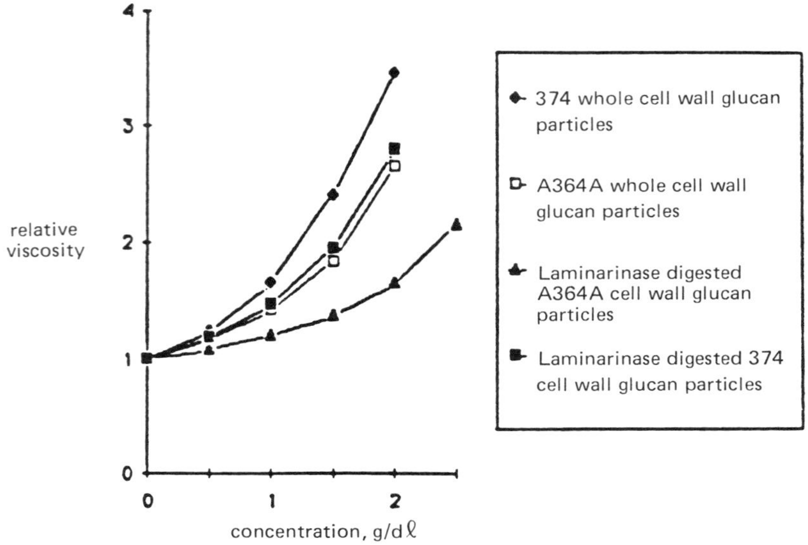

Figure 6-12. Whole Cell Wall Glucan Particles Digested with Laminarinase (44)

Table 6-8. Hydrodynamic Volume (dl/g) of Whole Cell Wall Glucan Matrix Particles After Laminarinase Digest (44)

S. cerevisiae Strain	Before Enzyme Digest	After Enzyme Digest	% Decrease
A364A	0.106	0.057	46
374	0.088	0.070	20
377	0.091	0.065	28
377 (AA)*	0.091	0.036	60

*Laminarinase digestion is preceded by acetic acid extraction.

To illustrate and prove that the β(1-6) component is responsible for the above observations, whole glucan particles from strain 377 were first extracted with hot acetic acid and then digested with laminarinase. Figure 6-13 clearly illustrates that the glucan matrix is hydrolyzed to a considerably greater degree and the viscosity profile of this sample is now significantly altered. Laminarinase digestion led to a 28% decrease in the hydrodynamic volume of glucan whole cell wall matrix particles. However, when acetic acid extraction preceeded the enzyme digestion, the decrease in hydrodynamic volume was 60%.

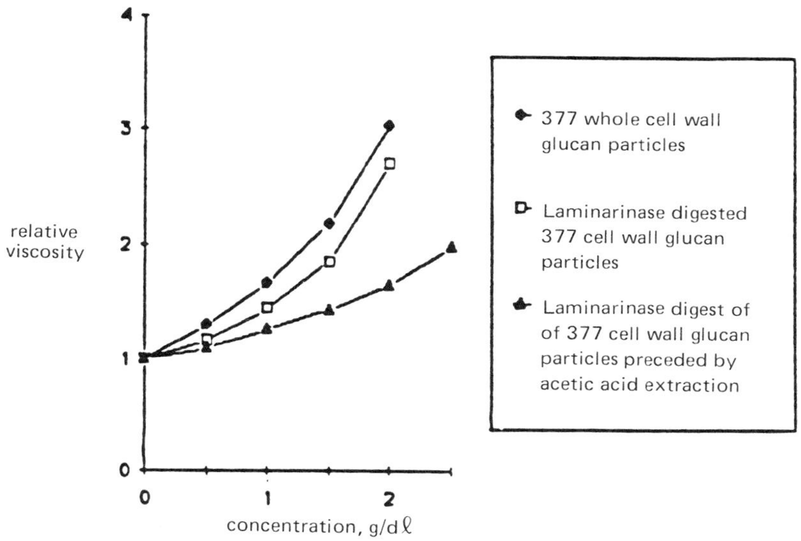

Figure 6-13. The Effect of β(1-6) Crosslinking on the Hydrolysis of Whole Cell Wall Glucan Particles by Laminarinase (44)

Isolation of Glucan Mutants: Rationale for the isolation of mutants with an altered β-glucan structure was derived from results of the rheological analysis. Digestion of the insoluble glucan particles with the enzyme laminarinase showed that glucan from the strains *Saccharomyces cerevisiae* 374 and 377 was significantly more resistant to the digestion. This was attributed to a higher degree of β(1-6)-D-glucosidic crosslinking. The structural component in the yeast cell wall is the β(1-3)-linked backbone chain, therefore hydrolysis of these linkages will break down the matrix. However, linkages adjacent to β(1-6) crosslinks or branches will not be susceptible to enzymatic hydrolysis. Laminarinase does have contaminating β(1-6) glucanase activities; however, breakdown of the β-glucan matrix structure will only be produced by disruption of the structure bearing backbone chains. This implied that 374 and 377 cells would be more resistant to osmotic lysis than A364A cells when their cell walls are digested with lamarininase.

This hypothesis was tested by comparing the viability of A364A and 377 cells under controlled conditions of growth, laminarinase digestion and regeneration. Strain 377 typically showed 10-100 fold higher resistance to laminarinase digestion than A364A indicating that it would be possible to utilize this approach as an enrichment procedure to differentiate cells with a higher frequency of β(1-6)-D-glycosidic linkages in their cell walls.

The criterion for resistance (R) was measured as the ratio of viable cells (V) after enzymatic digestion to the total initial cell count (T).

UV irradiated cells of *S. cerevisiae* A364A were taken through three consecutive laminarinase digestion procedures in order to isolate mutants with an increased β(1-6)-glucan content. Two mutants, R3 and R4, were isolated.

Table 6-9 indicates that mutants R3 and R4 have approximately 6000 and 16000 fold increased resistance to laminarinase over the strain A364A. It must be noted that R3 and R4 are not morphological mutants, and it is therefore not likely that they are *cdc* mutants. Resistance to the enzyme laminarinase, however, could also be attributed to an altered mannan envelope since mannan is located on the exterior of the yeast cell wall. Rheological analyses have been used to quantitate the hydrodynamic properties of the insoluble glucan fraction from mutant R4 and are discussed later.

Table 6-9. Laminarinase Resistance of Glucan Mutants (44)

	A364A	R3	R4
Initial Count (T) (cfu/ml)	5.1×10^{6}	9.8×10^{5}	7.4×10^{5}
Viable Count (V) (cfu/ml)	45	5.5×10^{4}	1.0×10^{5}
Viable Cells/Total Cells	8.8×10^{-6}	5.6×10^{-2}	1.4×10^{-1}

Complementation and Tetrad Analysis: Mutant R4 has been used in the subsequent genetic studies. Diploid strains were constructed with A364A and R4 and their resistance to laminarinase was determined using the standard procedure (Table 6-10).

All the α haploid strains with the exception of STX27-1C-1A have significantly lower resistance than R4. Therefore, the effect of the mutant allele in the diploid strains could be distinguished. The controls in this experiment are provided by the diploids constructed with the parental strain A364A. These diploids have low resistance levels comparable to that of A364A. The diploids constructed with the mutant R4, however, all show increased resistance levels similar to that of R4. In other words, the mutant allele in R4, responsible for the altered β-glucan structure, is dominant in the diploids and thus provides a selectible phenotype for this mutation.

With the knowledge that the mutant allele is dominant, a tetrad analysis was performed on the meiotic products from the diploid R4 x A2 which exhibited the highest resistance level. Within all the tetrads dissected, the resistance phenotype segregated 2:2 indicating that a single mutated gene was responsible for the altered glucan structure.

Table 6-10. Comparison of Resistance to Laminarinase (44)

Haploid Strains	Resistance V/T*
A364A	2.4×10^{-5}
R4	0.35
STX27-1C-1A	0.22
JX121	6.9×10^{-3}
A2	3.9×10^{-3}
Diploid Strains	
A364A x STX27-1C-1A	6.5×10^{-3}
A364A x JX121	6.8×10^{-4}
A364A x A2	7.9×10^{-3}
R4 x STX27-1C-1A	0.35
R4 x JX121	0.30
R4 x A2	0.454

*Resistance = viable cells/total cells

Cloning Strategies for β-Glucan Biosynthesis-Related Genes: The isolation of mutants with lesions in all the appropriate loci makes the strategy for isolation of the genetic loci straightforward.

DNA isolated from wild-type yeasts would be shotgun cloned into *E. coli*:yeast shuttle vectors, and then transformed into the glucan mutant strain by using standard techniques (62). Transformants containing the appropriate cloned gene can be selected by reversion from the mutant to the wild-type phenotype. Alternatively, DNA isolated from the mutant strains with increased resistance to a particular β-glucanase, for example, could be cloned into *E. coli*:yeast shuttle vectors and transformed into a wild-type strain. Transformants would then be selected by increased resistance to the β-glucanase relative to the wild-type level. Appropriate controls must be performed to ensure that the altered phenotype of the selected transformants is due to the presence of the desired gene and not to some other artifactual event.

An alternative cloning strategy is available through exploitation of the long half-life of mRNA's encoding glucan synthetases. By using temperature sensitive *rna*1 mutations (63), the transport of mRNA to the cytoplasm can be switched off at the beginning of the experiment. Growth would be permitted to continue for a predetermined time to allow mRNA's with short half-lives to be cleared from the cytoplasm, while mRNA's with long half-lives would survive. At this point, cytoplasmic mRNA would be isolated and radioactive complementary DNA synthesized by using the mRNA as a template. The radiolabelled cDNA can then be used as a probe in colony hybridizations to select clones from a yeast genomic library which contain the genes that correspond to mRNA's with long half-lives. A similar strategy involving differential screening with cDNA probes has been used to isolate the yeast heat shock genes (64).

The cloning of β-glucan biosynthesis-related genes on plasmid vectors which are maintained at high copy number in the host yeast cell will result in overproduction of the corresponding gene products. This in itself may influence the overall activity at the plasma membrane and hence influence the structure of the β-glucan synthesized. For example, cloning of the β(1-6)-glucan synthetase on a multicopy vector in yeast may result in an increased frequency of β(1-6)-branching in the β(1-3)-glucan. However, overproduction of these gene products may be tightly regulated or be lethal to the transformed cell, hence it may be necessary to use single copy vectors such as yeast replicon plasmids.

Rheological Characterization of Altered Glucan Matrix: In order to determine if the mutation is associated with glucan structure, viscosity measurements on the suspensions of R4 glucan particles were made. The results were analyzed using the linear model (Table 6-11).

Table 6-11. Hydrodynamic Properties of R4 Whole Cell Wall Glucan Particles (44)

Whole Cell Wall Matrix Particles of *Saccharomyces cerevisiae* R4	Shape Factor J	Hydrodynamic Volume v(dl/g)	Maximum Packing Fraction ϕ_m	%Δv
NaOH extracted	2.5	0.087	0.66	
NaOH and acetic acid extracted	2.5	0.103	0.44	+18
NaOH extracted and laminarinase digested	2.5	0.071	0.54	−18

It is interesting to note that the viscosity profile of R4 whole glucan particles is almost identical to that of A364A whole glucan. After the acetic acid extraction, however, the R4 glucan particles have an enhanced thickening behavior due to an increased hydrodynamic volume. Furthermore, laminarinase digestion of R4 glucan particles does not break down the network structure to the extent of A364A glucan. The hydrodynamic volume of the R4 glucan particles after the digestion decreased by 18% compared to 46% for the A364A sample.

These results support the hypothesis that the degree of β(1-6)-D-glycosidic linkages are a key element in controlling the hydrodynamic properties and morphology of yeast whole cell wall glucan matrix particles. The increase in thickening properties of R4 glucan particles after the acetic acid extraction attributed to the absorption of additional water by the glucan network. The mechanism for the increase in hydrodynamic volume may be the increased hydration or expansion of the glucan matrix resulting from the removal of the β(1-6)-D-glycosidic linkages which reduces the spacial constraint of the linear β(1-3) linked chains. This transition would be accompanied by a decrease in the storage modulus, G', of the glucan particles after the acetic acid extraction. In other words, the rigidity of the glucan particles should decrease. Therefore, the compression modulus of suspensions of glucan particles from strains A364A and R4 was determined in order to establish the effect of β(1-6) crosslinking on the rigidity of the glucan network structure.

Under two different centrifugation conditions the network-compression modulus increases rapidly in a critical volume fraction region and appears to approach an asymptotic value (Figures 6-14). In fact, if the critical volume fraction values are converted to concentration of the glucan suspensions using the relationship ϕ = Cv (v = hydrodynamic volume) we observed that the values obtained represent the critical concentration at which the viscosity of the suspenions increases asymptotically. Hence, the critical concentration region represents a transition from a dispersion consisting of discrete glucan particles to a solidly packed state.

The first important outcome of this analysis is the considerable difference in the compression modulus values between the wild-type A364A glucan and the mutant R4 glucan. Under both conditions the mutant R4 exhibits at least a 10 fold higher compression modulus than the wild type glucan. These results alone tie in very closely with the rheological characterization of the glucans. The mutant R4 was isolated on the basis of the higher β(1-6) component which makes the glucan envelope resistant to the lytic enzyme laminarinase. Furthermore, acetic acid extraction resulted in an increase in the thickening properties of R4 glucan which we have explained in detail in the preceeding section. A higher β(1-6) component would be expected to increase the rigidity of the glucan matrix, and this is exactly what is reflected in the compression modulus values.

The second outcome is the effect that is observed for acetic acid extracted glucans. In our previous rheological analysis we argued that the extraction of the β(1-6) crosslinked component allowed the backbone chains in the matrix to be less constrained and simultaneously occlude more water. Our results show that the compression modulus, k(ϕ), decreases after the extraction, thus supporting the proposed mechanism for absorption of water into the β(1-3) glucan matrix.

Conclusions of Work on Engineered Yeast Glucans: In this study we introduced a new approach to identify, describe, and control the morphology and matrix structure of yeast cell walls.

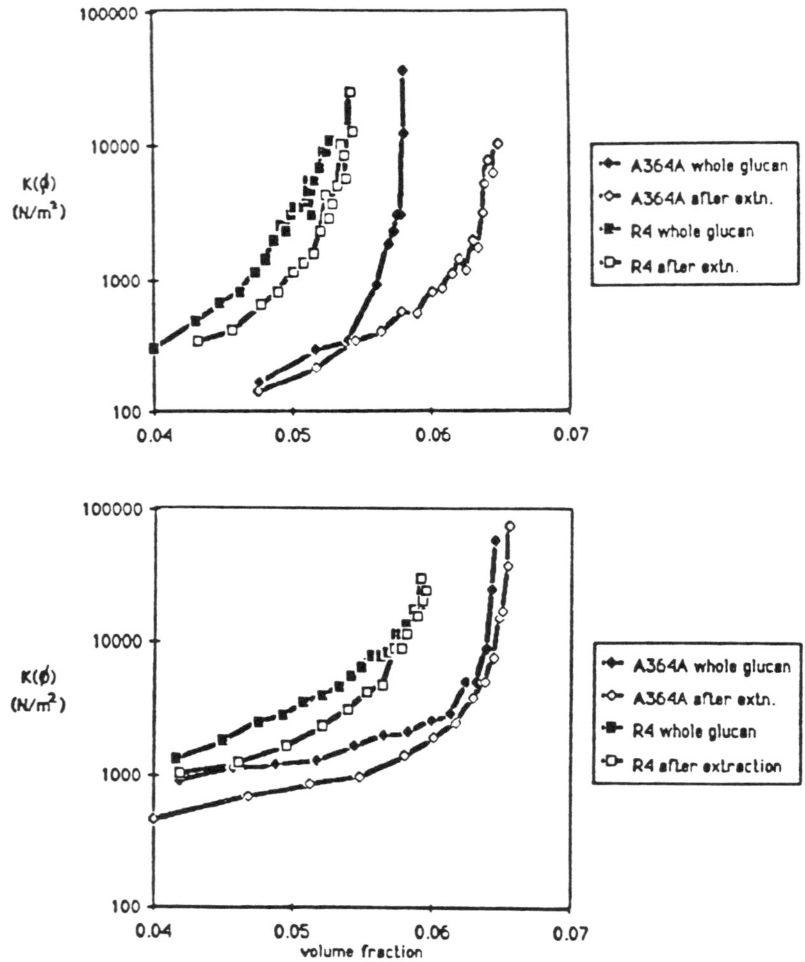

Figure 6-14. Compression Modulus, k(ϕ), versus Volume Fraction for Whole Cell Wall Glucan Suspensions (a. Low Centrifugal Field; b. High Centrifugal Field) (44)

The following conclusions resulted:

1) The amount of β(1-6)-D-glycosidic linkages in the cell wall glucan distinctly affect the morphology and matrix structure of yeast cell walls.

2) Acetic acid extraction of the β(1-6) component from cell walls increases their hydrodynamic volume and hence, enhances the thickening properties.

3) Strains of *Saccharomyces cerevisiae* with a high degree of β(1-6) glycosidic crosslinking in their cell wall glucan exhibit an increased resistance to the enzyme laminarinase.
4) The degree of β(1-6) glycosidic crosslinking can be used as the mechanism to control the extent of hydrolysis of β(1-3) glucan chains in the cell wall matrix by laminarinase. Hence, the morphology and matrix can be manipulated to yield desired rheological properties of the suspensions of glucan cell wall particles with unique physical-chemical properties.
5) Hydrodynamic analysis provides parameters of shape and volume which can quantitatively characterize the cell morphology and cell wall matrix structure.
6) The hydrodynamic analysis can be used to provide the criteria to isolate mutants with an altered cell wall glucan structure.
7) A genetic analysis shows that the altered glucan phenotype which is due to a single gene is selectable and introduction of this lesion into different hosts may be used to genetically modify the cell wall glucan.
8) The cell wall matrix particles obtained from *Saccharomyces cerevisiae* can be used to represent the morphology and structure of the original cell wall and have functional applications as particles of uniform size and shape.

The employment of rheology and enzymes with specific hydrolytic activities will be essential in characterizing the complex nature of polysaccharide matrices which cannot be achieved through chemistry alone. Furthermore, these enzymes can be used to provide the desired properties by cleaving junction points in a matrix to disrupt the ordered conformation of the polysaccharide chains thus controlling the hydrodynamic properties.

The potential utility of polysaccharide matrix structures in the food and high technology specialty industries has not been realized. The low toxicity and food-grade qualities of yeast glucans can be employed in the food industry for flavor entrapment/release and as non-caloric food thickeners. The ability to control the hydrodynamic properties (water content) of these particles can be used to satisfy the qualitative standards in food products such as consistency and texture. We can expect, however, that this technology will have a greater impact in high technology applications of the separations and biomedical industries. The economics of production of high value therapeutic compounds (hormones, enzymes) by chromatography dictate that throughput can be sacrificed for improved resolution. This is directing the separations industry to look at smaller particles for packings for which chemical synthesis is becoming increasingly difficult. Use of micron-sized glucan particles of controlled size and permeability therefore becomes an attractive cost-effective alternative.

Chromatographic resolution of optical isomers using polysaccharides is another area that requires investigation. Numerous publications report the separation of enantiomers on polysaccharide columns. For example, potato starch has been used to resolve chiral biphenyl compounds and amino acid derivatives (65), β- and γ-cyclodextrins have been used for separation of numerous enantiomers (66), and microcrystalline cellulose triacetate is currently used in HPLC separations of racemic mixtures (67,68).

Finally, the non-antigenic nature of yeast glucans in addition to the hydrodynamic properties cited above make this system a very attractive vehicle in controlled drug delivery.

The applications mentioned here merely outline the potential utility for polysaccharide ultrastructures and serve as an insight to the questions which research and development can answer. Strategies similar in principle to the yeast glucan project can be employed to investigate other interesting microbial products such as polysaccharides involved in flocculation and capsule formation.

Conclusions

The applications for polysaccharides can be grouped into three broad market categories. These are, the consumer applications which include food, cosmetics and detergents; the industrial market (paper, textile, petroleum); and the growing biomedical and high technology market.

In 1983 the consumer applications for polysaccharides accounted for approximately 30% of the total market volume or 43% by sales. It is interesting to note that the forecasts for U.S. consumption of polysaccharides in 1988 indicate a slight decrease of the consumer applications market with respect to the industrial market. This is accompanied by an almost stagnant average price associated with their polysaccharide products. It is clear from these observations that these industries have exhausted the utility of currently used polysaccharides. Additional value, which is related to performance, cannot be extracted from the current polysaccharide products and their production processes, hence, their value essentially reflects market trends and variations in production costs (cost of raw materials, availability). The contribution of biotechnology to these markets will not be reflected in market volume but in sales. The ability to design macromolecules with 'added value' is the overall objective. In this paper we have proposed the direction for research and development in this field.

ACKNOWLEDGEMENTS

The work on yeast glucans and *Z. ramigera* was supported by the Naval Air Development Center (Contract N62269-84-R-0253) and the B.F. Goodrich Company, respectively.

The authors thank Ms. Virginia Burr for her help in the preparation of the manuscript.

REFERENCES

1. Jansson, P.E., Keene, L. and Lindberg, B. Structure of the extracellular polysaccharide from *Xanthomonas campestris*. Carbohydr. Res. 45:275-282 (1975)
2. Sutherland, I.W. and Ellwood, D.C. In: "Microbial Technology - Current State and Future Prospects". (A.T. Bull, D.C. Ellwood and C. Ratledge, eds.) Society for General Microbiology Symposium 29, pp. 107-150. University Press, Cambridge (1979)
3. Souw, P. and Demain, A.L. Nutritional studies on xanthan production by *Xanthomonas campestris* NRRL B1459. Appl. Envir. Microbiol. 37:1186-1192 (1979)
4. Evans, C.G.T., Yeo, R.G. and Ellwood, D.C. Continuous culture studies on the production of extracellular polysaccharides by *Xanthomonas juglandis*. In: "Microbial Polysaccharides and Polysaccharases". (Berkley, W.C., Gooday, G.W. and Ellwood, D.C., eds.) Academic Press, New York (1979)
5. Cadmus, M.C., Knutson, C.A., Logoda, A.A., Pittsley, J.E. and Burton, K.A. Synthetic media for production of quality xanthan gum in 20 liter fermentors. Biotechnol. Bioeng. 20:1003-1014 (1978)
6. Daniels, M.J., Barber, C.E., Turner, P.C., Sawczyc, M.K., Byrde, Z.J.W. and Fielding, A.H. Cloning of genes involved in pathogenicity of *Xanthomonas campestris* pv. *campestris* using the broad host range cosmid pLAFR1. The EMBO J. 3:3323-3328 (1984)
7. Sutherland, I.W. Biosynthesis of microbial exopolysaccharides. Adv. Microbiol. Physiol. 24:79-150 (1982)
8. Davidson, I.W., Sutherland, I.W. and Lawson, C.J. Localization of O-acetyl groups of bacterial alginates. J. Gen. Microbiol. 98:603 (1977)
9. Pindar, D. and Bucke, C. The biosynthesis of alginic acid by *Azotobacter vinelandii*. Biochem. J. 152:617-622 (1975)
10. Deavin, L., Jarman, T.R., Lawson, C.J., Righelato, R.C. and Slocombe, S. The production of alginic acid by *Azotobacter vinelandii* in batch and continuous culture. In: "Extracellular Microbial Polysaccharides". pp. 14-26. (Sanford, P.A. and Larkin, A., eds.) American Chemical Society, Washington (1977)
11. Kelco Algin/hydrophilic derivation of alginic acid for scientific water control. 2nd ed. Kelco (Merck and Co., Inc.), Chicago, IL.
12. Goldberg, J.B. and Ohman, D.E. Cloning and expression in *Pseudomonas aeruginosa* of a gene involved in the production of alginate. J. Bacteriol. 158:1115-1121 (1984)
13. Darzins, A. and Chakrabarty, A.M. Cloning of genes controlling alginate biosynthesis from a mucoid cystic fibrosis isolate of *Pseudomonas aeruginosa*. J. Bacteriol. 159:9-18 (1984)
14. Darzins, A., Nixon, L.L., Vanags, R.I. and Chakrabarty, A.M. Cloning of *Escherichia coli* and *Pseudomonas aeruginosa* phosphomannose isomerase genes and their expression in alginate-negative mutants of *Pseudomonas aeruginosa*. J. Bacteriol. 161:249-257 (1985)
15. Robeson, J.P., Barletta, R.G. and Curtiss, R., III. Expression of a *Streptococcus mutans* glucosyltransferase gene in *Escherichia coli*. J. Bacteriol. 153:211-221 (1983)

16. Pucci, M.J. and Macrina, F.L. Cloned *gtf*A gene of *Streptococcus mutans* LM7 alters glucan synthesis in *Streptococcus sanguis*. Infect. Immu. 48: 704-712 (1985)
17. Srivastava, R.A.K., Nigan, J.N., Pillai, K.R. and Barauk, J.N. Purification, properties and regulation of amylase produced by a thermophilic *Bacillus* sp. Indian J. Exp. Biol. 18:972-976 (1980)
18. Haigler, C.H., Brown, R.M., Jr. and Benziman, M. Calcofluor white ST alters the *in vivo* assembly of cellulose microfibrils. Science 210:903-906 (1980)
19. Harada, T. In: "Microbial Extracellular Polysaccharides". (P.A. Sandford and A. Larkin, eds.) pp. 265-283. American Chemical Society, Washington (1977)
20. Metzler, D. "Biochemistry: The Chemical Reactions of Living Cells". p. 1129. Academic Press, New York (1977)
21. Troy, F.A., Frerman, F.E. and Heath, E.C. The biosynthesis of capsular polysaccharide in *Aerobacter aerogenes*. J. Biol. Chem. 246:118-133 (1971)
22. Bayer, M. and Thurow, H. Polysaccharide capsule of *Escherichia coli*: Microscope study of its size, structure and sites of synthesis. J. Bacteriol. 130:911-936 (1977)
23. Smit, J. and Nikaido, H. Outer membrane of gram-negative bacteria. XVIII. Electron microscopic studies on porin insertion sites and growth of cell surface of *Salmonella typhimurium*. J. Bacteriol. 135:687 (1978)
24. Anderson, N.S., Campbell, J.W., Harding, M.M., Rees, D.A. and Samuel, J.W.B. X-ray diffraction studies of polysaccharide sulphates: Double helix models for *k*- and *l*-carrageenans. J. Mol. Biol. 45:85-99 (1969)
25. Rees, D.A. and Smith, P.J.C. Polysaccharide conformation. Part. VIII. Test of energy functions by monte-carlo calculations for monosaccharides. J. Chem. Soc. Perkin Trans. 2 pp. 830-835 (1975)
26. Rees, D.A. "Polysaccharide Shapes". 2nd ed., Chapman and Hall, London (1977)
27. Flory, P.J. "Statistical Mechanics of Chain Molecules". Wiley, New York (1969)
28. Aspinall, G.O. "The Polysaccharides". Vol. 1, C5, Academic Press, New York (1982)
29. Flory, P.J. "Principles of Polymer Chemistry". Cornell Univ. Press, Ithaca (1953)
30. Tanford, C. "Physical Chemistry of Macromolecules". Wiley, New York (1961)
31. Mitchell, J.R. "Polysaccharides in Food". (J.M. Blanshard and J.R. Mitchell, eds.) Butterworths, p. 51-72 (1979)
32. Murphy, V.G. and French, A.D. The structure of V amylose dehydrate; a combined X-ray and stereochemical approach. Biopolymers 14:1487-1501 (1975)
33. Wu, H.C. and Sarko, A. The double-helical molecular structure of crystalline β-amylose. Carbohydr. Res. 61:7-25 (1978)
34. Powell, D.A. Structure, solution properties and biological interactions of some microbial extracellular polysacchraides. In: "Microbial Polysaccharides and Polysaccharases". (R.C.W. Berkeley, G.W. Gooday and D.C. Ellwood, eds.) Academic Press, New York (1979)
35. Goebel, C.V., Dimpfl, W.L. and Brant, D.A. The conformational energy of maltose and amylose. Macromolecules 3:644-654 (1970)
36. Smidrod, O. and Haug, A. Estimation of the relative stiffness of the molecular chain in polyelectrolytes from measurements of viscosity at different ionic strengths. Biopolymers 10:1213-1227 (1971)

37. Rodriguez-Sanchez, D. Kienzle-Sterzer, C.A. and Rha, C.K. Intrinsic viscosity of chitosan solutions as affected by ionic strength. In: "Proc. IInd Int. Conf. on Chitin-Chitosan". (S. Hirano and S. Takura, eds.) pp. 30-34 (1982)
38. Cleland, R.L. and Wang, J.L. Ionic polysaccharides. III. Dilute solution properties of hyaluronic acid fractions. Biopolymers 9:799 (1970)
39. Lang, E.R. Rheological analysis of hyaluronic acid and hyaluronic acid/protein solution. Ph.D. Thesis, MIT, Cambridge, MA (1982)
40. Pramanik, A.G. and Choudhury, P.K. Physiochemical and macromolecular properties of sodium amylose xanthate in dilute solution. J. Polym. Sci. Part A-1, 6:1121-1134 (1968)
41. Triredi, H.C. and Patel, R.D. Studies on carboxymethylcellulose: Estimation of the relative stiffness of the polyions. Makromol. Chem. Rapid Commun. 3:317-321 (1982)
42. Robinson-Lang, E., Kienzle-Sterzer, C.A., Rodriguez-Sanchez, D. and Rha, C.K. Rheological behavior of a typical random coil and polyelectrolyte: Chitosan. In: "Proc. IInd. Int. Conf. on Chitin-Chitosan". (S. Hirano and S. Takura, eds.) pp. 34-38 (1982)
43. Davidson, R.L. "Handbook of Water-Soluble Gums and Resins". McGraw Hill, New York (1980)
44. Jamas, S., Rha, C.K. and Sinskey, A.J. Morphology of yeast cell wall as affected by genetic manipulation of β(1-6) glucosidic linkages. Biotech. Bioeng. (In Press)
45. Easson, D.D., Jr., Peoples, O.P., and Sinskey, A.J. Genetic studies on *Zoogloea ramigera* and chemical characterization of a *Z. ramigera* exocellular polysaccharide. (Manuscript in preparation)
46. Easson, D.D., Jr., Pradipasena, P., Sinskey, A.J. and Rha, C.K. Rheological characterization of the exocellular polysaccharide of *Zoogloea ramigera* 115. (Manuscript in preparation)
47. Friedman, B.A. and Dugan, P.R. Concentration and accumulation of metallic ions by the bacterium *Zoogloea*. Dev. Ind. Microbiol. 9:381-388 (1968)
48. Norberg, A.B. and Enfors, S.O. Production of extracellular polysaccharide by *Zoogloea ramigera*. Appl. Env. Microbiol. 44:1231-1237 (1982)
49. Stauffer, K.R. and Leeder, J.G. Extracellular microbial polysaccharide production by fermentation on whey or hydrolyzed whey. J. Food Sci. 43:756-758 (1978)
50. Ikeda, T., Shuto, H., Saito, T., Fukui, T. and Tomita. An extracellular polysaccharide produced by *Zoogloea ramigera* 115. Eur. J. Biochem. 123:437-445 (1982)
51. Franzen, L.E. and Norberg, A.B. Structural investigation of the acidic polysaccharide secreted by *Zoogloea ramigera* 115. Carbohydr. Res. 128:111-117 (1984)
52. Friedman, A.M., Long, S.R., Brown, S.E., Buikema, W.J. and Ausubel, F.M. Construction of a broad host range cosmid cloning vector and its use in the genetic analysis of *Rhizobium* mutants. Gene 18:289-296 (1982)
53. Stauffer, K.R., Leeder, J.G. and Wang, S.S. Characterization of zooglan-115, an exocellular glycan of *Zoogloea ramigera* 115. J. Food Sci. 45:946-952 (1980)
54. Phaff, H.J. Structure and biosynthesis of the yeast cell envelope. In: "The Yeasts". (A.H. Rose and J.S. Harrison, eds.) Academic Press, New York (1971)
55. Sentendreu, R., Victoria-Elorza, M. and Villanuera, J.R. Synthesis of yeast wall glucan. J. Gen. Micro. 90:13-20 (1975)
56. Manners, D.J., Masson, A.J. and Patterson, J.C. The structure of a β(1-3)-D-glucan from yeast cell walls. Biochem. J. 135:19-30 (1973)
57. Jamas, S., Rha, C.K. and Sinskey, A.J. Glucan compositions and process for preparation thereof. U.S. Patent Application (1984)

58. Bluhm, T.L. and Sasko, A. The triple helical structure of lentinan, a linear α(1-3)-D-glucan. Can. J. Chem. 55:293-299 (1977)
59. Marchessault, R.H., Deslander, Y., Ogawa, K. and Sundararajan, P.R. X-ray diffraction data for β(1-3)-D-glucan. Can. J. Chem. 55:300-303 (1977)
60. Harada, T. In: "Extracellular Microbial Polysaccharides". (P.A. Sandford, ed.) Am. Chem. Soc. Symposium Series No. 45, pp. 265-283. Am. Chem. Soc., Washington, D.C. (1977)
61. Ogawa, K., Tsurugi, J. and Watanabe, T. The dependence of the conformation of a β(1-3)-D-glucan on chain length in alkaline solution. Carbohydr. Res. 29:399-405 (1973)
62. Botstein, D. and Davis, R.W. Principles and practice of recombinant DNA research with yeast. In: "Molecular Biology of the Yeast *Saccharomyces*". (J.N. Strattern, E.W. Jones and J.R. Broach, eds.) Cold Spring Harbor Lab., pp. 607-636 (1982)
63. Hutchinson, H.T., Hartwell, L.H. and McLaughlin, C.S. Temperature-sensitive yeast mutant defective in ribonucleic acid production. J. Bacteriol. 99:807-814 (1969)
64. Finkelstein, D.B., Strausberg, S. and McAlistor, L. Alterations of transcription during heat shock of *Saccharomyces cerevisiae*. J. Biol. Chem. 257:8405-8411 (1982)
65. Hess, H., Burger, G. and Musso, H. Complete enantiomer separation by chromatography on potato starch. Angew. Chem. Int. Ed. Engl. 17:612-614 (1978)
66. Armstrong, D.W. Chiral stationary phase for high performance liquid chromatographic separation of enantiomers. J. Liq. Chromatogr. 7(5-2):353-376 (1984)
67. Lindner, K.R. and Mannschrek, A. Separation of enantiomers by high-pressure liquid chromatography on triacetylcellulose. J. Chromatogr. 193:308-310 (1980)
68. Blaschke, G. Chromatographic resolution of racemates. Angew. Chem. Int. Ed. Engl. 19:13-24 (1980)

Chapter 7

Use of Microorganisms in the Production of Unique Ingredients

Nayan Trivedi

INTRODUCTION

The use of microorganisms in food preparation is an age old practice. Although the ancient Egyptians are credited with the idea of using yeast to leaven bread, historical records indicate that as early as 2300 B. C., bread was prepared in Babylonia. The food processing industry thus has the distinction of using microorganisms many centuries before any other industry. Today, production of baker's yeast is one of the largest fermentation industries, producing over 1.8 million tons of yeast, at 30% solids, annually. Besides yeast, other food products, such as cheese and pickles, employ microorganisms in their manufacture. In fact, many current food processing techniques take advantage of biochemical fermentations attributed to individual microorganisms, for it is these characteristic fermentation patterns and the production of various organic compounds that impart unique flavor and color to many food products. The emergence of rDNA technology in the 1970's has generated a great deal of interest in using biotechnology in the food industry. Besides several traditional uses, there are many areas where biotechnology may make a big impact. Since the basis of rDNA technology is the isolation and manipulation of specific genes which code for the production of a given protein, usually an enzyme, if specific proteins or enzymes responsible for the production of certain flavors and colors can be identified and isolated, then perhaps cheaper and more consistent production could be achieved. This chapter will be restricted to discussion of the production of flavor and color compounds by microorganisms.

MICROBIAL PRODUCTION OF FLAVOR COMPOUNDS

L-glutamate has been used to enhance the flavor of natural food products since the beginning of this century (1909). Today, over 600 million pounds of glutamate are produced worldwide (11). Ninety seven percent of this is consumed by the food industry. Glutamic acid is produced by the fermentation of molasses using *Corynebacterium qlutanicum*, *Corynebacterium lilium*, *Brevibacterium flavum* and *Micrococcus glutamicus*. Since these organisms produce glutamic acid in excess of 60% of the theoretical yield, this product is not considered a prime target for rDNA technology, although microbial cell immobilization could help reduce production cost.

A few years after the discovery of the flavoring function of glutamate, the flavoring compound in bonito fish was analyzed and found to be the histidine salt in inosinic acid (21). Inosinic acid together with guanadinic acid has subsequently become known as the 5' nucleotides in the food industry. Unlike glutamate, it took approximately 50 years to commercialize the use of 5' nucleotides. Kuninaka *et al.* (24) elucidated the relationship between the chemical structure and flavor potentiation of 5' nucleotides. The first step was to demonstrate that the 5' isomer of nucleotides possessed flavor enhancing activity. Hypoxanthine, inosine and ribose-5-phosphate were found to have no flavor activity. Ribosidic acid and 5' phosphomonoester linkages were found to be essential for the flavoring action of inosinate. In addition, the hydroxyl group at the 6th position was found to be essential in the structure of 5' inosinate. Replacement of the hydroxyl group by an amino group was found to sharply decrease the flavoring activity (23). Kuninaka also found that hydrogen at the 2-position could be replaced by a hydroxyl or amino group without a significant change in flavoring activity. Thus, not only 5' inosinate but 5' guanylate, and 5' xanthylate were found to impart flavoring activity. Qualitatively, the flavor producing effect of all of the above mentioned 5' nucleotides is similar and additive in nature.

Of the three 5' nucleotides, 5' xanthylate has been found to have the weakest flavoring activity and is not commercially produced. Honjo et al. (15) has reported that both hydroxyls of the phosphate group are essential, and primary and secondary disassociations are probably necessary for flavor activities of the 5' nucleotides. For a future research project, one might want to consider modifications of the purine moiety of both 5' inosinate and 5' guanylate since this might result in production of compounds which could show enhanced flavor activities.

Production of 5' Nucleotides

There are three ways to produce 5' nucleotides: 1) direct fermentation; 2) enzymatic breakdown of RNA; and, 3) chemical synthesis.

Direct Fermentation: 5' nucleotides can be produced directly by fermentation. Their yields in this process are generally low. An auxotroph of *Bacillus subtilis* has been reported to accumulate more than 7 g/L of inosine (1). There are several reports of the accumulation of somewhat larger quantities of nucleotides by microorganisms (2,47,48). In general, production of 5' nucleotides by microbial fermentation seems to be rather low yielding. Furuya et al. (10) has presented data showing fermentative production of 5' AMP by a wild strain of bacteria and actinomycetes. With the present knowledge of fermentation technology and genetics, it seems quite possible to increase fermentative production of nucleosides or nucleotides. If one could significantly improve the direct production of 5' nucleotides via fermentation using a GRAS organism such as *B. subtilis* growing in a clean medium (i.e. whey) it would provide a natural source of these substances. This could be a business with good profit potential.

Degradation of RNA by 5' Phosphodiesterase: Degradation of RNA by 5' phosphodiesterase is an alternate, and possibly the preferred route of production of 5' nucleotides. In this process, microbial (Torula yeast, baker's yeast, or bacteria) RNA can be separated and purified by chemical means. The extracted RNA is then treated with the enzyme 5' phosphodiesterase. The enzymatic hydrolysis of RNA results in simultaneous production of 5' AMP, 5' GMP, 5' CMP and 5' UMP (25,26,36). 5' IMP is produced either by enzyme action or chemically from 5' AMP. At the present time yeast is the source of choice for RNA since, after RNA extraction, yeast cells can be used as feed or for preparation of extracts and autolysates.

Penicillium citrinum, *Aspergillus* and *Streptomyces* species which produce 5' phosphodiesterase have also been isolated. Properties of the *Penicillium* enzyme have been studied in some detail. At first, this enzyme was called phosphodiesterase. Later it was recognized to be different from snake venom phosphodiesterase, and it was renamed nuclease P_1. The crude enzyme preparation of nuclease P_1 is used for the preparation of 5' nucleotides. The enzyme degrades single stranded DNA and RNA endo and exonucleolytically, but does not attack double stranded nucleic acids, especially in the presence of 0.4M sodium chloride. It cleaves substantially all 3'-5' phosphodiester linkages of single stranded polynucleotides and the 3' phosphomonoester linkages of mono and oligo nucleotides terminated by 3' phosphate. The optimum temperature of its reaction is 70°C and the optimum pH ranges from 4.0 to 8.5, depending on the kind of substrate used in the reaction (i.e. pH 4.0 for poly I and poly G; pH < 6 for DNA).

The crude preparation of nuclease P_1 is made from aqueous extracts of *Penicillium citrinum* grown on wheat bran medium. The crude extract is heated to inactivate other heat labile enzymes, including unwanted nucleases, without any loss of nuclease P_1 activity.

In commercial practice an aqueous solution of RNA is incubated with crude nuclease P_1 under optimal conditions. The 5' nucleotides resulting from enzymatic degradation are separated by an ion exchange process and then further purified. AMP is then converted to IMP by an adenyl deaminase enzyme preparation from an *Aspergillus* culture.

Chemical Synthesis: Several laboratories have evaluated the possibility of economically producing 5' nucleotides via chemical synthesis, for example, inosine has been economically phosphorylated to 5' IMP by chemical means (24). Since chemical synthesis is not widely used by the food industry, and is beyond the scope of this paper and is not discussed here.

Production of Esters

Microorganisms are known to produce a variety of volatile oils and flavoring compounds including esters, diacetyl, pyrazines, terpenes, L-menthol, and lactones. Esters impart a fruity, flowery flavor to foods. Since the 1800s, various workers have reported the isolation of microorganisms producing apple, pineapple, strawberry and muskmelon type flavors. Later, flavor production was attributed to ester formation by these microbes (38,39). Table 7-1 lists some microorganisms and the corresponding esters they produce.

Table 7-1. Organisms Producing Esters

Microorganism	Ester Produced
Pseudomonas fragi	Ethylbutyrate,
Pseudomonas fragi	Ethylisovalerate,
Pseudomonas fragi	Ethylhexanoate
Streptococcus diacetylactis ATCC 15346	} Ethylbutyrate Ethylhexanoate
S. lactis ML3	
S. cremoris TR	
Lactobacillus (No. 81)	
L. casei L323	
Pseudomonas strains No. 50 and 53	

Production of ethyl esters by cell free extracts of the five lactic bacteria and the two *Pseudomonas* species shown in the above table were studied by Hsono and Elliott (6). All of the extracts were found to be capable of esterifying butyric and caproic acids with alcohol. To further extend this research, one could immobilize the most active of the crude enzymes or its purified form and then study production of ethyl odanoate, ethyl butyrate or ethyl hexanoate under optimum reaction conditions to evaluate process economics.

Production of Diacetyl

Diacetyl or 2,3-butanedione is an essential ingredient in butter flavor formulations. *Leuconostoc dextranicus*, *Lactobacillus citrovorum* and *Streptococcus lactis* var. *diacetylactis* have been shown to produce butter flavor. The dairy industry uses diacetyl producing streptococci extensively to impart butter flavor to yogurt and other fermented products. All of the three above mentioned organisms need citrate for the production of diacetyl. Figure 7-1 depicts a pathway for diacetyl production from citrate. These organisms, when grown in milk media, obtain citrate readily since it is a constituent of milk. Citrate permease, an inducible enzyme, facilitates citrate permeation through the cell membrane where it is broken down to acetic acid and oxaloacetate. Oxaloacetate is decarboxylated to form pyruvate. Since an excessive intracellular concentration of pyruvate is believed to be toxic, the cells convert pyruvate to a non-toxic metabolite, diacetyl or acetoin. Usually acetolactate formed from pyruvate is decarboxylated to diacetyl under aerobic conditions and to acetoin under anaerobic conditions.

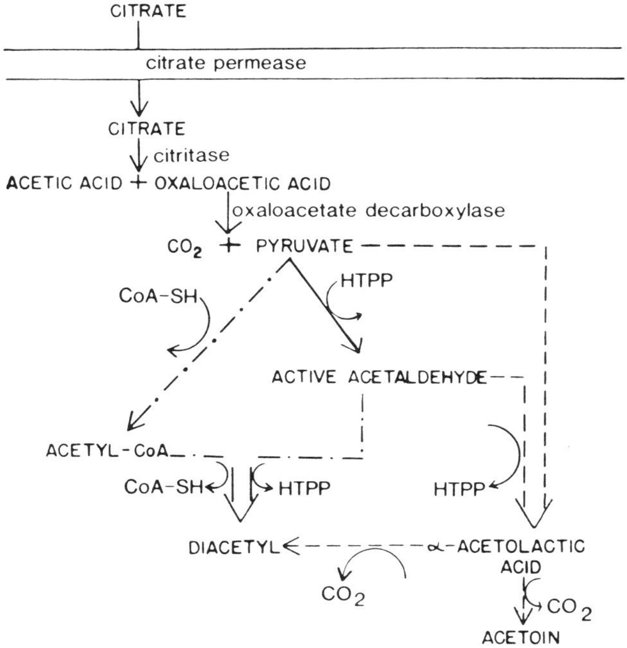

Figure 7-1. Pathway for Diacetyl Production from Citrate (20)

Some organisms do not use citrate as a source of energy for growth. In the case of *S. lactis* var. *diacetylactis*, it has been found that the addition of citrate resulted in a 35% increase in the specific growth rate in lactose containing media (13).

Several workers have successfully increased diacetyl production by mutating the parental strain of *S. lactis* var. *diacetylactis* (12,22). Kuila and Ranganathan irradiated *S. lactis* var. *diacetylactis* under ultraviolet light to get 99-99.9% kill and isolated higher diacetyl producing mutants by selection on citrate agar containing an insoluble citrate suspension. Mutants which could rapidly utilize citrate produced a clear zone around colonies after 18 hours of incubation. These were designated as Type I mutants. Others did not show clear zones on citrate agar for 72 hours, and these were designated as Type II. Parental strains showed a clear zone in 24-48 hours and were designated Type 0. Table 7-2 shows the production of diacetyl and acetoin by various mutants.

Table 7-2. Production of Diacetyl by UV Mutants of *S. lactis* var. *diacetylactis*

Cultures	No. of Isolates	Diacetyl and Acetoin (ppm)
Type I	100	138.0 (Range 94-170)
Type II	50	42.5 (Range 15-52)
Parent (Type 0)	20	30.7 (Range 20-40)

As shown in Table 7-2, the mutants produced a significantly greater amount of diacetyl than the parents. Since none of the mutants in this study showed increased citrate permease activity, one would speculate that the hyper-production of diacetyl by these mutants was probably due to increased activity of the decarboxylation step.

Another way to increase the accumulation of diacetyl is by increasing the amount of the enzyme citrate permease. Recently, the ability of *S. lactis* var. *diacetylactis* to ferment citrate, and consequently to produce diacetyl, was demonstrated to be a plasmid-mediated trait. Kempler and McKay (20) have found that the genes responsible for citrate fermentation are associated with plasmid DNA. This offers a possibility for creating a gene dosage effect by increasing the copy number of plasmids to boost the amount of enzyme. One would be safe in speculating that an increase in the amount of enzyme in the initial stage of the citrate fermentation system could result in hyper-production and accumulation of diacetyl. Thus it seems that there are several possibilities of manipulating *S. lactis* var. *diacetylactis* for hyper-production of diacetyl to the point of making this process commercially feasible.

Production of Pyrazines

Pyrazines are heterocyclic, nitrogen containing compounds which provide a roasted or nutty flavor note upon heating. In natural foods, it is produced by the browning reaction in which the carbonyl group of reducing sugars condenses with an amino group of protein when heated to temperatures of 100°C or higher.

Some of the microorganisms producing pyrazines are listed in Table 7-3.

Table 7-3. Microorganisms Producing Pyrazines

Microorganism	Pyrazine Produced
Bacillus subtilis	tetramethyl pyrazine
Corynebacterium glutanicum	tetramethyl pyrazine
Streptococcus lactis	2-methoxy-3-isopropylpyrazine
Pseudomonas perolens	2-methoxy-3-isopropylpyrazine
Pseudomonas taetrolens	2-methoxy-3-isopropylpyrazine
Streptomyces (from soil)	2-methoxy-3-isopropylpyrazine

Of these, a mutant of *Corynebacterium glutanicum* isolated by Demain et al. (7) deserves a closer look. These scientists mutated a biotin requiring strain of *C. glutanicum* with nitrosoguanidine. The mutation resulted in an auxotroph requiring isoleucine, valine, leucine and pantothenate for growth. This glutamic acid producing mutant also accumulated up to three grams/liter of tetramethyl pyrazine. This organism is grown in a complex seed medium followed by a defined production medium containing glucose, $NH_4H_2P0_4$, $(NH_4)_2SO_4$, urea, N-Z amine, inorganic salts, and trace minerals and vitamins. A shake flask culture of this organism has been reported to excrete pyrazine at up to 3 g/L in five days. At the end of the incubation period the whole broth can be centrifuged to remove cells and insoluble calcium carbonate. Pyrazines can then be crystallized by freezing the supernatant. The authors have postulated that tetramethyl pyrazine is derived from two moles of acetoin and two moles of ammonia. *C. glutanicum* is a gram positive bacterium used for glutamic acid production. Since its large scale propagation and fermentation has been well studied, this organism offers an opportunity to investigate the possibility of coproduction of glutamate and tetramethyl pyrazine in a fermentation process.

Another report on microbial production of tetramethyl pyrazine has been issued from Japan (44). This group recovered tetramethyl pyrazine from cultures of *Bacillus subtilis*. *B. subtilis*, being a GRAS organism, makes it an attractive candidate for extracellular production of flavors. Thus, both *B. subtilis* and *C. glutanicum* offer an opportunity to explore the possibility of microbial production of pyrazines since these organisms are routinely used for industrial production of amylases and glutamate, respectively.

Production of Terpenes

Terpenes are hydrocarbons built from a 5-carbon isoprene unit (2-methyl-1,3-butadiene) with an open chain, closed chain, cyclic, saturated or unsaturated structure. These compounds are of great interest to the food processing and wine industry since they are responsible for the characteristic odors of oils and grape aroma. Most of these compounds are produced by higher plants, but there are several reports on their production by microorganisms (4,5,6,8,27,30). Table 7-4 lists several microorganisms and the terpenes produced by them. Two of these are noteworthy. Production of terpenoids is a well known phenomenon within the *Ceratocystis* family. One member of this family, *C. variospora* has been reported to produce monoterpene alcohols at a gram per liter in a pilot plant scale experiment (41).

A second one is *Kluveromyces lactis*, a yeast, which can grow on lactose as a sole carbon source. This yeast is propagated on whey, a dairy industry waste product, for food or feed purposes and for alcohol production. An isolate of this yeast was found to accumulate up to 50 μg/L of linalool and citronellon in a shake flask culture. Drawert et al. (8) has optimized fermentation parameters, including composition of the medium, which resulted in improved yields of citronellon.

Table 7-4. Production of Terpenes by Microorganisms

Microbe	Terpenes Produced	Odor
Trametes odorata	Geraniol, d-limonene	
Phellinus spp.	α - pinene	
Kluyveromyces lactis	Citronellol, linalool, geraniol	fruity, flowery
C. moniliformis	Citronellol, geraniol, linalool	fruity, banana, peach
Ceratolystis spp.	Nerol, terpineol	
Ascoidea hylecocti	Citronellol, linalool	fruity, flowery
Ceratocystis coerulescens	Citronellol, citronellyl acetate	fruity
C. fimbriata	Citronellol, linalool, geraniol	sweet, fruity
C. variospora	Geraniol, citronellol, nerol	fragrant, banana
C. virescens	Linalool, geranylacetate	fruity
Lentinus lepideus	Linalool, several sesquiterpenes	aromatic fruity
Penicillium decumbens	Thujopsene, nerolidol	soapy perfume

With the proper genetic manipulations, followed by optimization of fermentation parameters, one could increase yields of citronellon and linalool in the medium and use the cells for food or feed purposes. Thus, one can use dairy industry waste products and improve the process economics of natural flavor production.

Production of L-Menthol

The market for fragrance chemicals used primarily in the cosmetics industry, and also in toiletries and household detergents, amounts to more than two billion dollars annually. L-menthol claims a big share of this market. World demand for L-menthol is over 3,000 tons every year. L-menthol is an essential oil produced by the plant *Mentha piperita*. During the early flowering stage, these plants produce L-menthone which is converted to L-menthol during full bloom. In this process yields of L-menthol are significantly reduced.

As shown in Figure 7-2, L-menthone conversion is catalyzed by two enzymes to two different end products (33). The final production of L-menthol in the natural process is only 40% based on L menthone. Microorganisms can be used to improve production economics in this process. A strain of *Pseudomonas putida*, YK-2, has been reported to catalyze conversion of L-menthone to L-menthol. One could harvest plants early enough to recover maximum L-menthone and then catalyze its conversion to L-menthol using *P. putida* to improve the yield of menthol production.

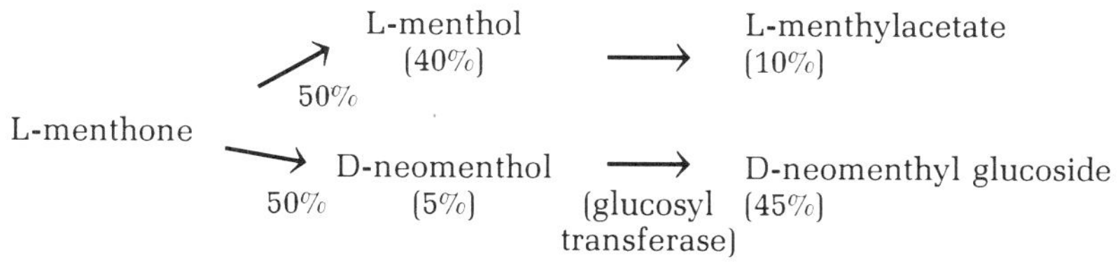

Figure 7-2. Pathway for Conversion of L-menthone to L-menthol

Another area in which microbes can help in L-menthol production is in obtaining high yields of a more desirable, purer form of L-menthol from its racemic mixture. Most work in this area has been done primarily by Japanese scientists, since Japan is historically the largest producer of menthol. Schindler and Schmid (42) have summarized literature on menthol production in their review. The best case is that of Omata et al. (34,37,46) who have demonstrated stereospecific hydrolysis of DL-menthol succinate by a yeast *Rhodotorula minuta* var. *texensis*. Cells of *Rhodotorula* were entrapped in a polyurethane matrix and immobilized. Using this immobilized yeast and DL-menthol succinate as a substrate and water saturated n-heptane as a solvent, a conversion ratio of 72.6% was achieved, yielding L-menthol of 100% optical purity. The half life of the cells was estimated to be 53-63 days. This is quite impressive work with a potential for commercialization. Some of the patent examples noted by Schindler and Schmid include raactions on the 800 Kg scale, clearly demonstrating that they were virtually ready for adaptations to large scale operations. These examples again show the potential for use of microbes in flavor production.

Production of Lactones

Lactones impart fruity, coconut-like, buttery sweet and nut-like flavor notes. At present these compounds are made via chemical synthesis from keto acids. At the present, microbial production of lactones would have to meet the economic reality of lower production cost to compete with chemically synthesized lactones. However, a clear advantage with a microbe based process is the production of optically pure forms and the fact that this route would produce "natural" flavors. Muys, van der Van and DeJonge (33) of the Unilever Research Laboratory, Netherlands, were probably the first to demonstrate microbial reductions of γ- and δ- keto acids to yield optically active lactones. They propagated species of *Candida* and *Saccharomyces*, *Penicillium notatum*, *Cladosporium butyricum*, *C. suaveolens*, and *Sarcina lutea*, on nutrient agar. In a qualitative test, performed by incubating a small quantity of cells of each organism with glucose and keto acid solution at 30°C for 48 hours, all the above organisms demonstrated lactone formation as evidenced by the characteristic odor of lactone in their respective filtrates. Several of these organisms were then grown on nutrient agar, and cell mass representing 12 g wet weight of each organism was incubated with 10 gms of glucose and 500 mg of δ-keto capric acid at 30°C for 48 hours. Following incubation, the hydroxy acids were extracted from the filtrate and lactonized at 130-140°C for one hour. Table 7-5 shows the yields of lactone produced by each organism:

Table 7-5. Production of Lactone from Keto Acids by Microorganisms

	δ-deca-lactone recovered		
Microorganism	**mg**	**% Yields**	**(a)D+**
Cladesporium butyricum	71	14.2	−15°
Saccharomyces cerevisiae	355	71	+48.2°
Saccharomyces fragilis	329	65.8	+48.5°
Candida pseudotropicalis	252	50.4	+48.2°
Candida globiformis	233	46.6	+55.8°
Sarcina lutea	302	60.4	−29.2°
	1400*	58.3	−36.8°

* from 2400 mg keto acid
\+ The odor of dextro and levorotatory δ-deca-lactones was identical.

The researchers (33,49) selected *S. cerevisiae* to test for reductions of the various keto-acids using the following system: 10% yeast slurry, 0.05 to 0.1% of the respective keto-acid, 5-10% glucose, incubation at 30°C for 48 hours. Table 7-6 shows the yield of lactones produced by baker's yeast from various keto acids.

This work clearly demonstrates the possibility for using baker's yeast in the production of γ and δ-lactones. Baker's yeast is a mass produced, GRAS organism which makes its use very practical. Also, with immobilization of yeast one may be able to improve process economics. This is yet another instance where microbial production of flavoring compounds may be practical.

Table 7-6. Lactone Production by Baker's Yeast (33)

Keto Acid	(g)	Yeast (g)	Glucose (g)	Volume (ml)	Lactone Yield (Percent)	(a)D
−C9	10	1,800	450	18,000	77	+50.4°
−C10	10	680	680	10,200	85	+46.1°
−C11	10	1,800	450	18,000	72	+44.1°
−C12	10	1,320	1,330	20,000	60	+40.4°
−C8	10	1,000	300	6,000	13	+58.4°
−C9	8.8	1,800	1,800	18,000	55	+58.2°
−C10	200	12,000	12,000	102,000	47	+55.6°
−C11	3	600	600	6,000	63	+47.9°
−C12	7	600	300	6,000	46	+48.8°

Debittering Citrus Fruit Juice

Certain citrus fruit juices that turn bitter soon after their extraction are found unsuitable for the food processing industry. In the state of California alone this problem causes an approximate loss of eight million dollars ($8,000,000) every year. The problem is most prevalent in some varieties of oranges, grapefruits and lemons, along with minor citrus crops. The bitterness is caused by limonoids and flavonoids, represented primarily by limonin, nomillin and naringin. The intact fruits, however, contain nonbitter precursors, for example the limonoate A-ring lactone of limonin. This precursor differs from limonin in having a free carbonyl group at C-18 and a free hydroxyl at C-17. Immediately upon extraction of juice the precursor lactonizes to its bitter form, limonin.

There are two main approaches whereby biotechnology may help remove bitterness: 1) use of enzymes or microbes which might convert bitter compounds to their nonbitter metabolic forms; and, 2) selective removal of bitter compounds by resins.

Use of microbes/enzymes to remove bitterness: *Aspergillus niger* possesses an enzyme which converts the naringin from grapefruit to naringenin, a nonbitter compound (3). Recently, Hasegawa et al. (14) of the USDA laboratory in California, have reported the isolation of five bacterial species from soil that convert limonoids to nonbitter compounds found in naval oranges. These scientists immobilized bacterial cells in an acrylamide gel and packed them in a column. Passing navel orange juice through the column reduced limonin content originally from 20 down to 5 ppm. In addition, the column could be used up to 15-20 times, before regeneration was needed.

Selective removal by resins: In another approach, a polystyrene divalent-benzene crosslinked polymeric adsorbent resin was used to selectively adsorb naringin and limonin from citrus fruits such as oranges, lemons, grapefruits, and tangerines to remove bitterness. This was accomplished without sacrifice to the overall quality of the juice or its solids content (40,50). This resin is commercially available as Duolite S-861 from Rohm and Haas (Philadelphia).

In still another approach, scientists from the USDA laboratory in Florida (43) polymerized water soluble cyclodextrins with epichlorohydrin and other crosslinking agents to form water insoluble polymers. Some of the polymers retained the ability of the cyclodextrin monomer to form inclusion compounds with limonin and naringin to remove them from the juice. The β-cyclodextrin polymers were found to selectively remove limonin from naval orange juice and limonin and naringin from grapefruit juice. Some of the lemonoids, limonin, nominlin and obacunone have shown some promise in laboratory trials as pest control compounds against insects such as cottonboll-worm and the fall armyworm. It is quite possible to use limonoids for insect control, if proven effective in field tests, to reduce process costs and thereby put biotechnology to work for the juice processing industry.

To summarize the microbial production of flavor compounds, there is agreement with a recent review in that, "in spite of the inherent advantages of using microbes in flavor production, only a few processes have made it out of the test tubes into the processing plant" (19). This is due to both regulatory issues and process economics. However, there are some rather good opportunities such as: terpene production by *K. lactis* as a coproduct of feed or food yeast from whey permeate, menthol production, production of lactones using baker's yeast, and coproduction of pyrazines and glutamate by *C. glutanicum*, where industrial microbiologists and fermentation technologists team up with a marketing expert and, with a little bit of luck, come up with a viable process for microbial production of flavoring compounds.

MICROBIAL PRODUCTION OF COLOR COMPOUNDS

The certified food color business is a relatively small volume business in the giant food processing industry. It commands a seventy million dollar annual market and approximately three and one half million dollars in profit. Although this is a small volume, the products using colors constitute an annual market of twenty five billion dollars, and many of these products would not exist without colors (35).

The production of pigments by microorganisms is not new to microbiologists. *Pseudomonas aeruginosa* species 196 Aa and Sol 20 produces both a brownish red and yellowish green extracellular pigment (45). Actinomycetes, yeast and other fungal, produce a vast variety of colored pigments under different growth conditions. All of these pigments are not naturally suitable for human consumption. However, many species of microbes are known to produce beta carotenes, astaxanthin and other edible pigmented compounds. Three examples seem to be ready for exploration and commercialization. These are: 1) *Phaffia rhodozyma*, a yeast which produces red pigment; 2) *Monascus purpureus*, a fungus which produces a red pigment; and, 3) fermentation of beet juice with the yeast *Candida utilis* to deplete carbohydrates and nitrites from the juice.

Phaffia rhodozyma

Phaffia rhodozyma is a relatively recently described yeast (32) which produces a carotenoid pigment, astaxanthin (3,3'-dihydroxy-β,β-carotene-4,4'-dione). Astaxanthin is commonly found in the animal kingdom, but is rarely produced by microorganisms. It is displayed in the plumage of many birds, including flamingos and scarlet ibis, in marine invertebrates such as lobsters, crabs and shrimp, and in fishes such as trout and salmon where astaxanthin is responsible for flesh color. These fish, when raised in pens, often lack desirable red color. Johnson et al. (17) have found that a preparation of *Phaffia* yeast is important for imparting red color to the flesh of pen grown rainbow trout. Johnson and Lewis have done extensive studies on optimization of astaxanthin production by *P. rhodozyma* by manipulating culture conditions (18). These authors investigated optimum temperature, pH, substrate concentration and carbon source for the growth and for pigment production by *Phaffia*. The authors used shake flasks and a 20 L bench top fermenter with a 14 L working volume in their investigations. The media they used were yeast extract and malt extract broth for shake flask experiments, Difco yeast nitrogen base without ammonium sulfate and amino acids for carbon assimilation studies in shake flasks, and a defined medium with yeast extract for bench top (14 L working volume) fermenter scale up work.

Growth and astaxanthin production experiments were conducted in a 20 L fermenter containing 14 L medium with 1.5% glucose as a carbon source. Figure 7-3 shows astaxanthin formation by *P. rhodozyma*.

Astaxanthin formation by Phaffia rhodozyma

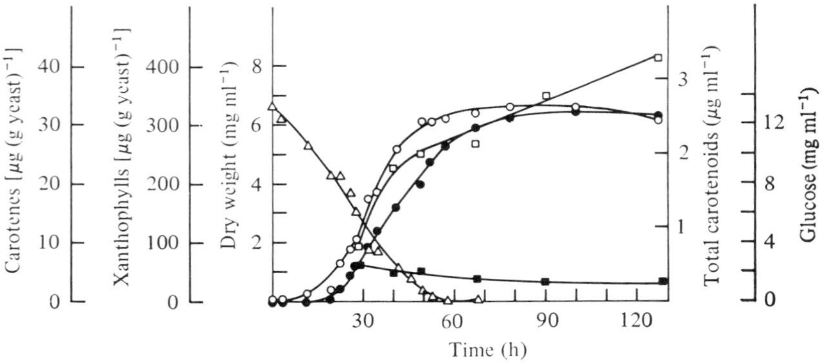

Figure 7-3. Production of Carotenoids by *P. rhodozyma* in a Batch Fermenter Culture
Yeast growth (○), total carotenoid formation (●), xanthophyll (astaxanthin) formation (□), carotene synthesis (■), and glucose utilization (△). The growth medium initially contained 1.5% (w/v) D-glucose (18).

In this study yeast growth had a 10 hour lag and the stationary phase was reached after about 80 hours. Astaxanthin was found to be the major xanthophyll in the yeast and was produced mainly during the exponential phase of growth. However, its production continued at low levels in the stationary phase of growth until the 128th hour of fermentation when the final concentration of 406 μg/g was obtained. Carotenes were produced at the very low level of 3-6 μg/g. The optimum pH was 5.0 and optimum temperature for growth and pigment production was 20-22°C.

Sucrose, glucose, maltose, cellobiose, succinate, mannitol, xylose, arabinose, and glucono-δ-lactones were tested for their effect on pigment production and growth of *P. rhodozyma*. Cellobiose supported more pigmentation than any other source, whereas glucose and sucrose promoted the most growth. The yields of astaxanthin were found to depend on the dissolved oxygen concentration at a low aeration rate. Thus a high aeration rate seems to be necessary for pigment production. Higher sugar concentrations were found to result in a greater aeration requirement for selective production of astaxanthin over carotenoids since, at a high glucose concentration, yeast have a tendency to have fermentative rather than aerobic metabolism. One may be able to further improve astaxanthin production by growing *Phaffia* under substrate-limiting conditions like that used in baker's yeast production.

Monascus purpureus

The second organism deserving consideration is *Monascus purpureus*. *Monascus* has been used in the orient for years for making red rice wine where it is traditionally cultivated in a solid-state fermentation on rice. The pigment produced is actually a mixture of red, yellow and purple pigmented polyketides. The pigment is insoluble in acid and is produced via a pathway similar to that for fatty acid biosynthesis. Typically, the pigment is produced intracellularly in the mold. The solid-state fermentation yield is 10-fold more than the submerged fermentation under other identical parameters (9,28,29). Recently Wong et al. (52) have studied the regulation of growth and pigment production by *M. purpureus*. These authors have found that the ratio of glucose to ammonium salts is important for pigment formation by this mold under submerged conditions in a synthetic medium. Since the *Monascus* pigment consists of at least six different compounds, the authors have suggested that culture conditions may be manipulated to produce a pigment with somewhat different shades. *Monascus* is known to produce up to tenfold more pigment when propagated under solid-state vs. submerged conditions. However, with enough microbial and biochemical understanding of the process, very much like citric acid production by *A. niger*, one should be able to successfully convert this fermentation to submerged culture. Furthermore, one might clone the genes responsible for pigment production and obtain a hyper pigment producer which would excrete pigment into the growth medium thereby making the process economically viable.

Candida utilis Red Beet Juice Fermentation

Some years ago an interesting approach was devised for the isolation of red beet color by Amber Laboratories in conjunction with von Elbe and Amundson of the University of Wisconsin (51). In their studies, *Candida utilis* yeast were used to ferment red beet juice. The fermentation produced alcohol, biomass for food or feed, and beet juice ready for concentration and use as a colorant (51).

In this work (51), beet juice was fermented in order to remove sucrose and most of the crude soluble protein. The inoculum level was one quarter pound yeast per pound of fermentable carbohydrate in the juice. The fermentation was run at 30 $\pm$ 2°C, and a pH of 5.0 $\pm$ 0.3 under aerated conditions. These conditions afforded optimum growth of yeast and maximum utilization of carbohydrates. The fermentation time was found to be 110 minutes for each percent of sucrose in the beet juice. At the end of fermentation, beet juice was centrifuged to remove yeast cells and then concentrated ten fold under a vacuum. The juice was acidified to pH 2.0 with HCl followed by a pass through a column packed with Sephadex G 25 resin to remove salts. The juice was then dried and found to have a 55% betacyanine content.

According to the patent (51), for beet juice with about 7-12% solids, and containing 80% sucrose on a dry basis, fermentation with *C. utilis* resulted in a five fold concentration of the beet color, 4% recoverable alcohol and 1.5% yeast solids. Besides this, the nitrate content and the earthy flavor, which makes use of beet juice as a colorant somewhat less attractive, were absent from the final product.

Production of color compounds by microorganisms may be made possible in these and other cases. However, it is quite reasonable to assume that in spite of optimizing production processes and the organism's performance, microbially produced colors may not be completely cost competitive with chemically synthesized colors. Despite this, since the cost of color ingredients in prepared products is usually so low, the food processing industry might accept a higher price for naturally produced colors.

REFERENCES

1. Aoki, R., Momose, H., Muramatsu, N., Kondo, Y. and Tsuchiya, Y. Recent studies of 5' nucleotides as new flavor enhancers. Flavor Chem. Adv. Chem. Series. 56:261 (1966)
2. Aoki, R., Momose, H., Tsunodas, T., and Motozaki, N. Japanese Patent SN 32731 (1959)
3. Chandler, B.V., Nicol, K.J. and Nicol, K. Debittering citrus products with enzymes. CSIRO Food Res. Q. 35(4):79 (1975)
4. Collins, R.P. and Halim, A.F. An analysis of the odorous constituents produced by various species of *Phellinus*. Can. J. of Microbiol. 18:65 (1972)
5. Collins, R.P. and Halim, A.F. Production of monoterpenes by the filamentous fungus *Ceratocystis variospora*. Lloydia 33:481 (1970)
6. Collins, R.P. and Halim, A.F. Essential oil composition of *Ceratocystis virescens*. Mycologia 69:1129 (1977)
7. Demanin, A., Jackson, and Tremer. Accumulation of tetramethyl-pyrazine. J. Bacteriol. 94:323 (1967)
8. Drawert, F. and Barots, H. Biosynthesis of flavor compounds by microorganisms 3. Production of monoterpenes by the yeast *Kluveromyces lactis*. J. Agri. and Food Chem. 26:765 (1978)
9. Evans, P.J. and Wang, H.Y. Pigment production from immobilized *Monascus* spp. utilizing polymeric resin adsorption. Applied Environ. Microbiol. 47(6):1323 (1984)
10. Furuya, A., Abe, S., and Kinoshitas, S. Food chemical studies on 5'- ribonucleotides 3. Flavor of sodium 5^1-ribonucleotides. Presented to Annual Meeting of the Agricultural Chemical Society of Japan (1963)
11. Genetic Engineering Applications for Industry. Chemical Technology Review No. 197:56. Noyes Data Corporation. (1979)
12. Gurovich, Sh. R., Grinevich, A.G. and Safiyazov, Zh. Free amino acids in cells of original and mutant strains of *Streptococcus acetoinicus*. Appl. Biochem. Microbiol. 3:92 (1967)
13. Harvey, R.J., and Collins, E.B. Role of citrate and acetoin in the milk metabolism of *Streptococcus diacetylactis*. J. Bacteriol. 86:1301 (1963)
14. Hasegawa, S. and Maier, V.P. New methods to remove citrus bitterness. Presentation to Food and Ag. Division, Annual Meeting of ACS, Miami Beach, FL, 1985 C&E News. 58 May 27, 1985

15. Honjo, M., Imai, K., Furukawa, Y., Moriyama, H., Yasumatsu, K., and Imada, A. Ann. Rpt. Takeda Res. Lab. 22:47 (1963) From Akira Kuninaka. Recent studies of 5' nucleotides as new flavor enhancers. Flavor Chem. Adv. Chem. Series. 56:261 (1966)
16. Hosono, A. and Elliott, J.A. Properties of crude ethyl ester-forming enzyme preparations from some lactic acid and psychrophillic bacteria. J. Dairy Sci. 57:1432 (1974)
17. Johnson, E.A., Conklin, D.E. and Lewis, M.J. The yeast *Phaffia rhodozyma* as a dietary pigment for salmonids and crustaceans. J. of Fisheries Research Board of Canada 34:2417 (1977)
18. Johnson, E.A. and Lewis, M.J. Astaxanthin formation by the yeast *Phaffia rhodozyma*. J. Gen. Microbiol. 115:173 (1979)
19. Kempler, G.M. Production of flavor compounds by microorganisms. Advances in Appl. Microbiol. 29:29 (1983)
20. Kempler, G.M. and McKay, L.L. Biochemistry and genetics of citrate utilization in *Streptococcus lactis* subspecies *diacetylactis*. J. Dairy Sci. 64:1527 (1981)
21. Kodama, S.J. Tokyo. Chem. Soc. 34, 751 (1913) from Akira Kuninaka, Recent studies of 5'-nucleotides as new flavor enhancers. Flavor Chem. Adv. Chem. Series 56:261 (1966)
22. Kuila, R.K. and Ranganathan, B. Ultraviolet light-induced mutants of *Streptococcus lactis* subspecies *diacetylactis* with enhanced acid or flavor producing abilities. J. Dairy Sci. 61:379 (1978)
23. Kuninaka, A. Recent studies of 5' nucleotides as new flavor enhancers. Flavor Chem. Ad. Chem. Series 56:261 (1966)
24. Kuninaka, A., Kibi, M., and Sakaguchi, K. History and development of flavor nucleotides. Food Technol. l8:287 (1964)
25. Kuninaka, A., Otsukas, S., Kobasyashi, Y. and Sakaguchi, K. Japanese Patent SN 16141 (1957)
26. Kuninaka, A., and Sakaguchi, K. Japanese Patent SN 23698 (1957).
27. Lanza, E., Ko, K.H., and Palmer, J.K. Aroma production by cultures of *Ceratocystis moniliformis*. J. Agr. Food Chem. 24:1247 (1976)
28. Lin, C. Isolation and culture conditions of *Monascus* spp. for the production of pigment in submerged culture. J. Ferment. Technol. 51:407 (1973)
29. Lin, C. and Suen, S.J. Isolation of hyperpigment-productive mutants of *Monascus* spp. F-2. J. Ferment. Technol. 51:757 (1973)
30. Manssen, H.P. Sesquiterpene hydrocarbons from *Lentinus lepideus*. Phytochemistry 21:1159 (1982)
31. Martinkus, C. and Croteau, R., Metabolism of terpenes. Plant Physiol. 68:99 (1981)
32. Miller, M.W., Yoneyama, M. and Soneda, M. *Phaffia*, a new yeast genus in *Deutromyotina* (Blastomycetes). Int. J. of Systematic Bacteriol. 26:286 (1976)
33. Muys, G. Tuynenburg, Van, B. der van, and DeJonge, A.P. Synthesis of optically active γ and δ-lactones by microbiological reduction. Nature 194:995 (1962)
34. Nippon Terpene Chemical Co. Ltd. Japanese Patent No. 77122690, 1979 from Chem. Abstracts. 1978, 87656g.
35. Noonan, J. An analysis of the factors affecting the future of FD and C colors. Cereal World 30(4):265 (1985)
36. Ogata, K., Nakao, Y., Igarashi, S. Omura, E., Sugino, Y., Koneda, M. and Suhara, I. Degradation of nucleic acids and their related compound by microbial enzymes. I. On the distribution of extracellular enzymes capable of degrading ribonucleic acid into 5'-mononucleotides in microorganisms. Agr. Bio. Chem. 27:110 (1963)

37. Omata, T., Iwamoto, N., Kumara, T., Tanaka, A. and Fukui, S. Sterioselective hydrolysis of DL-methyl succinate by gel-entrapped *Rhodotorula minuta* var. *texensis* cells in organic solvent. Eur. J. Appl. Microbiol. Biotechnol. 11:199 (1981)
38. Pereira, J.N. and Morgan, M.E. Identity of esters produced in milk cultures of *Pseudomonas fragi.* J. Dairy Sci. 41:1201 (1958)
39. Perry, K.D. A comparison of the influence of *Streptococcus lactis* and *S. cremoris* starters on the flavor of cheddar cheese. J. Dairy Res. 28:221 (1961)
40. Puri, A. New methods remove citrus bitterness. C&E News. 58-59. May 27, 1985. Presentaton to Food and Ag. Division, Annual Meeting of ACS, Miami Beach, FL, 1985.
41. Schindler, J. and Bruns, K. German patent No. 2,840,143 (1980)
42. Schindler, J. and Schmid, R.S. Fragrance or aroma chemicals - microbial synthesis and enzymatic transformation - A Review. Process Biochemistry 17(5):2 (1982)
43. Show, P.E. and Wilson, C.W. III. New methods to remove citrus bitterness. C&E News. 58-59 (May 27, 1985) Presentation to Food and Ag. Division. Annual Meeting of ACS, Miami Beach, FL, 1985.
44. Takuo, K., and Hiroko, K. Discovery of a pyrazine in a natural product: Tetramethylpyrazine from cultures of a strain of *Bacillus subtilis.* Nature 193:776 (1962)
45. Trivedi, N.B. n-alkane metabolism by bacteria. Ph.D. Thesis, University of Southwestern Louisiana, Lafayette, LA.
46. Ube Industries Ltd. Japanese Patent 8048396, 198O in Chem. Abst. 94:14031p (198l)
47. Uchida, K., Kuninaka, A., Yoshino, H. and Kibi, M. Japanese Patent SN36392 (1959)
48. Uchida, K., Kuninaka, A., Yoshino, H. and Kibi, M. Fermentative production of hypoxanthine derivatives. Agr. Biol. Chem. 25:804 (1961)
49. Unilever, N.V. Belgian Patent 592,593.
50. U.S. Patent 4,439,458 (Issued to Coca Cola Foods)
51. von Elbe, J. and Amundson, C.H. U.S. Patent 4,027,042, (1977)
52. Wong, H.C., Lin, Y.C. and Koehler, P.E. Regulation of growth and pigmentation of *Monascus purpureus* by carbon and nitrogen concentrations. Mycologia 73:649 (1981)

Chapter 8

Potential Applications of Plant Cell Culture

David Evans and William Sharp

INTRODUCTION

For many branded food products the raw material is purchased as a commodity. Food companies provide added-value based on processing technology or marketing. The tools of plant biotechnology, which shorten the time for crop improvement, offer the food industry the opportunity to modify their raw materials. Such modifications will permit development of plant varieties specifically selected for traits with added-value for the processor or the consumer. Biotechnology developed varieties offer food companies the opportunity to have proprietary raw material for use with specific brand names.

In recent years, the food industry has become more active in the application of plant biotechnology. Table 8-1 is a summary of published food industry programs aimed at using biotechnology to modify raw materials. The biotechnology tools of clonal propagation, somaclonal variation, gametoclonal variation, and protoplast fusion permit new variety development on an accelerated schedule, making them attractive for food industry applications. When integrated with conventional breeding, these intermediate term technologies will permit modification of raw materials to meet food industry specifications during the next several years.

CLONAL PROPAGATION

Clonal propagation provides potential for the large-scale production of genetic carbon copies of superior genetic varieties for commercial use. A wide variety of plant species can be clonally propagated from plant tissue using carefully defined culture media (1). This has resulted in a

Table 8-1. Plant Cell Culture Effort of Selected Food Companies

Food Company	Biotechnology Company	Major Interests
Campbell Soup Company	DNAP	Tomato improvement
General Foods Corporation	DNAP	Unpublished
Unilever	In-House	Palm oil
Hunt	IPRI	Tomatoes
Nestle	Calgene	Soybeans
American Home Foods	DNAP	Popcorn
Kraft	DNAP	Vegetables
Frito-Lay	In-House	Potatoes
Heinz	Plant Cell Research Institute	Tomatoes
Hershey	DNAP	Cacao
McCormick	NPI	Unpublished
Kirin Beer	Plant Genetics, Inc.	Unpublished

worldwide $30 million business since 1960. Examples of commercially tissue culture propagated crops include strawberry, asparagus, and oil palm (2).

Future clonal propagation advances currently being researched focus on propagation using bioreactor technology for industrial scale clonal propagation. Potential applications of clonal propagation include development of cloned varieties, propagation of elite plants and hybrids difficult or time-consuming to propagate using conventional seed technology, virus-free nursery stock, and propagation of certain perennials, estate crops, fruit trees, and forest trees.

The generation of clones can be achieved at both the organ and cellular level. For micropropagation (or mericloning) the apical meristem, the mass of cells at the growing tip of shoots, is removed and induced to develop many additional shoots. These can be separated and rooted, each one a clonal copy (3). Alternatively, cells taken from the donor plant can be induced to divide and multiply in culture and then the plants are regenerated. The advantage with this second approach is that many cells, and, therefore, many plants, can be grown in a shorter period of time. From these cells, there are two routes to plant regeneration: 1) organogenesis in which the cells first form shoots and are then separated and rooted; and, 2) somatic embryogenesis in which each cell develops into an embryo, identical to the seed embryo but arising from body or somatic cells. Each somatic embryo has a shoot and a root; thus in one step, both growing regions necessary for plant development are formed. Somatic embryos can be grown in liquid medium in large numbers (4).

Clonal propagation is proving particularly useful in plants where the unique combination of traits would be lost during seed production or where standard methods of asexual propagation are inefficient or lacking. Clonal propagation is also valuable in the reproduction of certain high value hybrids that are too expensive to produce commercially due to the absence of male sterility, or where seed production is inefficient or seeds show poor germinability. It is especially useful in crops that must be propagated asexually due to natural infertility, such as of the root and tuber crops.

In asparagus, an elite line with high yield and resistance to both *Fusarium* and *Verticillium* (fungal diseases), has been propagated and field yields have doubled. Clonally propagated strawberries have proven hardier with substantially better yields. These technologies have been applied to potato for the propagation of elite, virus-free plants and to sugar cane for the introduction of specific variants arising in tissue culture. By manipulating development, microtubers have been formed in potato cultures, providing numerous units that can be efficiently delivered. A similar system is being developed for the tropical yams (5).

Clonal propagation may also allow for the rapid development of improved plant varieties and replace the time-consuming conventional breeding procedures currently used for certain perennial varieties, such as estate crops, fruit trees, and forest trees. Clonal propagation may also permit the scale-up of production of certain superior perennial breeding lines selected either in the field or from biotechnologically altered plants for the rapid introduction of cloned varieties. Certain cloned varieties may have the potential to become competitive with seed-propagated varieties. For example, cloned estate crops may be an important potential commercial application of clonal propagation. Plant breeding of perennial crops has lagged far behind conventional crop breeding because of the time requirement for maturity. Using tissue culture it is possible to produce uniform, high yielding, and disease-resistant perennial crops. The methodology for cloning perennials has been developed in coffee, cacao, and oil palm trees.

Clonal propagation via somatic embryogenesis offers the potential of truly large-scale production. Being structurally similar to seed embryos, somatic embryos offer the prospect of efficient storage and dissemination. Work is in progress to adapt bioreactor technology to produce millions of plant cells and embryos continuously or on demand. For somatic embryos to be manipulated in bioreactors then mechanically handled, their growth must be regulated (6). As shown by Ammirato (4), embryo development can be controlled to prevent premature germination or abnormal maturation.

There are a wide range of potential delivery systems. One current technology, techniculture transplantation, involves direct planting of culture-derived embryos and plants into seedling trays; machines deliver the plants to the field. One potential delivery system is the adaptation of the fluid-drilling technique developed in the United Kingdom for pregerminated seeds. Another area of future technology is the development of artificial seeds through encapsulation or incorporation into seed tapes. Clonal propagation technologies are currently providing the means for large-scale propagation and mechanical delivery systems with the potential for more different and efficient technologies in the short term.

SOMACLONAL VARIATION

In contrast to clonal propagation which faithfully produces genetic carbon copies, evidence has accumulated suggesting that regeneration of plants from callus, leaf tissue explants, or plant protoplasts (walless cells) results in recovery of somaclonal variants (7). This allows the researcher to genetically modify genes in cultured cells of existing plant varieties for development of new breeding lines for use in development of improved varieties. Regeneration of plants from callus culture of existing commercial varieties has been associated with recovery of genetic variants. In tomato (7), potato (8), sugarcane (9) and other crops (10), somaclonal variants have been produced and selected for development of new breeding lines that have moved plant breeding a step forward with new agronomic and processing benefits.

The genetic variability recovered in regenerated plants probably reflects both preexisting cellular genetic differences and tissue culture-induced variability. Skirvin and Janick (11) systematically compared plants regenerated from callus of five cultivars of geranium. Plants obtained from geranium stem cuttings *in vivo* were uniform, whereas plants from *in vivo* root and petiole cuttings and plants regenerated from callus were all quite variable. Changes were observed when regenerated plants were compared to parent plants for plant and organ size, leaf and flower morphology, essential oil constituents, fasciation, pubescence, and anthocyanin pigmentation. Long-term cell cultures result in tissue culture-induced variability with regard to chromosome number that is expressed in both callus and in plants regenerated from callus (12). This results in regeneration of aneuploid plants that are commercially useless in sexually propagated species. On the other hand, aneuploidy may not interfere with productivity of asexually propagated crops such as sugar cane and potato.

Somaclonal variation depends on the occurrence and recovery in plants regenerated from cell culture of Mendelian and non-Mendelian genetic variation. These changes in the integrity of the genome are attributed to mutant induction, mitotic crossing-over, and organelle mutation. Two important selection steps of the procedure serve as sieves to recover a population of plants produced by self-fertilization of first generation tissue culture-derived plants most suitable for a breeding program. These are: 1) the culture medium provides a sieve for singling out cells from the foundation cell populations which possess genome competence for plantlet regeneration; and, 2) greenhouse selection identifies those plants regenerated from cell cultures of the donor plant with normal development that are capable of undergoing flower and fruit formation and which set seed. This permits elimination of plants with deleterious genetic backgrounds. The regenerated plants are selfed and the resulting seed is used for field trial evaluation and selection of breeding lines.

Evans and Sharp (7) made the first detailed genetic evaluation of the plants regenerated from cell culture of a sexually propagated crop plant. In evaluating the plants regenerated from tissue cultures of tomato, they have provided evidence for recovery of single gene changes using somaclonal variation. This discovery demonstrates that the biotechnology tool of somaclonal variation can result in stable changes in the genetic traits of tissue culture regenerated plants. Several of these single gene changes of tomato have been well characterized and localized to specific regions on the chromosome map of tomato. Among the variants of tomato observed using this procedure were changes in fruit color, plant architecture, and characteristics to improve the ease of mechanical harvesting. This new tool provides scientists with the ability to develop new breeding lines for the generation of improved varieties for food and industrial products from existing varieties within a reasonable time frame. This compresses the time requirement for modification of existing varieties.

From work at DNA Plant Inc. on somaclonal variation on tomatoes (13), one can conclude that: 1) the recovery of several single gene mutations in breeding lines from different tomato varieties; 2) somaclones including dominant, semi-dominant and recessive nuclear mutations; 3)single gene mutations using the somaclonal variation procedure with a frequency in the neighborhood of one mutant in every 20-25 regenerated plants; 4) evidence suggesting that new single gene mutants not previously reported using conventional mutagenesis have been recovered using somaclonal variation; 5) the occurrence of 3:1 ratios for single gene mutants in plants produced by self-fertilization of first generation tissue culture-derived plants which suggests that mutants are of clonal origin and that the mutation occurred prior to shoot regeneration (i.e., no mosaics are detected); 6) evidence suggests that mitotic recombination (reciprocal or non-reciprocal) may also account for some somaclonal variation; and, 7) evidence suggests that mutations in chloroplast DNA (detected by both maternal inheritance and restriction enzyme analysis) can also be recovered.

Variation was previously observed in plants regenerated from long-term cell culture. However, in most cases, these changes have been attributed to abnormalities in chromosome number, due to variations that accumulate in these cells. In a few cases, the plants regenerated from cell culture were evaluated for useful variation. Shepard and his colleagues (8) have used this method to isolate new breeding lines of potato with disease resistance. Although detailed genetic analysis has not been published, it is apparent that regenerated clones are highly variable. There is evidence that some of the variability is due to chromosomal rearrangements (8). As in sugarcane, lines have been identified that are resistant to diseases (late blight and early blight) to which the parent line, Russet Burbank, is susceptible (8). However, potato which is grown from tubers rather than seed, is one of the few asexually propagated crops. The work reported by Evans and Sharp (7) supports this earlier work on potato and extends it to seed propagated crops by establishing a genetic basis for variation from cell culture. It is expected that during the next few years, this new biotechnology tool of somaclonal variation will allow breeders to introduce new useful characteristics into existing plant varieties, thereby developing new, improved crop varieties in a short period of time.

GAMETOCLONAL VARIATION

Gametoclonal variation is recovered by regenerating plants from cultured microspore or pollen cells or pollen still contained within the anther. This genetic variation is the result of both recombination occurring during meiosis and tissue culture induced mutation. The gametes are products of meiosis, governed by Mendel's laws of independent assortment and segregation and contain half the amount of genetic information found in normal somatic cells. The most exciting opportunity is to culture anthers of hybrid plants that contain genetic information that represents a mixture of two parents. The resulting haploid plants can be used to produce new true-breeding recombinant lines that express the best characters of each parent.

To recover gametoclones, anthers containing immature microspores are removed from plants grown in the greenhouse or field and are transferred to a culture medium suitable for recovery of haploid plants. These haploid plants are then treated in some fashion (usually treatment with the chemical colchicine) to double the chromosome number to produce true-breeding lines with two copies of all genes. These doubled haploid plants are extremely valuable as breeding parents since a breeder can be certain that no masked genetic traits are present and can predict the genetic behavior of these plants in subsequent generations (14).

Gametoclonal variation has exciting commercial opportunities in two cases (13). First, many crops, particularly some cereals, can be regenerated more easily from anthers than from somatic tissue. Hence, anthers of rice and wheat have been cultured to recover new variants using a procedure resembling somaclonal variation. This approach has resulted in new breeding lines of rice with disease resistance. Secondly, anther culture can be used to achieve hybrid sorting. In this method, anthers of hybrid lines are cultured to recover haploid plants that express the best characteristics of each parent line. This method of hybrid sorting has been used extensively and quite successfully by Chinese workers to produce several new crop varieties. For example, two rice varieties, Hua Yu No. 1 and No. 2 were developed that express high yield resistance to bacterial blight and wide adaptability. These and other rice varieties developed via gametoclonal variation are currently grown on more than 100,000 hectares in China. Similarly new varieties of wheat (Jingdan 2288) and tobacco (Danyu No. 1) that simultaneously express multiple, desirable characters have been developed using gametoclonal variation (15).

The most exciting potential for gametoclonal variation has not yet been fully realized. Based on the success of the Chinese workers, it appears feasible to use gametoclonal variation to "sort out" commercial F_1 hybrid seed. In one step it will be possible to recover true-breeding lines that express the most desirable characteristics of new commercial hybrids.

PROTOPLAST FUSION

Protoplast or plant cell fusion is a technology that permits the combination of different plant cells and leads to the production of new hybrid plants. In some cases hybrids can be produced that cannot be obtained using conventional breeding methods, resulting in new breeding lines.

Plant cells are surrounded by a cell wall that protects the cell's contents. This wall can be enzymatically digested to produce protoplasts, wall-less cells, that are easier to manipulate. Mixed protoplasts isolated from plant cells of two parents can be fused by using a multistep chemical treatment that includes treatment with polyethylene glycol. Fused protoplasts are then grown in culture medium suitable for plant regeneration to permit recovery of new hybrid plants.

This procedure is limited to those species that are capable of plant regeneration from protoplasts. Nonetheless, several somatic hybrid plants have been recovered that express genetic information from each parent line. This procedure has been particularly useful for recovering hybrids between tobacco and closely related species, but recently has been extended to several food crops (16).

In most cases, somatic hybrids produced by protoplast fusion contain a fairly uniform mixture of genetic information from each parent line. Hence, it is necessary to integrate these hybrids into a conventional breeding program. The process is time-consuming but opens the door to the transfer of new traits that are not accessible using conventional breeding approaches.

At this point in time, no new plant varieties have been produced using protoplast fusion. However, new breeding lines have been produced between cultivated tobacco and several wild tobacco species with disease resistance (17). The disease resistance, particularly resistance to tobacco mosaic virus, has been documented in the new hybrids as well as backcrossed breeding lines developed from these new hybrids. Use of these new hybrids offers a novel source of resistance to several economically important diseases.

Another use of protoplast fusion is the transfer of cytoplasmic genetic information. During fertilization, the male cytoplasm is preferentially excluded from the developing embryo. Thus, in most crops, conventional sexual hybrids contain and express only maternally derived cytoplasmic genes. As protoplast fusion combines both parent cytoplasms into a single protoplast, fusion can be used to produce unique nuclear-cytoplasmic combinations not possible using conventional breeding (18). In some cases, even though protoplasts fuse, the two parental nuclei do not fuse. In these cases, cybrids (cytoplasmic hybrids) are produced that contain novel mixtures of chloroplast and mitochondrial genes. By creating cybrids, small blocks of DNA can be transferred between breeding lines using a one-step process. Several agriculturally important traits, such as disease resistance, herbicide resistance, and male sterility for hybrid seed production are all encoded in chloroplasts and mitochondria. Hence, use of protoplast fusion to transfer organelles will open the door to development of many new plant varieties.

Evidence from a number of somatic hybrids suggests that although two types of cytoplasms are initially mixed during protoplast fusion, resulting in heteroplasmons, eventually one parent type cytoplasm predominates resulting in cytoplasmic segregation. Following segregation, which may not occur until after meiosis, genetic information from the second cytoplasm is irreversibly lost. This phenomenon has been reported by examination of several traits encoded in cytoplasmic genes, including fraction-1 protein in *N. glauca* + *N. langsdorffii* somatic hybrids (19) and male sterility versus male fertility (20). Cosegregation has been reported for several cytoplasmic traits (21). For example, Flick and Evans (22) have found that F-I-P and tentoxin sensitivity cosegregated in *N. glauca* + *N. tabacum* somatic hybrids. Cosegregation implies that following cytoplasmic segregation, the cytoplasm contains all characters from one parent or the other, and, therefore, precludes widespread cytoplasmic gene recombination. Evidence has been presented that mitochondria and chloroplasts sort independently in somatic hybrids (23).

Despite reports of cosegregation of cytoplasmic genes, biochemical evidence using restriction analysis of mitochindrial DNA of some cybrids has suggested that recombination occurs between parental mitochondrial DNAs prior to cytoplasmic segregation in somatic hybrids (24). In these reports, mitochondrial recombination following protoplast fusion produced a wide range of morphological phenotypes controlling male sterility in the hybrid plants. Similarly, Nagy et al. (25) detected recombination in mitochondrial DNA of *N. tabacum* + *N. knightiana* somatic hybrids resulting in novel mitochondrial DNA restriction enzyme patterns. Recombination of chloroplast DNA in higher plants has not yet been documented. Most likely these intriguing results of cytoplasmic segregation and recombination predate an era of extensive research on cytoplasmic genetics with higher plants. Newly established cytoplasmic markers for antibiotic resistance (26) should facilitate research in this area. The production of unique nuclear-cytoplasmic combinations using protoplasts will aid in the development of new breeding lines not available using conventional methods.

This technology has greatest application to those crops capable of plant regeneration from protoplasts. This list presently includes tobacco, tomato, potato, carrot, rapeseed, lettuce, and alfalfa and is expanding rapidly.

PROTECTION OF NEW DEVELOPMENTS

The above techniques are all aimed at the development of new genetic variants of crop plants. However, in order to attract corporate interest, it is also necessary to protect new germplasm developed by these tools. The two approaches that are available to protect new plant breeding lines are male sterility for production of hybrid seed and molecular fingerprinting.

Male sterility is under genetic control and permits inexpensive and efficient production of hybrid seed. Protoplast fusion opens the door to efficient transfer of genes controlling male sterility, especially since these genes are encoded in mitochondria. Hence, cellular geneticists can work with the plant breeder to first identify superior parent lines to permit inexpensive production of the hybrid seed. Commercial F_1 hybrids produced using male sterility are protected since growers that save seed of the F_1 hybrid no longer have the uniform hybrid characteristics in their seed; F_2 seed segregates and produces a non-uniform planting containing many undesirable plants. Hence, when growers use F_1 seed, they must return year after year to purchase new F_1 seed.

Molecular fingerprinting is the use of molecular methods to precisely characterize the genetic information of new breeding lines. DNA can be isolated from a breeding line and can be treated with restriction enzymes to generate a precise banding pattern that uniquely reflects the DNA present in a new breeding line produced by genetic engineering. Such precise characterization will aid in identification of varieties covered by the Plant Variety Protection Act. In addition, a recent patent office decision (*Ex parte* Hibberd) reverses an earlier position and opens the door to obtain utility patents on plant varieties, plants, and plant cell cultures. In following an earlier decision by the U.S. Supreme Court, Diamond v. Chakrabarty (1980), this new patent office ruling affords the same rights to patent developments in plant biotechnology as currently exists for microbial research.

CONCLUSIONS

Genetic engineering related to the development of improved plant varieties for the food processor and the consumer is the big research opportunity of the decade. This makes it possible to breed the raw plant material according to specifications prescribed by the processor or to develop fresh market products geared to the consumer. Clonal propagation is in use for scale-up of special parents for breeding programs, mass propagation of elite plants, and multiplication of virus-free germplasm. Somaclonal variation and gametoclonal variation are keys to genetic engineering technology presently being used by the food industry to develop new breeding lines for variety development. Protoplast technology is being commonly applied to a few crops but is still predominantly in the research laboratory. During the next few years, regeneration procedures will be developed to allow plantlet regeneration from protoplast fusion products on a large-scale that will enlarge the portfolio of food crops amenable to protoplast fusion mediated improvement.

It should be noted that plant biotechnology is moving rapidly, and it is clear that food companies with a leadership position in the technology will retain their product leadership positions as biotechnology moves into the mainstream of raw material modification for the food industry.

REFERENCES

1. Hu, C. Y. and Wang, P. J. Meristem, shoot tip and bud culture. In: "Handbook of Plant Cell Culture" (D. A. Evans, W. R. Sharp, P. V. Ammirato, and Y. Yamada, eds.) pp. 177-227, MacMillan Press, NY (1983)
2. Morris, C. Developments in genetic engineering. Food Engineering 55:18-38 (1983)
3. Murashige, T. The impact of plant cell culture on agriculture. In: "Frontiers of Plant Tissue Culture" (T. A. Thorpe, ed.) pp 15-26, Univ. Calgary Press, Calgary, Canada (1978)
4. Ammirato, P. V. The regulation of somatic embryo development in plant cell cultures: Suspension culture techniques and hormone requirements. Bio/Technology 1:68-74 (1983)
5. Ammirato, P. V. In: "Handbook of Plant Cell Culture" (P. V. Ammirato, D. A. Evans, W. R. Sharp, and Y. Yamada, eds.) Vol. 3, pp. 327-354, Macmillan Press, NY (1983)
6. Styer, D. J. Bioreactor technology for plant propagation. In: "Tissue Culture in Forestry and Agriculture" (R. Henke, K. Hughes, M. Constantin, and A. E. Hollaender, eds.) pp. 117-130, Plenum Press, NY (1985)
7. Evans, D. A., and Sharp, W. R. Single gene mutations in tomato plants regenerated from tissue culture. Science 221:949-951 (1983)
8. Shepard, J. The regeneration of potato plants from leaf cell protoplasts. Scientif. Amer. 246:154-166 (1982)
9. Larkin, P. J., and Scowcroft, W. R. Somaclonal variation—a novel source of variability from cell cultures for plant improvement. Theoret. Appl. Genet. 60:197-214 (1981)

10. Evans, D. A. and Sharp, W. R. Somaclonal and gametoclonal variation. In: "Handbook of Plant Cell Culture" (D. A. Evans, et al, eds.) Vol. 4 Macmillan Press, NY (1985)
11. Skirvin, R. M., and Janick, J. Tissue culture-induced variation in scented *Pelargonium* spp. J. Amer. Soc. Hort. Sci. 101:281-290 (1976)
12. D'Amato, F. Cytogenetics of differentiation in tissue cell cultures. In: "Plant, Cell, Tissue, and Organ Culture" (J. Reinert and Y. P. S. Bajaj, eds.) pp. 343-357, Springer-Verlag, Berlin (1977)
13. Evans, D. A., Sharp, W. R., and Medina-Filho, H. P. Somaclonal and gametoclonal variation. Amer. J. Bot. 71:759-774 (1984)
14. Sharp, W. R., Reed, S. M., and Evans, D. A. Production and application of haploid plants. In: "Contemporary Bases for Crop Breeding" (P. B. Vose and S. Blixt, eds.) pp. 347-381, Pergamon Press, London (1984)
15. Chu, C. C. Haploids in plant improvement. In: "Plant Improvement and Somatic Cell Genetics" (I. K. Vasil, W. R. Scowcroft, and K. J. Frey, eds.) pp. 129-158, Academic Press, NY (1982)
16. Bravo, J. E. and Evans, D. A. Protoplast fusion for crop improvement. In: "Plant Breeding Reviews" (J. Janick, ed.), Vol. 3 pp. 193-218, AVI Press, Westport, CN (1985)
17. Evans, D. A., Flick, C. E., and Jensen, R. A. Incorporation of disease resistance into sexually incompatible somatic hybrids of the genus *Nicotiana*. Science 219:907-909 (1981)
18. Gleba, Yu. and Evans, D. A. Hybridization of somatic plant cells: Genetic analysis. In: "Genetic Engineering: Principles and Methods" (A. Hollaender and J. K. Setlow, eds.) Vol. 6, pp 175-209, Plenum Press, NY (1984)
19. Chen, K., Wildman, S.G., and Smith, H.H. Chloroplast DNA distribution in parasexual hybrids as shown by polypeptides composition of fraction-1 protein. Proc. Nat. Acad. Sci., Wash. 74:5109-5112 (1979)
20. Flick, C. E., Kut, S. A., Bravo, J. E., Gleba, Yu., and Evans, D. A. Segregation of organelle traits following protoplast fusion in *Nicotiana*. Bio/Technology 3:555-560 (1985)
21. Aviv, D., Fluhr, R., Edelman, M., and Galun, E. Progeny analysis of the interspecific somatic hybrids: *Nicotiana tabacum* (cms) + *Nicotiana sylvestris* with respect to nuclear and chloroplast markers. Theoret. Appl. Genet. 56:145-150 (1980)
22. Flick, C. E., and Evans, D. A. Evaluation of cytoplasmic segregation in somatic hybrids in the genus *Nicotiana*: Tentoxin sensitivity. J. Hered. 73:264-266 (1982)
23. Galun, E. Somatic cell fusion for inducing cytoplasmic exchange: A new biological system for cytoplasmic genetics in higher plants. In: "Plant Improvement and Somatic Cell Genetics" (L K. Vasil, W. R. Scowcroft, and K. J. Frey eds.), pp. 205-220, Academic Press, New York (1982)
24. Belliard, G., Vedel, F., and Pelletier, G. Mitochondrial recombination in cytoplasmic hybrids of *Nicotiana tabacum* by protoplast fusion. Nature 281:401-403 (1979)
25. Nagy, F., Torol, L, and Maliga, P. Extensive rearrangements in the mitochondrial DNA in somatic hybrids of *Nicotiana tabacum* and *Nicotiana knightiana*. Molec. Gen. Genet. 183:437-439 (1981)
26. Fluhr, R., Aviv, D., Galun, E., and Edelman, M. Efficient induction and selection of chloroplast-encoded antibiotic-resistant mutants in *Nicotiana*. Proc. Nat. Acad. Sci. 82:1485-1489 (1985)

Chapter 9

Application of Genetic Engineering Techniques for Dairy Starter Culture Improvement

Larry McKay

INTRODUCTION

Due to the importance of genetics in strain improvement programs, the discussion presented here will concentrate on the potential application of genetic engineering techniques to the industrially important dairy streptococci. This chapter is not intended to be a comprehensive review of the literature on plasmid biology and genetics of dairy streptococci. This topic has been eloquently reviewed by Gasson and Davies (1).

Many properties in mesophilic streptococci are unstable and may be spontaneously lost, resulting in strains that are unsuitable for manufacturing purposes. Other strains produce undesirable off-flavors, and most strains are susceptible to attack by bacteriophage. Because of such deficiencies the number of suitable strains for some processes is rather limited and there is a continual need to obtain new or modified strains. Furthermore, as advanced technology (large mechanized cheese operations and use of retentates for cheese manufacture) is introduced into the dairy fermentation industry, a wider range of strains with a greater array of metabolic capabilities will be required. The question concerns how these strains will be developed.

STRAIN SELECTION AND DEVELOPMENT

The potentiality of a microorganism is determined by the nature of the genetic material possessed in the individual cell. Access to the large number of possible variations in the expression of that genotype can be obtained in several ways (2). One method is the isolation and selection of strains from ecological sources such as green plant material or raw milk.

In this process, which could be referred to as natural genetic engineering, nature develops the strains and industry employs an extensive screening program in order to isolate variants having the necessary metabolic properties for use in dairy fermentation processes. This is the technique which the dairy industry has primarily relied upon since the turn of the century. It is archaic in contrast to modern strain improvement programs especially with respect to available technology in molecular biology. A second way to develop a new strain is the artificial mutagenesis of existing strains for a particular purpose. This could be referred to as the classical genetic approach to strain improvement, but it too has been more or less ignored by the dairy fermentation industry. The third approach, and the one that has sparked excitement in the dairy and food industries, is the use of recombinant DNA technology (r-DNA) to actually "tailor-make" a strain for a particular purpose. The application of this approach to dairy or food fermentation microorganisms is likely to be slow. One explanation is that the fundamental genetic knowledge of the microorganisms involved in food fermentations has not been adequately developed (3). These industries are now playing catch-up so that they too may exploit this new technology. Application of r-DNA technology requires an advanced understanding of the genetic make-up and plasmid biology of the microorganisms of interest, as well as the development of gene transfer systems which are functional with these organisms.

Plasmid Biology

Plasmid biology has become an exciting area of research with respect to the lactic acid bacteria involved in dairy, meat, and vegetable fermentation processes, as well as in those strains used for silage inocula and as probiotics. Plasmids can be defined as small circular pieces of DNA which exist in the bacterial cell and replicate independent of the chromosome. A failure of plasmid duplication during parental cell division results in two daughter cells, one that possesses the plasmid and one that is devoid of that particular plasmid. The latter cell is said to be "cured" and will have lost those metabolic properties dictated by the plasmid DNA.

It is now well established that the mesophilic group N streptococci used in dairy fermentation processes harbor plasmids of diverse sizes (4). The average range of distinct plasmid types per cell appears to be from 4 to 7, although the actual range may be from 2 to more than 10. Most of these plasmids are cryptic (function unknown), but some have now been shown to carry identifiable properties. These are listed in Table 9-1.

Lactose Metabolism: In order to grow in milk, dairy streptococci must have the capacity to ferment lactose to lactic acid. This reaction furnishes the cell with energy in the form of ATP, necessary for growth

Table 9-1. Metabolic Functions That Have Been Linked or Tentatively Linked to Plasmid DNA in Mesophilic Dairy Streptococci.

Function	Selected References
1. Sugar utilization	
a. Lactose	1, 4
b. Galactose	5, 6
c. Sucrose	7, 8, 9
d. Glucose and mannose	7
e. Xylose	7
2. Proteinase activity	1, 4, 10
3. Citrate utilization	1, 4, 11
4. Production of antagonistic compounds	
a. Nisin	1, 4, 7, 8, 9
b. Diplococcin	12
c. Bacteriocins	13
5. Nisin resistance independent of nisin production	14
6. Resistance to UV	15
7. Bacteriophage resistance	
a. Restriction/modification systems	16, 17
b. Reduced phage adsorption	18, 19
c. Temperature-dependent mechanism	14, 20
8. Copper resistance	21
9. Kanamycin resistance	22

and cell maintenance, and lactic acid which accumulates and contributes to the flavor and physical properties of the finished product. In addition, the low pH achieved aids in preventing the growth of spoilage and pathogenic organisms (23). The ability of group N streptococci to ferment lactose is an unstable property in many strains and the genes responsible have now been linked to plasmid DNA. The specific enzymes coded by the plasmid are EII^{lac}, $FIII^{lac}$, and phospho-β-galactosidase of the phosphoenolpyruvate (PEP)-dependent phosphotransferase system (24). Although the ability to metabolize lactose is a plasmid-mediated characteristic within these bacteria, cloning the genes for lactose metabolism on a high copy-number plasmid could increase the gene dosage and thus conceivably increase the rate of acid production. The cloning of these genes on a high copy-number plasmid has occurred naturally in a derivative of *S. lactis* (25). Although the phospho-β-galactosidase activity was about twice that found in the strain carrying the normal lactose plasmid, it had no effect upon growth or rate of acid production when the cells were propagated in milk. In this particular case, the linkage of phospho-β-galactosidase to a high copy-number plasmid did not increase the organism's ability to produce acid; however, other metabolic traits may react differently when associated with a high copy-number plasmid.

Unlike the group N streptococci, *S. thermophilus* has received considerably less research attention. This organism hydrolyzes lactose via a β-galactosidase to yield glucose and galactose. The β-galactosidase gene is assumed to be chromosomally-encoded since strains with no detectable plasmids ferment lactose and exhibit β-galactosidase activity (26,27). The hydrolysis of lactose to its constituent monosaccharides by β-galactosidase is of interest for several reasons. Lactose is not digested very well by a large proportion of the world's population (28). Lactose also has low solubility in water which leads to problems in concentrating whey and in preparation of certain food items (29). When compared to sucrose, lactose has a relatively low level of sweetness. These problems can be overcome to a large extent by hydrolysis of lactose to its monosaccharides which are sweeter, more soluble, and more digestible than lactose (29). Investigators have thus searched for microorganisms capable of producing high levels of β-galactosidase (30,31,32,33). Although a number of sources of β-galactosidase now exist, as suggested by Greenberg and Mahoney, *S. thermophilus* is a promising source for the production of this enzyme because it is a food approved organism and also because the *S. thermophilus* β-galactosidase is more heat stable than the β-galactosidase from other sources used commercially (34). These reports suggest that genetic engineering techniques, including r-DNA technology could not only be used to improve upon the organism's ability to ferment lactose, but could also be used to construct strains, for commercial purposes, which overproduce β-galactosidase.

Proteinase Activity: When grown in milk, the group N streptococci are also dependent on their proteinase enzyme system to degrade casein and thus acquire the necessary nitrogenous compounds, in the form of amino acids and peptides (23). Casein degradation may also be involved in cheese ripening. Proteinase activity is an unstable property and has been linked to plasmid DNA in some of these strains (1,4,10). As suggested by Gasson and Davies (1), the genes for different proteinases could be exchanged between strains, and control over the level of their expression should be possible through the application of genetic engineering techniques. This could possibly lead to strains capable of accelerating the ripening of cheese, either by manipulating existing starter proteinases or through the introduction of new ones from non-dairy sources (1). This approach is nearing reality due to the recent cloning of a piece of DNA coding for the proteinase enzyme system from *S. cremoris* Wg2 (10).

Citrate Metabolism: A third characteristic vital for certain dairy fermentation processes is the ability of *Streptococcus lactis* subspecies *diacetylactis* to ferment citrate producing diacetyl (butter aroma). This property has been linked to a 5.5 Mdal [Mega Dalton] plasmid coding for the transport of citrate into the cell (1,4,11). Work on this plasmid, as well as attempts to develop genetic techniques applicable to the dairy streptococci should eventually lead to the construction of strains that more efficiently utilize citrate and subsequently produce the flavor compound diacetyl (11). It may also be possible to use the "classical genetic approach" to isolate mutants of *S. diacetylactis* defective in lactic dehydrogenase. Such mutants may be capable of producing sufficient quantities of diacetyl in the absence of citrate (35).

Dairy streptococci are undeniably indispensable for the manufacture of many fermented dairy products and, as indicated above, a number of these critical cheese-making functions are encoded by genes carried on unstable plasmid DNA. These findings suggest that genetic engineering techniques could be used to stabilize these important functions by integrating the plasmid or the desired gene into the chromosome. This has been accomplished for lactose utilization and for the proteinase enzyme system from *S. lactis* C2 and *S. lactis* 712 (1,36). Such an approach could lead to "stabilized" dairy starter cultures.

Production of Antagonistic Compounds: The ability of dairy streptococci to produce antagonistic compounds, other than hydrogen peroxide and lactic acid, is well documented. These substances include nisin produced by some *S. lactis* strains, diplococcin produced by several *S. cremoris* strains and bacteriocins produced by various group N streptococci. Diplococcin production in *S. cremoris* has been linked to a 54 Mdal plasmid (12) and bacteriocin production in *S. diacetylactis* WM4 has been linked to an 88 Mdal plasmid (13). In addition, the conjugal transfer of the genetic determinants for nisin (8,9), diplococcin (12), and bacteriocin (13) have been demonstrated. Nisin is of particular interest since it is a polypeptide antibiotic that has potential as a food preservative. The possible linkage of nisin-producing ability to plasmid DNA may lead to the construction of a "super-nisin"-producing derivative through genetic manipulations (8); such a strain may have commercial value. In addition, the ability to transfer the genetic factors controlling inhibitory substance production to other lactic acid bacteria may ultimately lead to construction of strains with enhanced antagonistic properties against food spoilage organisms and food-borne pathogens. Such strains could extend the shelf-life of fermented dairy products as well as other food products.

Bacteriophage Resistance: The other major microbiological problem facing the dairy fermentation industry is the presence of bacteriophage. The isolation of phage-resistant mutants as a means of circumventing the problem was first attempted in 1934 in New Zealand. The process was more or less abandoned in the 1950's, because many of the resistant strains isolated were slow acid producers, reverted to phage sensitivity, or were lysed by a new group of phages (37). In the 1960's workers in New Zealand again attempted to isolate phage-resistant strains for commercial purposes. Appropriate strains were isolated that did, for a time at least, extend the number of strains available. As phage began to appear against these strains, the process was again abandoned (37). As stated by the New Zealand workers (38), the problem facing the starter culture technologist was how to select for strains that would remain resistant long enough to make their introduction into cheese plants worthwhile. Then in 1976 two reports were published, one from England (39) and the other from New Zealand (40), documenting the successful isolation of phage-resistant mutants. The New Zealand workers subsequently developed a defined starter system consisting of a blend of phage-resistant mutants (41). This blend of strains is used without rotations, but the whey is regularly monitored for the presence of phage. If phage appears against any of the strains, that strain is removed from the blend and a phage-resistant mutant is developed to replace it. This system has been studied by several U.S. laboratories (42,43) and is now also used commercially in the U.S.

The question is whether phage-resistant mutants can be constructed on a rational basis. The answer is positive for several reasons. It is now recognized that certain plasmids in the group N streptococci code for restriction-modification (R/M) systems. Sanders and Klaenhammer have shown that in *S. cremoris* KH, 2 to 5% of the colonies isolated after a heat challenge were partially deficient in R/M capacity and a 10 Mdal plasmid was missing in these variants (16). Chopin et al. in France (17), examined a strain of *S. lactis* that contained 9 plasmids, and provided evidence that R/M systems were related to a 19 Mdal and a 21 Mdal plasmid in this strain. Genetic evidence was also provided through conjugal transfer of the 19 Mdal plasmid to a plasmid-free recipient. Sanders and Klaenhammer have presented evidence that a 30 Mdal plasmid in *S. lactis* ME2 coded for a property that retarded phage adsorption (18). Work by de Vos, et al. also showed that variants of *S. cremoris* SK11 which contained a 34 Mdal plasmid were resistant to a given phage, whereas those variants missing the 34 Mdal plasmid were sensitive to the phage (19). The 34 Mdal plasmid was shown to confer resistance by reducing phage adsorption. It was also shown by McKay and Baldwin (14) that a 40 Mdal plasmid isolated from *S. diacetylactis* DRC3 coded for phage resistance at 21°C and 32°C but not at 37°C. The plasmid was also effective against a variety of phages and was conjugally transferred to *S. cremoris* B1 and *S. lactis* H1 (McKay and Baldwin, unpublished data). Both strains acquired resistance to their host phage. The mechanism of phage resistance is currently unknown, but does not appear to involve R/M systems or inability of the phage to adsorb. Steenson and Klaenhammer reported the presence of a conjugal plasmid in *S. lactis* ME2 that confers lactose-fermenting ability and a phage-resistance mechanism which is heat sensitive (20). Unlike *S. lactis* transconjugants, *S. cremoris* ME2 transconjugants carrying the conjugal plasmid did not demonstrate phage resistance inactivation at high temperature and appeared to stably maintain the plasmid.

Clearly, the increasing evidence for the association of plasmids with phage resistance in lactic streptococci provides a genetic mechanism to explain the rapid development of phage-sensitive dairy starter cultures. The manipulation of these plasmids by genetic engineering techniques is one approach for obtaining phage-resistant mutants for commercial purposes. It may also be possible to develop phage-resistant strains by conjugally transferring the appropriate plasmid to selected phage-sensitive strains of dairy streptococci, as well as to other lactic acid bacteria. Additionally, it may be of value to combine the different genetic loci for phage resistance into a single strain and to stabilize the phage resistance phenotype by integrating the phage-resistance genes into the chromosome or into a high copy-number plasmid.

Other Potential Strains

Further efforts on the application of genetics to dairy streptococci could be in the development of strains more suitable for mass culturing techniques, in obtaining a larger number of strains for use in preparing frozen culture concentrates, or in developing strains for use as freeze-dried or spray-dried culture concentrates. Although there are many alterations that could be made in strains for dairy fermentation processes, it must be kept in mind that such variants should retain their proper acid-producing abilities as well as their ability to bring about the desired physical change and to produce the desired flavor characteristics in the final product.

GENE TRANSFER SYSTEMS

To facilitate the manipulation of desired plasmid or chromosomally-linked genes, the development of gene transfer systems in dairy streptococci was required. Conjugation and transduction are well documented and transfection and protoplast fusion have also been reported (1). High-frequency conjugative plasmids may be important in constructing strains for industrial purposes, since it will be necessary to have efficient systems for moving genes from one strain to another. Anderson and McKay (44) have isolated a recombinant plasmid containing a transfer region which correlates with high-frequency conjugation. It may be possible to move this region onto other non-conjugative plasmids, in order to get rapid mobilization of the plasmid into a desired strain. Alternatively, high-frequency conjugative plasmids might be developed into cloning vectors applicable to dairy fermentation organisms. Although the above gene transfer systems have aided in studying the genetics and plasmid biology of these organisms, the development of an efficient plasmid transformation system was vital for further genetic studies and for the use of recombinant DNA technology for strain improvement. One method was to utilize existing transformation systems in *Streptococcus sanguis* Challis and *Escherichia coli* in order to clone in genes from *S. lactis* for further analysis (45,46,47). The alternative approach was to develop a transformation system within dairy streptococci. Kondo and McKay described polyethylene glycol-induced transformation of *S. lactis* protoplasts using plasmid DNA (48).

The frequency was variable, but upon optimizing various parameters, transformation was increased to about 104 transformants per μg DNA. The feasibility of using this *S. lactis* protoplast transformation system in molecular cloning was then demonstrated by cloning a DNA fragment coding for the lactose-metabolizing genes into *S. lactis* (48). The development of a transformation system in dairy streptococci was a key to performing recombinant DNA experiments for strain improvement.

CONCLUSIONS

A high frequency transformation system, coupled with the development of other gene transfer systems, will open new avenues of investigation for the study of gene expression, gene regulation, and plasmid analysis in the lactic streptococci. One could speculate endlessly on the possibilities for future application of genetic engineering to dairy starter cultures. Some of the more realistic, and potentially beneficial applications include: 1) developing naturally-produced flavor compounds (pineapple, citrus, grape-like, peach, fruity, banana) through the isolation of the responsible genes and transfer of these genes to lactic acid bacteria; 2) developing strains capable of converting whey into a marketable end product through flavor and textural alterations; 3) cloning genes coding for the production of proteins with intrinsic sweetness qualities so that addition of the altered starter culture would lessen the need for supplementary sweeteners; 4) developing food grade organisms into hosts used to produce medicinals for human use (growth hormones, for example), thereby circumventing the need for extensive purification when *E. coli* is used as the host; and, 5) cloning genes beneficial to human or animal health including enzymes and nutritive additives.

It is now well established that mesophilic streptococci used in starter cultures harbor plasmids of diverse sizes and that some of these plasmids code for properties vital for successful dairy fermentation processes. Amplification of desired plasmid genes should be possible through the isolation of copy-number mutants, which may increase the efficiency of the fermentation process as well as the quality of the final product. Stabilization of plasmid-mediated traits by integrating the genes into the chromosome may also prove beneficial. Clearly, the study of plasmid biology in dairy starter cultures has become a prerequisite for future strain improvement programs. This knowledge, coupled with the developing plasmid-transfer systems applicable to dairy streptococci is essential for gene cloning work within this group of bacteria.

REFERENCES

1. Gasson, M. J., and Davies, F. L. In: "Advances in the Microbiology and Biochemistry of Cheese and Fermented Milk" (F. L. Davies and B. A. Law, eds.) pp. 99-126, Elsevier Applied Science Publishers, New York (1984)
2. Ralph, R. J. In: "Dairy Fermentation Technology" (P. M. Linkletter, ed.) pp. 1-9. The University of New South Wales, Kensington (1969)
3. Legates, J. E. In: "Impact of Applied Genetics (Microorganisms, Plants, and Animals)" pp. 107-114. U.S. Office of Technology Assessment. U.S. Government Printing Office, Washington, D.C. (1981)
4. McKay, L. L. Functional properties of plasmids in lactic streptococci. Antonie van Leeuwenhoek 49:259-274 (1983)
5. Park, Y. H., and McKay, L. L. Distinct galactose phosphoenol-pyruvate-dependent phosphotransferase system in *Streptococcus lactis*. J. Bacteriol. 149:420-425 (1982)
6. Crow, V. L., Davey, G. P., Pearce, L. E., and Thomas, T. D. Plasmid linkage of the D-tagatose-6-phosphate pathway in *Streptococcus lactis*: Effects of lactose and galactose metabolism. J. Bacteriol. 153:76-83 (1983)
7. LeBlanc, D. J., Crow, V. L., and Lee, L. N. In: "Plasmids and Transposons: Environmental Effects and Maintenance Mechanisms" (C. Stuttard and K. R. Rozee, eds.) pp. 31-41, Academic Press, New York (1980)
8. Gasson, M. J. Transfer of sucrose fermenting ability, nisin resistance, and nisin production in *Streptococcus lactis* 712. FEMS Microbiol. Lett. 21:7-10 (1984)
9. Gonzalez, C. F., and Kunka, B. S. Transfer of sucrose-fermenting ability and nisin production phenotype among lactic streptococci. Appl. Environ. Microbiol. 49:627-633 (1985)
10. Kok, J., Maarten van Dijl, J., van der Vossen, J. M. B. M., and Venema, G. Cloning and expression of *Streptococcus cremoris* proteinase in *Bacillus subtilis* and *Streptococcus lactis*. Appl. Environ. Microbiol. 50:94-101 (1985)
11. Kempler, G. M., and McKay, L. L. Biochemistry and genetics of citrate utilization in *Streptococcus lactis* subsp. *diacetylactis*. J. Dairy Sci. 64:1527-1539 (1981)
12. Davey, G. P. Plasmid associated with diplococcin production in *Streptococcus cremoris*. Appl. Environ. Microbiol. 48:895-896 (1984)
13. Scherwitz, K. M., Baldwin, K. A., and McKay, L. L. Plasmid linkage of a bacteriocin-like substance in *Streptococcus lactis* subsp. *diacetylactis* WM4 and its transferability to *Streptococcus lactis*. Appl. Environ. Microbiol. 45:1506-1508 (1983)
14. McKay, L. L., and Baldwin, K. A. Conjugative 40-megadalton plasmid in *Streptococcus lactis* subsp. *diacetylactis* DRC3 is associated with resistance to nisin and bacteriophage. Appl. Environ. Microbiol. 47:68-74 (1984)
15. Chopin, M. C., Moillo-Batt, A., and Rouault, A. Plasmid-mediated UV-protection in *Streptococcus lactis*. FEMS Microbiol. Lett. 26:243-245 (1985)
16. Sanders, M. E., and Klaenhammer, T. R. Evidence for plasmid linkage of restriction and modification in *Streptococcus cremoris* KH. Appl. Environ. Microbiol. 42:944-950 (1981)
17. Chopin, A. M., Chopin, C., Moillo-Batt, A., and Langella, P. Two plasmid-determined restriction and modification systems in *Streptococcus lactis*. Plasmid 11:260-263 (1984)

18. Sanders, M. E., and Klaenhammer, T. R. Characterization of phage-sensitive mutants from a phage-insensitive strain of *Streptococcus lactis*: Evidence for a plasmid determinant that prevents phage adsorption. Appl. Environ. Microbiol. 46:1125-1133 (1983)
19. de Vos, W. M., Underwood, H. M., and Davies, F. L. Plasmid encoded bacteriophage resistance in *Streptococcus cremoris* SK11. FEMS Microbiol. Lett. 23:175-178 (1984)
20. Steenson, L. R., and Klaenhammer, T. R. Heat stable phage resistance in *Streptococcus cremoris* by a high frequency conjugal plasmid. Abstr. of the Annual Meeting of the American Society for Microbiol. p. 126 (1985)
21. Otto, R., de Vos, W. M., and Gavrieli, J. Plasmid DNA in *Streptococcus cremoris* Wg2: Influence of pH on selection in chemostats of a variant lacking a protease plasmid. Appl. Environ. Microbiol. 43:1272-1277 (1982)
22. Dobrzanski, W. T., Bardowski, J., Kozak, W., and Zajdel, J. In: "Microbiology - 1982" (D. Schlessinger, ed.) pp. 225-229, American Society for Microbiology, Washington, D.C. (1982)
23. Sharpe, M. E. Lactic acid bacteria in the dairy industry. J. Soci. Dairy Technol. 32:9-18 (1979)
24. McKay, L. L. In: "Developments in Food Microbiology - I" (R. Davies, ed.) pp. 153-182, Applied Science Publishers Ltd., Essex, England (1982)
25. Anderson, D. G., and McKay, L. L. In vivo cloning of *lac* genes in *Streptococcus lactis* ML3. Appl. Environ. Microbiol. 47:245-249 (1984)
26. Herman, R. E., and McKay, L. L. Isolation and partial characterization of plasmid DNA from *Streptococcus thermophilus*. Appl. Environ. Microbiol. 50:1103-1106 (1985)
27. Hutkins, R., Morris, H. A., and McKay, L. L. Galactokinase activity in *Streptococcus thermophilus*. Appl. Environ. Microbiol. 50:777-780 (1985)
28. Hourigan, J. A. Nutritional implications of lactose. Austr. J. Dairy Technol. 39:114-120 (1984)
29. Smart, J. B., Crow, V. L., and Thomas, T. D. Lactose hydrolysis in milk and whey using β-galactosidase from *Streptococcus thermophilus*. N. Z. J. Dairy Sci. Technol. 20:43-56 (1985)
30. Blankenship, L. C., and Wells, P. A. Microbial beta-galactosidase: A survey for neutral pH optimum enzymes. J. Milk Food Technol. 37:199-202 (1974)
31. Rao, M. V., and Datta, S. M. Lactase activity of microorganisms. Folia Microbiol. 23:210-215 (1978)
32. Sorensen, S. G., and Crisan, E. V. Thermostable lactase from thermophilic fungi. J. Food Sci. 39:1184-1187 (1974)
33. Wierzbicki, L. E., and Kosikowski, F. V. Lactase potential of various microorganisms grown in whey. J. Dairy Sci. 56:26-32 (1973)
34. Greenberg, N. A., and Mahoney, R. R. Production and characterization of β-galactosidase from *Streptococcus thermophilus*. J. Food Sci. 47:1824-1835 (1982)
35. McKay, L. L., and Baldwin, K. A. Altered metabolism in a *Streptococcus lactis* C2 mutant deficient in lactic dehydrogenase. J. Dairy Sci. 57:181-186 (1974)
36. McKay, L. L., and Baldwin, K. A. Stabilization of lactose metabolism in *Streptococcus lactis* C2. Appl. Environ. Microbiol. 36:360-367 (1978)
37. Lawrence, R. C., and Pearce, L. E. Cheese starters under control. Dairy Industries 37:73-78 (1972)
38. Heap, H. A., and Lawrence, R. C. Selection of starter strains for cheesemaking. N. Z. J. Dairy Sci. Technol. 11:16-20 (1976)
39. Marshall, R. J., and Berridge, N. J. Selection and some properties of phage-resistant starters for cheese-making. J. Dairy Res. 43:449-458 (1976)

40. Limsowtin, G. K. Y., and Terzaghi, B. E. Phage resistant mutants: Their selection and use in cheese factories. N. Z. J. Dairy Sci. Technol. 11:251-256 (1976)
41. Limsowtin, G. K. Y., Heap, H. A., and Lawrence, R. C. Multiple starter concept for cheesemaking. N. Z. J. Dairy Sci. Technol. 12:101-106 (1977)
42. Thunell, R. K., Sandine, W. E., and Bodyfelt, F. W. Phage-insensitive, multiple-strain starter approach to Cheddar cheese making. J. Dairy Sci. 64:2270-2277 (1981)
43. Richardson, G. H., Hong, G. L., and Ernstrom, C. A. Defined single strains of lactic streptococci in bulk culture for Cheddar and Monterey cheese manufacture. J. Dairy Sci. 63:1981-1986 (1980)
44. Anderson, D. G., and McKay, L. L. Genetic and physical characterization of recombinant plasmids associated with cell aggregation and high-frequency conjugal transfer in *Streptococcus lactis* ML3. J. Bacteriol. 158:954-962 (1984)
45. Harlander, S. K., and McKay, L. L. Transformation of *Streptococcus sanguis* Challis with *Streptococcus lactis* plasmid deoxyribonucleic acid. Appl. Environ. Microbiol. 48:342-346 (1984)
46. Harlander, S. K., McKay, L. L., and Schachtele, C. F. Molecular cloning of the lactose metabolizing genes from *Streptococcus lactis*. Appl. Environ. Microbiol. 48:347-351 (1984)
47. Harlander, S. K., McKay, L. L., and Schachtele, C. F. Construction of a gene bank of *Streptococcus lactis* plasmid pLM2001 deoxyribonucleic acid in *Escherichia coli*. J. Dairy Sci. 68:1337-1345 (1985)
48. Kondo, J. K., and McKay, L. L. Plasmid transformation of *Streptococcus lactis* protoplasts: Optimization and use in molecular cloning. Appl. Environ. Microbiol. 48:252-259 (1984)

Chapter 10

Production of L-Ascorbic Acid from Whey

Theodore Cayle, John Roland, David Mehnert, Robert Dinwoodie, Russell Larson, Jeremy Mathers, Maura Raines, Warren Alm, Samir Ma'ayeh, Sarah Kiang and Richard Saunders

INTRODUCTION

For each pound of cheese manufactured there are approximately nine pounds of whey by-product created. In 1983 there were 4.82 billion pounds of cheese produced in the United States. This translates to approximately 2.75 billion pounds of whey solids that had to be disposed of in some manner. While there are a number of uses that have been found for whey, the cost of disposing of this waste product stream is staggering. Large cheese manufacturers can spend millions of dollars per year in this endeavor.

In addition, as environmental controls become more stringent, fewer municipalities tolerate the combining of the high BOD whey with other effluents processed by the local sewage disposal facilities. Similar concerns are also preventing traditional land spreading practices from continuing in many parts of the country.

Ultrafiltration has provided the cheesemaker with a means of channeling the component of whey with the greatest value into the marketplace. The whey protein concentrate (WPC) is an item of worldwide commerce, with significant dollar value. In excess of 86 million pounds of WPC, worth $40 million, were sold in the United States in 1983.

That is the bright side. The dark side is that for each pound of WPC produced, 2.3 pounds of permeate solids are also produced, with a lactose content of approximately 85%. It is projected that 400 million pounds of permeate solids will be available in the United States in 1985.

A number of options have been proposed to convert permeate to a value added product: lactose sweetener, single cell protein, baker's yeast, alcohol and methane are examples of commercially viable products currently being produced. Other more profitable alternatives are constantly being evaluated.

Whereas, commercial production of vitamin C is from D-glucose via the "Reichstein synthesis" (1), the reports in the literature disclosing that plants and yeasts are capable of synthesizing L-ascorbic acid from galactose (2-9) provided the impetus to investigate this route as a means of obtaining the vitamin from whey.

The final step in the biological synthesis is the oxidation of L-galactono-1,4-lactone to L-ascorbic acid. In higher plants this is catalyzed by galactonolactone dehydrogenase (EC 1.3.2.3) which requires cytochrome C as the electron acceptor. However, yeasts employ L-galactonolactone oxidase (EC 1.1.3.24), an enzyme which uses oxygen as the electron acceptor. It was for this reason that it was desirable to evaluate a yeast bioconversion as the final step in a commercially feasible synthesis of L-ascorbic acid from whey. This chapter will present the steps in the selection and modification of a suitable yeast, and its use in the bioconversion of L-galactono-1,4-lactone to L-ascorbic acid.

METHODS

The following steps, also shown in Figure 10-1, summarize the total synthesis employed, for the most part using conventional catalysis to produce the lactone from whey.

1) Whey is ultrafiltered and both WPC and permeate are recovered.
2) Permeate is continuously hydrolyzed via an immobilized lactase reactor (Damrace Hydrolysis System, Damrow Co., Madison, WI).
3) The D-glucose/D-galactose stream is fed to a continuous fermenter containing a mutant, flocculating strain of *Saccaromyces cerevisiae*, where the glucose is converted to ethanol.
4) The beer is continuously withdrawn and the alcohol is recovered by distillation.
5) The galactose is recovered by ion exclusion chromatography (Dowex 1-X8, bisulfite, Dow Chemical U.S.A., Functional Products and Systems Department, Midland, MI).
6) The dried galactose is oxidized with conventional inorganic catalysts to D-galacturonic acid.
7) The D-galacturonic acid is reduced with conventional inorganic catalysts to L-galactonic acid.
8) The L-galactonic acid is concentrated to remove water, thereby forming L-galactono-1,4-lactone.
9) The L-galactono-1,4-lactone is converted to L-ascorbic acid via a mutant strain of *Candida norvegensis*, grown on the ethanol produced in step 4.

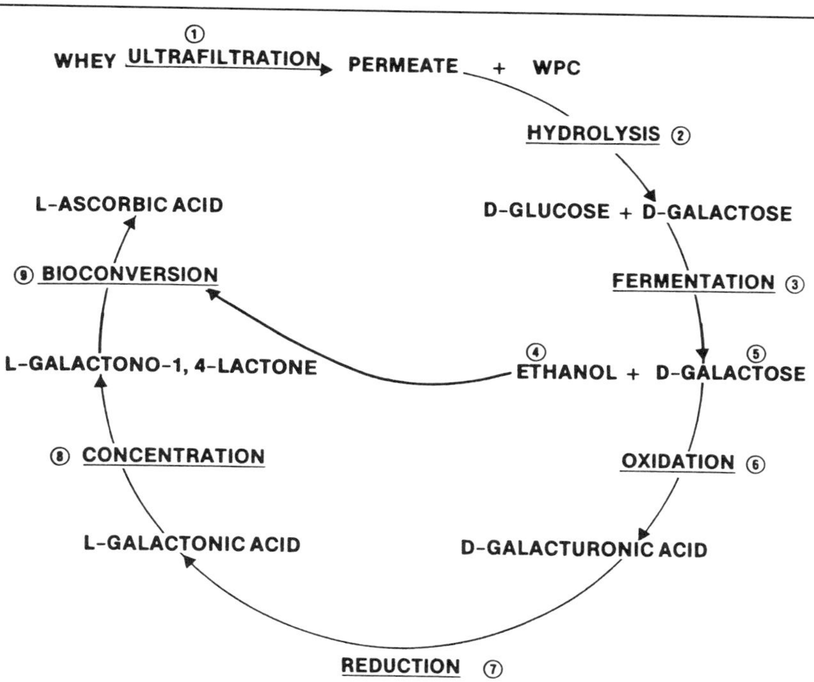

Figure 10-1. Steps in the Synthesis of L-Ascorbic Acid from Whey

The chemical steps in the synthesis of L-ascorbic acid from D-galactose are illustrated in Figure 10-2.

The L-galactone-1,4-lactone can be enzymatically converted to L-ascorbic acid by the following reaction:

$$\text{L-galactono-1,4-lactone} + O_2 \xrightarrow{\text{oxidase}} \text{L-ascorbic acid} + H_2O_2$$

Screening of Mutants

Candida norvegensis CBS 2145 served as the parent in a series of mutations that were made using conventional mutagenic agents, such as ethyl methane sulfonate, N-methyl-N^1-nitro-N-nitrosoguanidine, nitrous acid, ethidium bromide, ultra-violet light and gamma irradiation.

A minimum of 2000 isolates from each treatment were plated on agar containing the standard medium minus glycine, and powdered calcium carbonate. The relative quantity of ascorbic acid produced by each isolate could be determined by the size of the clear zone around each colony.

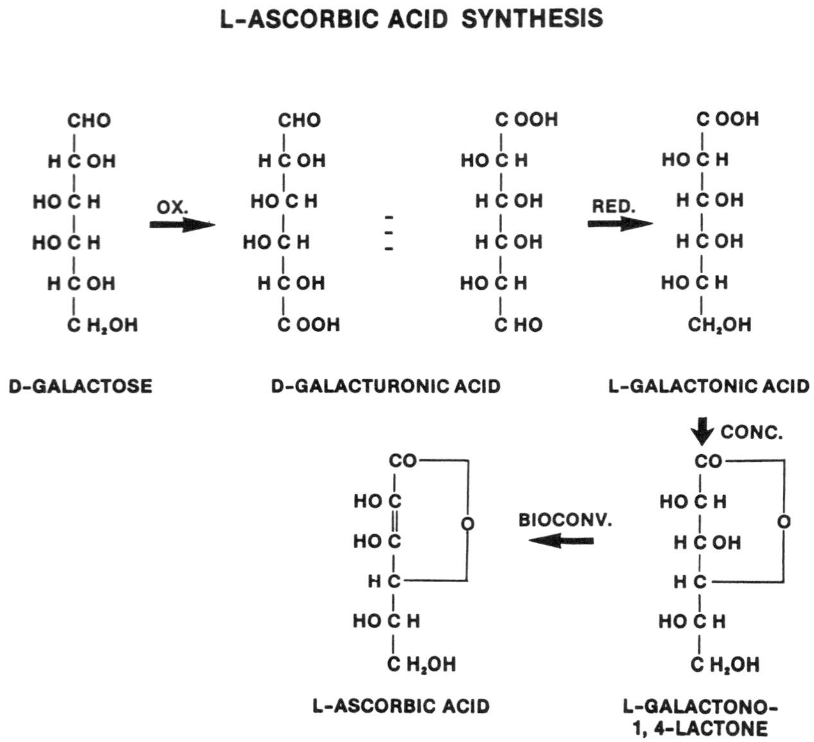

Figure 10-2. Synthesis of L-Ascorbic Acid from D-galactose

Five hundred of the highest producing isolates were selected for the next stage of screening, a mini-fermentation in 96 well microtiter plates. The plates were shaken at 400 RPM, 30°C for 48 hours. The quantity of L-ascorbic acid produced in each well was determined by titrating with 2,6-dichlorophenol-indophenol.

Fifty of the highest producing microtiter isolates were then cultured for shake flask trials. The yeast inoculum was cultivated at 30°C for 24 hours on G-agar slants (glucose 0.5%, trypticase-peptone 0.2%, yeast extract 0.5%, agar 1.5%). One slant was used as the inoculum for each 500 ml flask containing 50 ml of glycine medium shown in Table 10-1. Incubation was for 48 hours at 30°C, 400 RPM. The course of the shake flask conversions was monitored by appropriate analytical procedures at 24 and 48 hours. Quantitative assays of L-ascorbic acid and analogs were carried out by redox-titration using 2,6,-dichlorophenol-indophenol (10). The values were confirmed using an HPLC procedure (11). A summary of the mutation screening protocol can be seen in Figure 10-3.

Table 10-1. Glycine Medium Used for Cell Culture

Medium (pH adjusted to 4.0)	Grams Per Liter
Ethanol	20.0
Glycine	7.0
Corn steep	5.0
L-galactono-1,4-lactone	5.0
Monosodium glutatmate	2.0
NH_4Cl	1.0
$MgSO_4{\cdot}7H_2O$	0.5
Mineral mix	2.0 ml
Mineral Mix	
EDTA (2Na)	5.0
$ZnSO_4{\cdot}7H_2O$	0.22
$CaCl_2{\cdot}2H_2O$	0.735
$MnSO_4{\cdot}H_2O$	0.6725
$FeSO_4{\cdot}7H_2O$	0.915
$(NH_4)_6{\cdot}Mo_7O_{24}{\cdot}4H_2O$	0.10
$CuSO_4{\cdot}5H_2O$	0.25
$CoCl_2{\cdot}6H_2O$	0.25

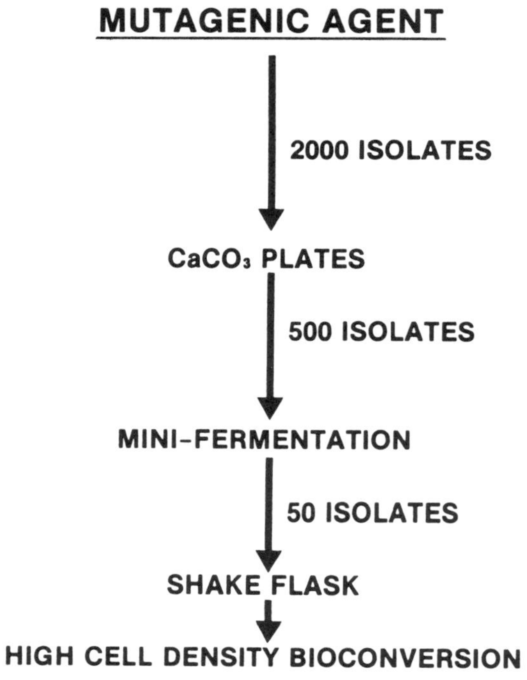

Figure 10-3. Mutation Screening Protocol

High Cell Density Bioconversions

High cell density biomass conversions (25-37.5 g DW/L) were carried out in conventionally stirred, aerated fermenters, such as New Brunswick Scientific 7, 14 or 30 liter fermenters. Product formation was monitored and control of the fermenter environment was achieved using physical and chemical sensors linked to a microcomputer. The pH, dissolved oxygen and off-gas (oxygen and carbon dioxide) levels were monitored and logged continuously throughout the bioconversion, as were the concentrations of ethanol, lactone and ascorbic acid. Substrate levels were controlled using a computer-linked automated HPLC system (12). The HPLC system consisted of a model 6000 solvent delivery system, Model 721 programmable system controller, Model 730 data module, a WISP 710B automatic injector and a Model 401 refractive index detector (Waters Assoc. Milford, MA). Separations were achieved on a 0.4 x 25 cm Aminex HPX-85 alcohol analysis column which was protected by a microguard ion-exclusion pre-column cartridge (Bio-Rad Labs., Richmond, CA). The mobile phase was 3mM HNO_3. Separations were carried out at a flow rate of 0.5 mL per minute at 21°C. Ascorbic acid levels were determined using an inline electrochemical detector (13).

Because of the highly aerobic nature of the bioconversion, pure oxygen was either gassed directly to the fermenter, or combined with air to achieve the desired dissolved oxygen level. This was typically maintained at 20-30% of saturation. The exit gas can be scrubbed through a potassium hydroxide trap to remove carbon dioxide and recycled to the fermenter.

Upon completion of the bioconversion, ascorbate, unconverted lactone, and alcohol were recovered and the lactone and alcohol recycled. The clarified bioconversion mixture was passed through a Duolite S-761 resin (Diamond Shamrock Ion Exchange Functional Polymers Division, Redwood City, CA) column where color and other interfering compounds are removed. Alcohol was recovered during concentration and the pH-adjusted ascorbate-lactone mixture was passed through a Dowex 1-X8 Cl- column. The lactone was recovered with a water wash, concentrated and returned to the fermenter where it was used to maintain the lactone level during bioconversion. The L-ascorbic acid was recovered from the resin column with 0.05 M HCl, followed by concentration and crystallization from a 1:2 mixture of ethanol:ethyl acetate (Figure 10-4).

Isolation of Mitochondria

Mitochondria were isolated from *Candida* cultures by disrupting the cells in liquid nitrogen according to Tzagoloff (14). To determine substrate specificity, 2 mM of L-galactono-1,4-lactone or the compound being evaluated as a potential substrate, were incubated with the mitochondria in 50 mM citrate buffer, pH 6.8 at 37°C for 30 minutes in a total volume of 3 ml. The reaction was terminated with 0.3 ml 50% TCA, the mixture centrifuged, and the supernatant assayed for L-ascorbate by HPLC as described above.

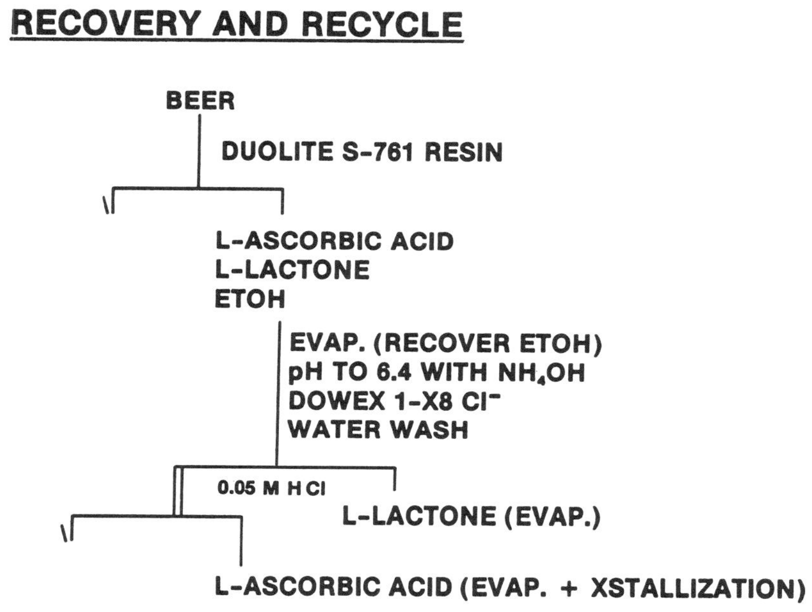

Figure 10-4. Scheme for Recovery/Recycling of Ascorbate, Lactone and Ethanol

RESULTS

Early in the screening program it became evident that the genus *Candida* included species that were capable of converting the lactone to ascorbate, though the initial results were not very encouraging. Figure 10-5 summarizes the results of the original survey of 87 wild type *Candida* cultures grown in shake flasks in the presence of lactone for 24 hours. It can be seen that 39 of the isolates, or 45% of the cultures produced a maximum of 10 mg per liter of L-ascorbic acid. At the other extreme, one culture produced 90 mg per liter. It should also be noted that very little of the ascorbate product was exported into the medium by these cultures.

The screening program uncovered a strain of *Candida norvegensis*, CBS 2145, which produced 50 mg per liter of ascorbate, and which exported a minimum of 90% of its product into the medium. It also was capable of growing on ethanol, a carbon source which was not converted into any other isomer of L-ascorbic acid.

Figure 10-6 illustrates the genealogy of the *Candida norvegensis* mutants developed from the CBS 2145 parent. During this period of strain improvement, the product yield went from 0.30 grams per liter for CBS 2145 to 1.77 grams per liter for MF-78. This is graphically represented in Figure 10-7.

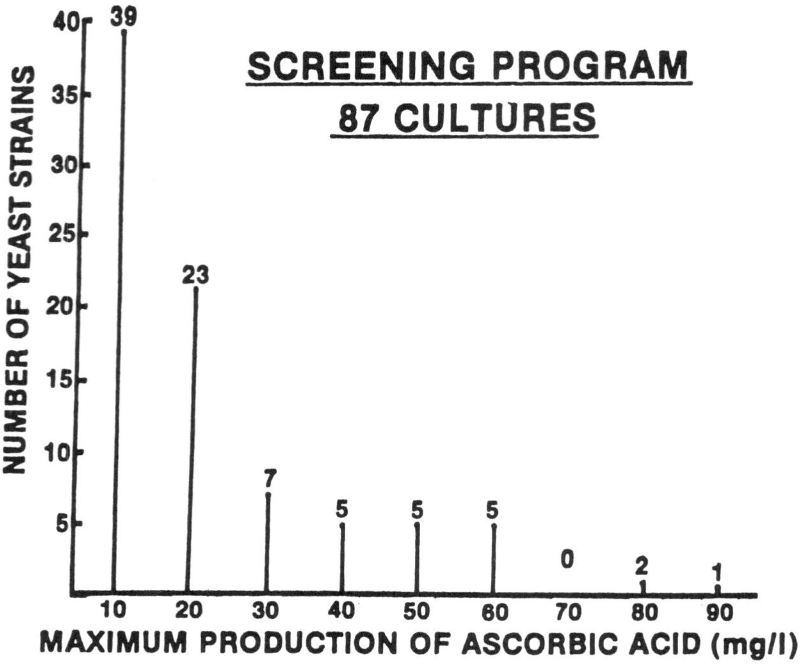

Figure 10-5. Maximum Production of L-Ascorbic Acid by Wild Type *Candida* sp.

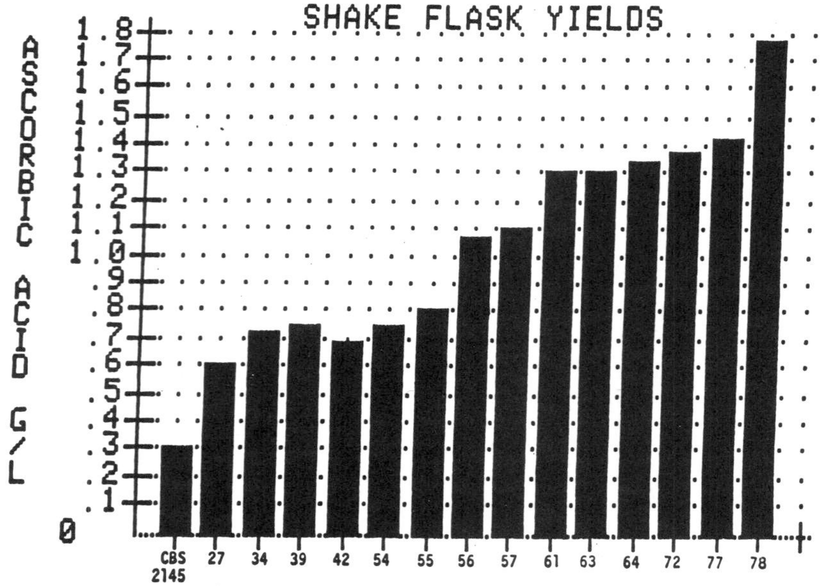

Figure 10-7. Shake Flask Yields of *Candida norvegensis* Mutants

MUTATION SEQUENCE

	YIELD - G/L
CBS 2145	0.30
I EMS	
MF-27	0.60
I UV	
MF-34	0.72
I UV/CAF	
MF-39	0.75
I UV	
MF-42	0.69
I NTG	
MF-54	0.75
I NA	
MF-55	0.80
I Ni^{+2} Res	
MF-56	1.07
I UV	
MF-57	1.10
I UV/Vn^{+2} Res	
MF-61	1.30
I UV	
MF-63	1.31
I Ce^{137}	
MF-64	1.34
I EtBr	
MF-72	1.38
I UV	
MF-77	1.43
I UV	
MF-78	1.77

Figure 10-6. Ascorbate Yield as a Function of Mutation Sequence

EMS = eythyl methane sulfonate, UV = ultra violet light,
CAF = caffine, NA = nitrous acid, EtBr = ethidium bromide
NTG = N-methyl-N^1-nitro-N-nitrosoguanidine

The high cell density bioconversion yields also increased significantly, reaching 7.51 grams per liter with MF-78 (Figure 10-8).

The specific conditions employed in the MF-78 bioconversion can be seen in Table 10-2. Modifications of the earlier conditions included

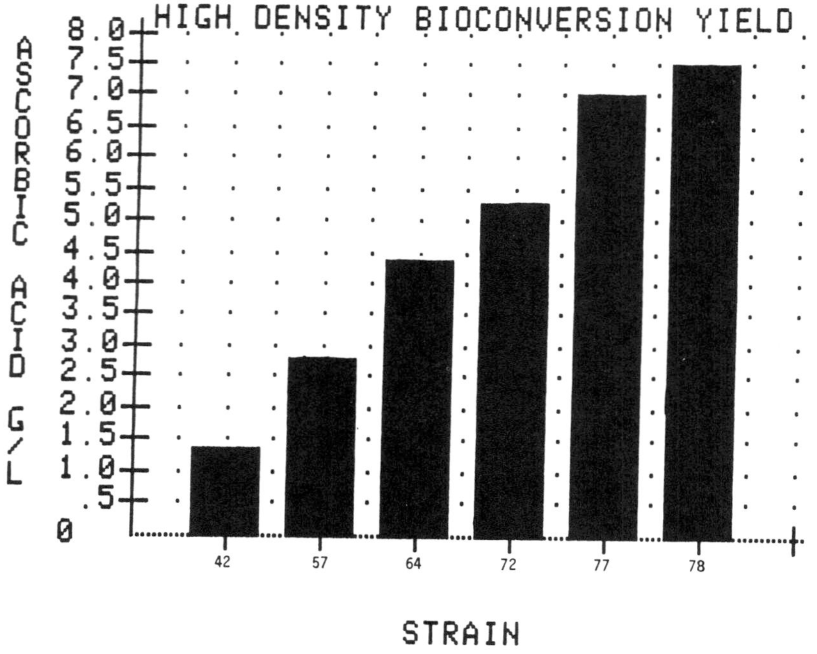

Figure 10-8. High Cell Density Bioconversion Yields of *Candida norvegensis* Mutants

maintaining the lactone concentration at 1.4%, starting the ethanol concentration at 1.5% and allowing it to fall to 0.5% at which level it was maintained, and including the spent yeast from the alcohol recovery at a level of 0.19% in place of the mineral mix.

Table 10-2. Bioconversion Conditions for Production of L-Ascorbic Acid by *Candida norvegensis* MF-78

Candida norvegensis, MF-78	37.5g dw/L
Temperature	28°C
D.O. (L/L air:oxygen 1:1)	20-30%
pH (initial)	5.0
Medium	
Lactone	1.4% (Maintained)
Ethanol	1.5→0.5% (Maintained)
Spent yeast	0.19%
Time	34 Hours
Fermenter Geometry	3:1
Agitation	300 RPM
L-Ascorbic Acid Titer	751 g/L

The rate of L-ascorbic acid production is shown in Figure 10-9. The 34 hour conversion resulted in an average production rate of 0.221 grams per liter per hour.

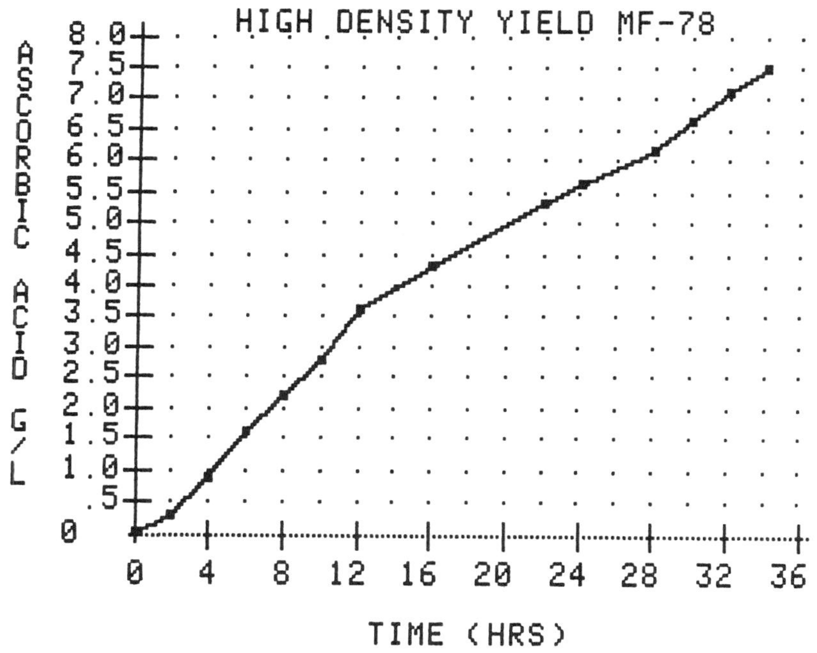

Figure 10-9. Rate of L-Ascorbic Acid Production by *Candida norvegensis* MF-78

The stoichiometry of this run can be seen in Figure 10-10. The reaction is substrate driven and requires an excess of lactone. However, this

STOICHIOMETRY
CANDIDA NORVEGENSIS MF 78
BIOCONVERSION 34 HOURS

LACTONE	+ ETOH	+ (O_2 / AIR) →	ASCORB.	+ LACTONE	+ ETOH	+ O_2
20g	84mL	(408L / 408L)	7.5g	12.49g	5mL	367L

PER LITER

Figure 10-10. Stoichiometry of the Bioconversion of L-Galactono-1,4-lactone to L-Ascorbic Acid by *Candida norvegensis* MF-78 component can be recycled, as can excess oxygen and ethanol, with an overall recovery of 90% of the lactone and oxygen and 95% of the ethanol.

The step yields and the cumulative yields are shown in Table 10-3. Fifty-five percent of the starting galactose is recovered as L-ascorbic acid, which translates to a yield of 24% based on permeate solids.

Table 10-3. Step and Cumulative Yields During the Conversion of D-Galactose to L-Ascorbic Acid

		YIELD		
		Step	Cumulative	From Galactose
Permeate (43% galactose)		100		
D-galactose		89	38	89
D-galacturonic acid		75	29	67
L-galactono-1,4-lactone		96	28	64
L-ascorbic acid		86	24	55
Fermentation	95			
Recovery	90			

The substrate specificity of the *C. norvegensis* L-galactonolactone oxidase, both in intact cells and with isolated mitochondria, was determined. A comparison with some literature values for *S. cerevisiae* and cauliflower mitochondria can be seen in Table 10-4. Both yeast L-galactonolactone oxidases and the cauliflower L-galactonolactone dehydrogenase are quite specific for the L-galactono-1,4-lactone.

Table 10-4. Substrate Specificity of L-Galactonolactone Oxidase and Related Enzymes

	RELATIVE ACTIVITY		
SUBSTRATE	*C. norvegensis*	*S. cerevisiae*[2,3]	Cauliflower[2,3]
L-galactonolactone[1,2]	100	100	100
L-gulonolactone[1]	5	0	0
D-galacturonic methyl ester[1,2]	<4	—	—
D-galactonolactone[2]	0	—	—
L-galactonic acid[2]	0	—	—

1. Bioconversion with intact cells.
2. Bioconversion with isolated mitochondria.
3. Reference 6.

DISCUSSION

A typical cheddar plant in the United States produces approximately 25 million pounds of cheese per year, though many have twice this capacity. If the whey from such a facility is ultrafiltered to recover the WPC, 10.75

million pounds of permeate solids would also be generated, with a negative market value. The procedure described here could be used to convert this whey to 2.58 million pounds of L-ascorbic acid with a significant dollar value.

It should be possible to amplify the L-galactonolactone oxidase either within the genus *Candida* or within another genus of microorganism with faster generation time, and/or a faster rate of bioconversion capability, thereby, even further enhancing the value of the process.

REFERENCES

1. Reichstein, T., and Grussner, A. A high-yield synthesis of L-ascorbic acid (Vitamin C). Helv. Chem. Acta. 17: 311-328 (1934)
2. Isherwood, F.A., Chen, Y.T., and Mapson, L.W. Synthesis of L-ascorbic acid in plants and animals. Biochem. J. 56: 1-15 (1954)
3. Mapson, L.W., Isherwood, F.A., and Chen, Y.T. Biological synthesis of L-ascorbic acid: the conversion of L-galactono-γ-lactone into L-ascorbic acid by plant mitochondrion. Biochem. J. 56: 21-28 (1954)
4. Mapson, L.W., and Isherwood, F.A. Biological synthesis of ascorbic acid: the conversion of derivatives of D-galacturonic acid into L-ascorbic acid by plant extracts. Biochem. J. 64: 13-22 (1956)
5. Mapson, L.W., and Breslow, E. Properties of partially purified L-galactono-γ-lactone dehydrogenase. Biochem. J. 65: 29 (1957)
6. Mapson, L.W., and Breslow, E. Biological synthesis of ascorbic acid: L-galactono-γ-lactone dehydrogenase. Biochem. J. 68: 395-406 (1958)
7. Bleeg, H. L-ascorbic acid in yeast and isolation of L-galactono-γ-lactone oxidase from the mitochondrion. Enzymologia 31: 105-112 (1966)
8. Nishikimi, M., Noguchi, E. and Yagi, K. Occurrence in yeast of L-galactonolactone oxidase which is similar to a key enzyme for ascorbic acid biosynthesis in animals, L-gulonolactone oxidase. Arch. Biochem. Biophys. 191: 479-486 (1978)
9. Bleeg, H., and Christensen, F. Biosynthesis of ascorbate in yeast. Eur. J. Biochem. 127: 391-396 (1982)
10. Bunton, N.G., Jennings, N., and Crosby, N.T. The determination of ascorbic acid and erythorbic acids in meat products. J. Assoc. Pub. Analysts 17: 105-110 (1979)
11. Bui-Nguyen, M.H. Application of high-performance liquid chromatography to the separation of ascorbic acid from isoascorbic acid. J. Chrom. 196: 163-165 (1980)
12. Dinwoodie, R.C., and Mehnert, D.W. A continuous method for monitoring and controlling fermentations using an automated HPLC system. Biotechnol. Bioeng. 27:1060-1062 (1985)
13. Pachia, L.A. Determination of ascorbic acid in foodstuffs, pharmaceuticals, and body fluids by liquid chromatography with electrochemical detection. Anal. Chem. 48: 364-367 (1976)
14. Tzagoloff, A. Assembly of the mitochondrial membrane system. J. Biol. Chem. 244: 5020-5026 (1969)

Chapter 11

The Genetic Modification of Brewer's Yeast and Other Industrial Yeast Strains

Igne Russell, Rena Jones and Graham Stewart

INTRODUCTION

The yeast genus *Saccharomyces* has often been referred to as "the oldest plant cultivated by man". Indeed, the history of beer, wine and bread making, with the fortuitous use of yeast, is as old as the history of man himself; however, recent years have seen the transition of yeast from being solely the "workhorse" in the traditional food and beverage industry to being in addition one of the organisms of choice for gene manipulation involving the "new genetics".

It is the opinion of some biotechnologists, an opinion not shared by these authors, that the future importance of this group of yeast will diminish due to the fact that they produce few secondary metabolites of commercial interest. However, many species of yeast (particularly of the genus *Saccharomyces*) are of GRAS (generally recognized as safe) status and produce two very important primary metabolites—ethanol and carbon dioxide. The ethanol is used in both beverages and for industrial or fuel purposes. The carbon dioxide is employed for leavening in baked goods, for carbonation of beverages, as a solvent in the supercritical state and for the culturing of vegetables in greenhouses under controlled environmental conditions.

In the production of fermentation ethanol (be it for beer, wine, potable spirit or industrial ethanol), the microorganism being employed should possess a number of important characteristics: 1) rapid and relevant carbohydrate fermentation ability; 2) appropriate flocculation and sedimentation characteristics; 3) genetic stability; 4) osmotolerance (i.e., the ability to ferment concentrated carbohydrate solutions); 5) ethanol tolerance and the ability to produce elevated concentrations of ethanol;

6) high cell viability for repeated recycling; and, 7) temperature tolerance. Most of the developments that are currently occurring in industrial yeast strains, are to improve their efficiency to produce ethanol eg., broadened substrate specificity, increased fermentation efficiency, greater ethanol, osmo- and temperature tolerance and appropriate sedimentation characteristics (20). In addition, cells of *Saccharomyces cerevisiae* have been genetically transformed to produce a number of important non-yeast proteins and peptides; such as, the antiviral protein interferon (9), and the acid protease chymosin, used in the milk-clotting step in cheese production (13).

With the development of novel genetic techniques such as recombinant DNA and cell fusion, the genetics of *Saccharomyces cerevisiae* and related species has been transformed from a purely classical field to one that includes classical genetics and molecular biology. The distinction does not concern the rational for employing genetic analysis to study biological problems, but rather procedural matters which nevertheless have profoundly influenced knowledge about yeast cells. The manipulations of classical genetics involve the intact organism. Mutations are identified by selecting yeast strains possessing new properties and then characterized by analyzing the progeny of sexual matings. The life cycle of yeast involves the fusion of haploid germ cells of opposite mating type (a and α) to form diploid zygotes that, under appropriate conditions, undergo a fairly typical meiosis. It is a particular advantage that both haploid and diploid cells can be propagated vegetatively in defined medium. With the advent of recombinant DNA technology, classically defined yeast genes have been isolated as cloned DNA by virtue of their ability to be functionally expressed in *Escherichia coli* such that they complement defects in a specific bacterial gene (22). Such cloned DNA segments have been introduced into yeast cells where they caused heritable changes. Indeed, almost all the techniques of the new yeast genetics are analogous to those that have been employed with *Escherichia coli* for many years. However, since yeast is a eukaryote it is invaluable for studying many basic questions in eukaryotic biology, including gene expression, DNA replication, recombination, transposition, chromosome segregation, chromatin structure, secretion and excretion, the cell cycle and the control of cell type (22).

INDUSTRIAL YEAST STRAINS

Much literature exists describing the genetics and biochemistry of laboratory strains of *Saccharomyces cerevisiae* but there is a general lack of knowledge regarding the genetics and biochemistry of industrial *Saccharomyces* strains. The haploid strain that the molecular biologist employs in the university laboratory as the organism of choice is usually totally unsuitable for use in the industrial world. Brewing yeasts, and many other industrial yeast, have been selected over time for exactly those characteristics which make them so unamenable to easy genetic manipulation in the laboratory. They are usually polyploid or aneuploid, lack a mating type characteristic, sporulate poorly if at all and the spores that do form are usually not in fours and exhibit poor viability, rendering tetrad analysis difficult (19). It would appear, however, that the widespread use of polyploid yeasts for industrial purposes is no accident. Owing to their multiple gene structure, polyploids are genetically more stable and less susceptible to mutational forces and thus can be used routinely with a higher degree of confidence than either haploid or diploid strains.

THE BREWING PROCESS

Beer is a dilute solution of ethanol, its characteristic flavor arises from the use of malt as the predominant source of fermentable carbohydrates and other yeast nutrients, and of hops as a source of bitter components. These characteristics make beer very different in flavor compared to products derived from grapes, apples or other fruits. It also separates beer from distilled liquids which are flavored either with added botanical products, for example gin, or which contain only the volatile constituents of fermented broths such as whiskey. Brewing was one of the earliest processes to be undertaken on the commercial scale. Of necessity, it became one of the first processes to be developed from an art into a technology. The beer industry has the largest dollar value of all the biotechnology industries. It is produced by an industry which has made a rapid transition from a craft organized in small units, to an industry making its products in large complexes. In a craft industry the usual way to produce a constant product is to change nothing. The modern brewing technologist must produce a constant product even though raw materials, types of plant and scale of operation are changing radically.

Beer production is divided into quite clear-cut processes: malting, mashing, fermentation, and storage. The first two processes together produce a medium known as wort; to this yeast is added, and fermentation is allowed to proceed. Wort is essentially an aqueous extract of malted barley, the primary raw material in the manufacture of beer. In principle, the production of beer is a simple process. Yeast cells are added to a nutrient medium (wort) and the cells take up nutrients and utilize them so as to increase the yeast population. In doing this, the cells excrete major end-products such as ethanol and carbon dioxide into the medium, together with a host of other minor metabolites. The final medium, after yeast removal, is the product beer. It contains some wort constituents, as well as all the non-volatile and many of the volatile substances produced by the metabolism of the yeast. The objective of individual breweries is to bring about the appropriate degree of metabolism so that the product contains the required mixture of by-products and to achieve this in a reasonable time.

BREWER'S YEAST STRAINS

The two main types of beer are lager and ale. These are fermented with strains of *Saccharomyces uvarum* (*carlsbergensis*) and *Saccharomyces cerevisiae*, respectively. Traditionally, lager is produced by bottom-fermenting yeasts at fermentation temperatures between 7-15°C. This means that at the end of the fermentation, the yeasts flocculate and collect at the bottom of the fermenter. Top-fermenting yeasts, used in the production of ales at fermentation temperatures between 18-22°C tend to be somewhat less flocculent and loose clumps of cells are carried to the fermenting wort surface adsorbed onto carbon dioxide bubbles. Consequently, top yeasts are collected for reuse from the surface of the fermenting wort (a process called skimming) whereas, bottom yeasts are collected (or cropped) from the fermenter bottom. The differentiation of lagers and ales on the basis of bottom and top cropping is becoming less and less distinct with the advent of vertical conical bottom fermenters and centrifuges. With centrifuges, non-flocculent yeast strains are required for both lagers and ales, where, as soon as fermentation is complete and before the yeast has had the opportunity to sediment, the fermented medium is passed through a centrifuge in order to separate the yeast from the "green" beer. With conical bottomed fermenters, a more sedimentary yeast (lager or ale) settles into the cone of the fermenter at the completion of fermentation. It is then removed and a portion reused as an inoculum for a subsequent fermentation.

Taxonomically the two species *Saccharomyces cerevisiae* and *Saccharomyces uvarum* (*carlsbergensis*) have been distinguished on the basis of their ability to ferment the disaccharide melibiose (11). Strains of *Saccharomyces uvarum* (*carlsbergensis*) possess the *MEL* gene(s) and consequently produce the extracellular enzyme α-galactosidase (melibiase) allowing utilization of melibiose whereas strains of *Saccharomyces cerevisiae* are unable to produce α-galactosidase and therefore cannot utilize melibiose. However, in two recent texts on yeast taxonomy (2, 10) these two groups of yeasts have been consolidated into one species, *Saccharomyces cerevisiae*. Indeed it has been proposed that the number of species within the genus *Saccharomyces* be reduced from 41 as described by Lodder (11), to either six (2) or seven (10). Whether or not these proposals will receive acceptance by scientists active in the field of yeast research and by fermentation technologists employing yeasts at the industrial level will have to await the test of time. Indeed, it is at the strain level that interest in brewing yeast centers. At the last count there were at least 1000 separate strains of *Saccharomyces cerevisiae*—these strains may be brewing, baking, wine, distilling or laboratory cultures. There is a problem classifying such strains in the brewing context; the minor differences between strains that the taxonomist dismisses are vitally important to brewers.

The behavior, performance and quality of a yeast strain is influenced by two sets of determining factors, collectively called, nature-nurture effects. The nurture effects are all the environmental factors (i.e., the phenotype), to which the yeast is subjected from inoculation onwards. On the other hand the nature influence is the genetic make-up (i.e., the genotype) of a particular yeast strain.

The requirements of an acceptable brewer's yeast strain can be defined as follows: "In order to achieve a beer of high quality, the yeast culture must be effective in removing the desired nutrients from the growth medium (i.e., the wort), it must impart the required flavor to the beer and finally, the microorganisms themselves must be effectively removed from the fermented wort after they have fulfilled their metabolic role" (18). This definition, although wide ranging and very general, allows the major activities of a brewing yeast strain during the fermentation of wort to be divided into a number of stages: 1) nutrition—sugar and amino acid uptake, etc.; 2) by-product formation—the excretion of compounds that contribute to the flavor of the product; and, 3) removal of the culture from the fermented wort by flocculation or centrifugation.

In terms of the development and selection of yeast strains for use in a brewery there are a number of parameters that are important as scale-up criteria: 1) fermentation rate; 2) decrease in specific gravity (°Plato of the wort); 3) taste and flavor match of the final product; 4) fusel oil, ester and organo-sulphur production; 5) consistent high cell viability; 6) ethanol tolerance and production; and, 7) inoculation (pitching) rate. More specifically, there are a number of factors that will affect fermentation rate *per se* and these include: 1) inoculation (pitching) rate; 2) yeast cell viability; 3) fermentation temperature; 4) wort dissolved oxygen concentration at pitching; 5) wort soluble nitrogen concentration; 6) wort fermentable carbohydrate concentration; and, 7) yeast storage conditions, eg., the influence of intracellular glycogen levels.

GENETIC MANIPULATION TECHNIQUES

There are a number of methods that are employed in the genetic research and development of brewer's yeast strains. These include hybridization, mutation and selection, rare mating, spheroplast fusion, and transformation. Transformation can be carried out using native DNA, recombinant DNA or by liposome-mediated DNA transfer. Hybridization cannot be used directly as a means to manipulate brewer's yeast strains for the reasons already discussed in this chapter. Nevertheless, hybridization is a technique that has made an invaluable contribution to the field of yeast genetics and is by no means obsolete today.

Hybridization has been used in conjunction with more novel genetic techniques to verify the genetic composition of recombinants. It can also be employed to provide a great deal of relevant genetic information about traits that are very germane to brewing fermentation systems. Hybridization has been used to study the genetic control of flocculation, phenolic-off flavor production and the uptake of wort sugars and dextrins.

Techniques that have greatest potential and promise as aids in the genetic manipulation of industrial yeast strains are: rare mating, mutation and selection, spheroplast or protoplast fusion (also called somatic fusion) and transformation (usually associated with recombinant DNA techniques). All of these methods have a total disregard for ploidy and mating type and thus have great applicability to brewing yeast strains. Rare mating has been used in conjunction with the *kar* (karyogamy defective) strains to introduce zymocidal (killer) activity into brewing strains (27). Mutation and selection has been used to induce auxotrophs and to select easily recognizable characteristics of brewing strains in order that such strains can be employed as spheroplast fusion partners and as recipients for transformation experiments (4). Mutation and selection has also been used to isolate derepressed mutants of brewing strains such that these strains possess the ability to metabolize maltose in the presence of glucose and thus have increased wort fermentation rates. Spheroplast fusion has been employed to fuse strains constructed by hybridization with brewing stains in order to introduce the novel capabilities of the hybridized strain into the brewing strains whilst still maintaining all the characteristics of the latter (19). Finally, transformation is being employed in this laboratory to introduce genes from non-*Saccharomyces* yeast strains into brewing strains.

RARE MATING AND ZYMOCIDAL ACTIVITY

When non-mating strains are mixed together at a high cell density, a few true hybrids with fused nuclei form which can usually be selectively isolated. An even more useful variation of this technique employs a yeast strain which harbors a specific nuclear gene mutation, designated *kar*. When this strain hybridizes with another strain the nuclei will not fuse and this permits the formation of cell lines with mixed cytoplasmic contents (heteroplasmons). Rare mating has been employed to transfer "zymocidal" or "killer" factor from laboratory haploid strains to brewing yeast strains. Some strains of *Saccharomyces* spp. (and other yeast genera) secrete a proteinaceous toxin called a "zymocide" or "killer" toxin which is lethal to certain other strains of *Saccharomyces* (27). Toxin producing strains are termed "killers" and susceptible strains are termed "sensitives". However, there are strains that do not kill and are not themselves killed and these are called "resistant" (Figure 11-1).

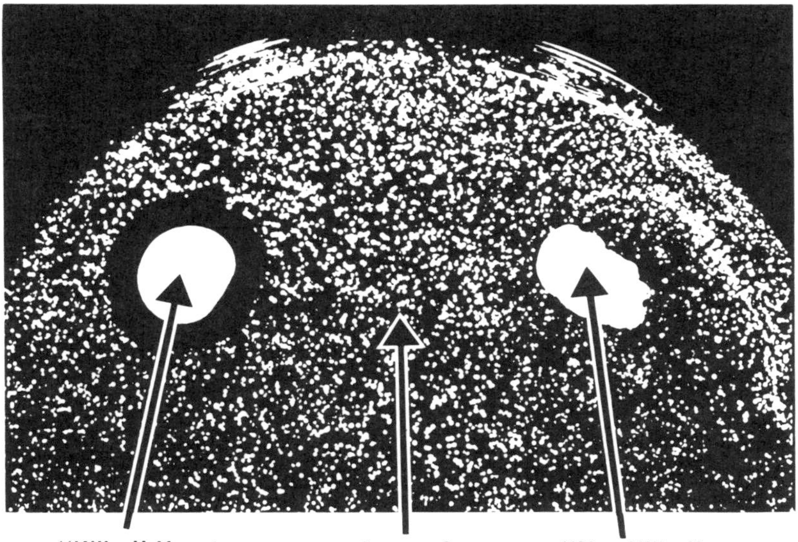

Figure 11-1. *Saccharomyces* Brewing Yeast with and without Zymocidal ("killer") Activity

The "killer" factor has been renamed "zymocide" to indicate that it is only lethal towards yeast and not towards bacteria or cells of higher organisms. Zymocidal yeasts have been recognized to be a serious problem in both batch and continuous fermentation systems (12). An infection of as a little as 1% of the cell population can completely eliminate all the brewing yeast from the fermenter. The brewer can protect the process from this occurrence in one of two ways: 1) maintain vigorous standards of hygiene to ensure that contamination with a wild yeast possessing zymocidal activity is prevented; or, 2) genetically modify the brewery yeast so that it is not susceptible to the zymocidal toxin. The first method is the one most brewers to date have been invoking to protect their process; however, genetic manipulation can also be used to produce a brewing strain that is less vulnerable to destruction by a zymocidal yeast infection.

The killer character of *Saccharomyces* spp. is determined by the presence of two species of cytoplasmically located dsRNA (7). M dsRNA (killer plasmid), which is killer strain specific, codes for killer toxin and also for a protein or proteins which render the host immune to the toxin (3). L dsRNA, which is also present in many non-killer yeast strains, specifies a capsid protein which encapsulates both forms of dsRNA thereby yielding virus-like particles. Although the killer plasmid is contained within these virus-like particles, the killer genome is not naturally transmitted from cell to cell by any infection process. The killer plasmid behaves as a true cytoplasmic element and requires at least 29 different chromosomal genes (*mak* for its maintenance in the cell). In addition, three other chromosomal genes (kex1, kex2 and rex) are required for toxin production and resistance to toxin (26).

In recent years, the technique of rare mating has been employed to produce hybrids with enhanced fermentation ability (8,27). A modification of the technique of rare mating, employing the *kar* mutation (karyogamy defective), has been successfully used by a number of research laboratories. The *kar* mutation offers a significant advantage in that it prevents nuclear fusion and hybrids from such a rare mating can be selected that contain only the brewing strain nucleus. However, such hybrids contain the cytoplasm of both parental cells, thereby permitting the introduction of cytoplasmically transmitted characteristics such as killer toxin production into brewing strains without altering the nucleus of the brewing strain. A brewing polyploid lager yeast strain was rare mated with a laboratory "killer" haploid strain and a number of rare mating products isolated (Figure 11-2).

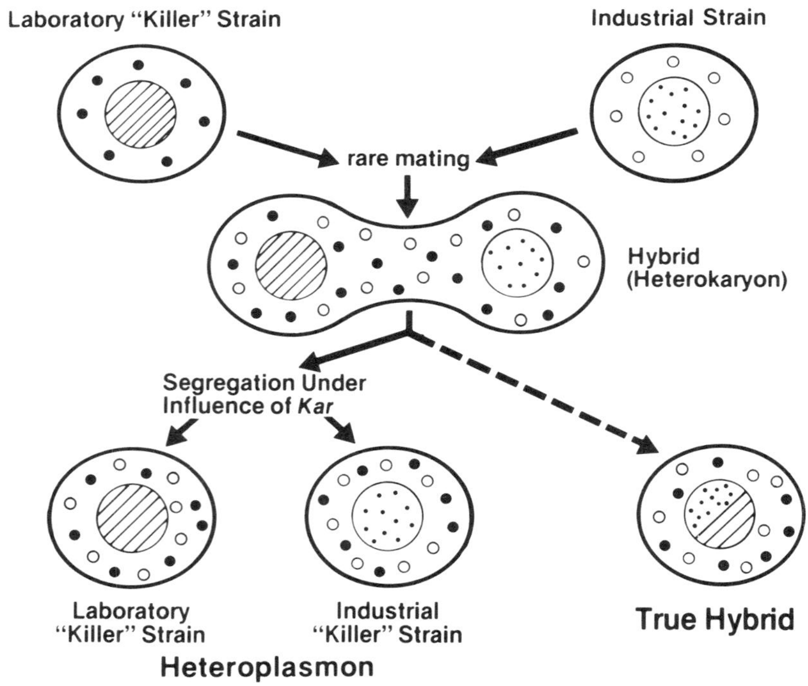

Figure 11-2. Construction of a Brewing "Killer" Strain Employing Rare Mating and the *kar* Mutation.

In addition to biochemical tests to characterize the rare mating products, agarose gel electrophoresis demonstrated that some rare mating products contained not only the 2 micron plasmid (from the parental brewing lager strain) but also the L and M dsRNA plasmids (from the haploid mating partner), which code for killer toxin production (Figure 11-3).

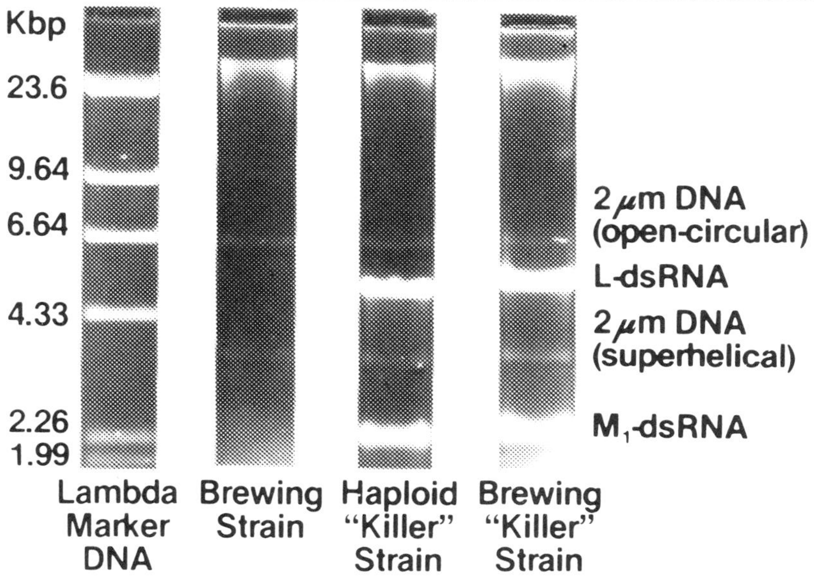

Figure 11-3. Plasmid Profile Illustrating the dsRNA Bands of a "Killer" Lager Yeast

To determine the effect of the zymocidal lager strain on a typical brewery fermentation, "killer" lager yeast was mixed at a concentration of 10% with an ale brewing strain (Figure 11-4A). The control was the ale strain mixed with 10% "non-killer" lager. The yeast was sampled throughout the fermentation and viable cells determined by plating onto nutrient agar plates incubated at 37°C, a temperature which inhibits the growth of lager yeast but allows the growth of ale yeast. Within 10 hours the "killer" lager strain had almost totally eliminated the ale strain. When the concentration of killer yeast was reduced to 1%, the ale yeast again was eliminated within 24 hours (Figure 11-4B).

The speed at which this occurs may well make a brewer apprehensive about employing such a yeast in the fermentation cellar particularly where several yeasts are employed for the production of different beers. An error on an operator's part in keeping lines and yeast tanks separate could result in serious consequences. In a brewery with only one yeast strain this would not be a cause for concern.

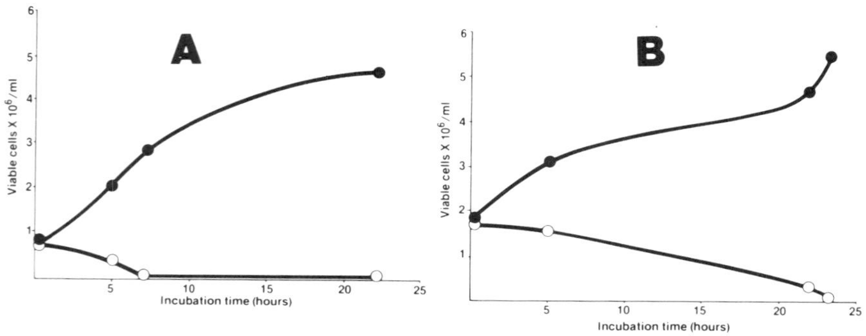

Figure 11-4. A) Control-Growth of Ale Yeast from Mixture of 90% Ale Yeast Plus 10% "Non-killer" Lager Yeast (●-●).
Test-Growth of Ale Yeast from Mixture of 90% Ale Yeast Plus 10% "Killer" Lager Yeast (O-O).
B) Control-Growth of Ale Yeast from Mixture of 99% Ale Yeast Plus 1% "Non-killer" Lager Yeast (●-●).
Test-Growth of Ale Yeast from Mixture of 99% Ale Yeast Plus 1% "Killer" Lager Yeast (O-O).

An alternative to the "killer" strain would be to produce a yeast strain that does not "kill" but is "killer resistant". That is to say, it has received that genetic complement that renders it immune to zymocidal activity. The construction of such a yeast would perhaps be a good compromise, since it does not itself kill, it allays the fear of the brewer that this yeast might kill all other production strains in the plant and at the same time it is not itself killed by a contaminating yeast with killer activity.

WORT SUGAR UPTAKE

Brewer's wort contains the sugars, sucrose, fructose, glucose, maltose and maltotriose together with dextrin material. In the normal situation, brewing yeast strains are capable of utilizing sucrose, glucose, fructose, maltose and maltotriose in this approximate sequence, although some degree of overlap does occur, leaving maltotetraose and the other dextrins unfermented. A major limiting factor in wort fermentation rates is the repression of maltose and maltotriose uptake by glucose (21). Only when approximately 50% of the wort glucose has been taken up by the yeast will the uptake of maltose commence. In other words, in most strains of *Saccharomyces cerevisiae* and related species, maltose utilization is subject to control by carbon catabolite repression, such that even in the presence of maltose, the maltose-utilizing system will be inactivated by high concentrations of glucose. In order to illustrate this phenomenon two brewing strains of *Saccharomyces cerevisiae* (coded 154 and 3001) have been studied. In both instances (Figures 11-5 and 11-6), the presence of glucose repressed the uptake of maltose. In an attempt to isolated spontaneous mutants of these strains in which the presence of glucose in the medium did not repress maltose uptake, mutants capable of growth in the non-metabolizable glucose analogue 2-deoxyglucose (2-DOG), were isolated (Figure 11-7).

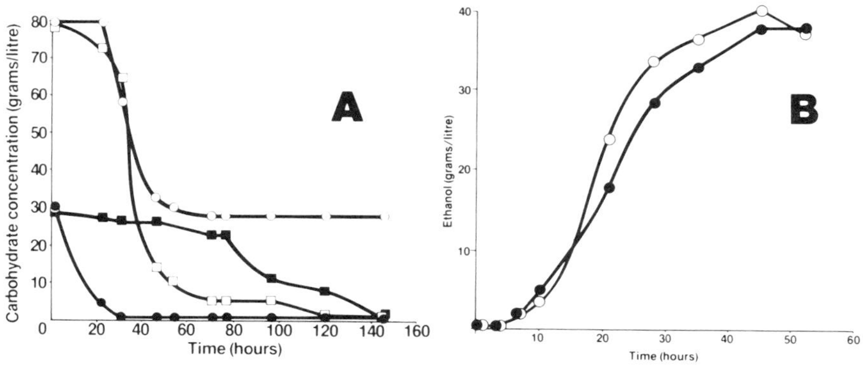

Figure 11-5. Fermentation Characteristics of a Brewing Strain of *Saccharomyces cerevisiae* Strain 154 and Its Derepressed Maltose 2-DOG Mutant.
A) Carbohydrate Uptake in an 8% (w/v) Maltose /3% (w/v) Glucose-peptone-yeast Extract Medium:
Glucose Uptake: Parent (●-●) and Mutant (■-■)
Maltose Uptake: Parent (O-O) and Mutant (□-□)
B) Ethanol Production in a 12°P Wort: Parent (●-●) and Mutant (O-O)

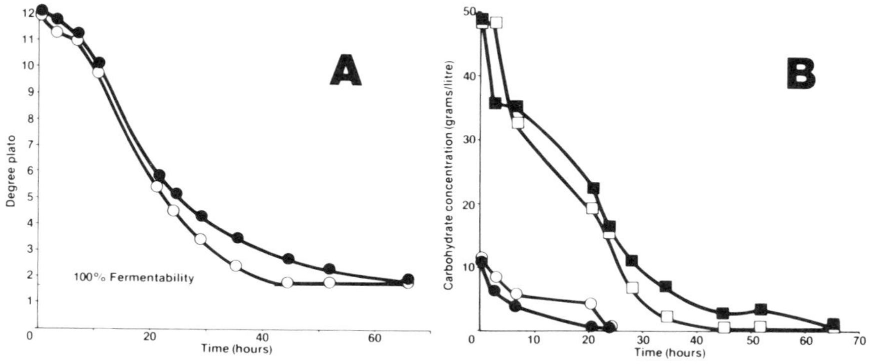

Figure 11-6. Fermentation Characteristics of a Brewing Strain of *Saccharomyces cerevisiae* (Strain 3001) and Its Derepressed Maltose 2-DOG Mutant in a 12°Plato Wort
A) Fermentation Rate: Parent (●-●) and Mutant (O-O)
B) Glucose Uptake: Parent (●-●) and Mutant (O-O)
Maltose Uptake: Parent (■-■) and Mutant (□-□)

A number of stable 2-DOG mutants were found to be capable of utilizing maltose in the presence of significant concentrations of glucose. For example, 2-DOG resistant mutants of *Saccharomyces cerevisiae* (strain 154) in a synthetic medium containing both maltose [8% (w/v)] and glucose [3% (w/v)] were able to completely metabolize the maltose (Figure 11-5A) whereas in the same medium, maltose uptake was slow with the parental strain and only 60% complete when fermentation ceased. Fermentation and ethanol formation rates in 12°P wort were also increased in the 2-DOG mutants when compared to the parental strain (Figure 11-5B).

determine glucose repression by screening on PY-maltose plates containing 2-deoxyglucose

↓

pregrow the sensitive strains in 10 mL of 10% PY-glucose medium for 24 hours

↓

pitch into 200 mL PY-maltose media containing 2-deoxyglucose

↓

shake at 21°C for 14 days

↓

streak onto PY-maltose plates containing 2-deoxyglucose

↓

pick a large single colony

↓

repitch this into 200 mL of PY-maltose media containing 2-deoxyglucose

↓

shake at 21°C and repeat as above for 2 more generations

Figure 11-7. Method for Isolation of Derepressed Mutants

Similar results were obtained with the other strain of *Saccharomyces cerevisiae* (strain 3001) studied; maltose uptake in 2-DOG mutants was not repressed by glucose. In 12°P wort, the overall fermentation rate was significantly faster in the 2-DOG mutants (Figure 11-6A) with complete fermentation being achieved in 45 hours compared to 65 hours in the parental strain. This increased fermentation rate was due to an increased maltose uptake rate in the 2-DOG mutants and compared to the parent strain, with glucose have little influence upon maltose uptake in the mutants (Figure 11-6B). A trained taste panel, using the triangle test method, determined that the beer produced with 2-DOG mutants of both strains (i.e., 154 and 3001) was not significantly different from that produced using the parental yeast strain; all beers were produced under similar pilot plant brewing conditions. The mechanism by which 2-DOG resistant mutants are derepressed is far from understood but such mutants have been reported to possess diminished levels of hexokinase (5,14). Possibly hexokinase is associated with a general regulatory system in yeast, involving overall repression.

Bailey and Woodword (1) have described a mutant allele (designated *grr-1*) in *Saccharomyces cerevisiae* that is characterized not only by 2-DOG resistance, but insensitivity to glucose repression for invertase, maltase and galactokinase, as well as the mitochondrial enzyme cytochrome *c* oxidase. The levels of hexokinase activity towards both fructose and glucose were also approximately three-fold higher in mutant clones. The molecular mechanism by which catabolite repression is effected in yeast is still unknown although many theories have been discussed. Nevertheless derepressed mutants of industrial yeast strains will be of significant economic importance particularly with respect to increased control of fermentation rate.

LOW CARBOHYDRATE (LOW CALORIE) "LITE" BEER

Low carbohydrate (usually low calorie) beer in North America represents a significant share of the beer volume; it constitutes 13% of the market or 2.2 billion liters per year in the U.S.A. and 8% or 170 million liters per year in Canada.

Several procedures for producing low carbohydrate beers are being employed or contemplated, most of them center on one of the following techniques: 1) dilution of regular strength beer with water; 2) addition of fungal α-amylase or glucoamylase and bacterial pullulanase to the wort during fermentation; 3) use of glucose, fructose and sucrose as an adjunct; 4) use of brewing yeast strains with amylolytic activity; and, 5) use of a malt enzyme preparation during mashing or fermentation.

As previously discussed, typical brewing yeast strains are capable of utilizing sucrose, glucose, fructose, maltose and maltotriose, leaving the maltotetraose and the larger dextrins unfermented. However, yeasts of the species *Saccharomyces diastaticus* (6) have been classified as a distinct species from that of *Saccharomyces cerevisiae* due to the fact that the former produces the extracellular enzyme glucoamylase, also called amyloglucosidase (α-1,4 glucan glucanohydrolase, E.C.3.2.1.3.). This enzyme possesses the ability to cleave α-1,4 and in some cases α-1,6 linkages, releasing glucose from the non-reducing end of starch chains. On the other hand, α-amylase (α-1,4-glucan-4-glucanohydrolase, E.C.3.2.1.1) is an enzyme which hydrolyzes random α-1,4 glucosidic linkages by-passing α-1,6 bonds. Three genes have been identified that are associated with glucoamylase production of *Saccharomyces diastaticus*, *DEX1*, *DEX2* and *STA3* (6). Using classical hybridization techniques a diploid strain, containing the *DEX* and *STA* genes in the homozygous condition. has been constructed and its fermentation rate studied in brewer's wort under static fermentation conditions (Figure 11-8).

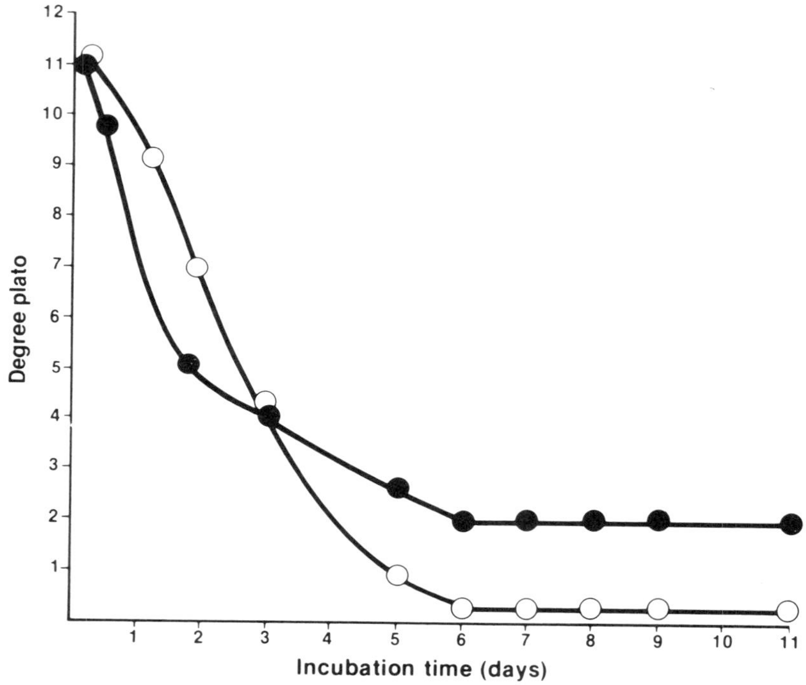

Figure 11-8. Static Fermentation of a Brewer's Wort (40 Liter Scale) by a *Saccharomyces cerevisiae* Brewing Strain (dex) (●-●) and a *Saccharomyces diastaticus* diploid (*DEX1/DEX1*, *DEX2/DEX2*, *STA3/STA3*) (O-O)

The initial fermentation rate of this strain was slower than a production ale brewing strain, however, the *DEX*-containing strain fermented the wort to a greater extent that the brewing strain due to the partial hydrolysis of the dextrins by the action of glucoamylase. Thus, *Saccharomyces diastaticus* strains possess the capacity to produce beer which has been fermented to a high degree and is desirable in the production of low carbohydrate beer; however, it should be noted that beer produced by these strains has a characteristic phenolic-off-flavor. Phenolic-off-flavors in beer are due on many occasions to the presence of 4-vinyl guaiacol (4-VG) which arises by the enzymatic decarboxylation of ferulic acid, a wort constituent (24). It has been found by Tubb et al. (23) that a single dominant nuclear gene designated *POF* (*p*henolic-*o*ff-*f*lavor) codes for the ferulic acid decarboxylation enzyme. Therefore, strains possessing the *POF* gene can produce the enzyme capable of decarboxylating ferulic acid. Whereas, brewing *Saccharomyces* strains normally can not decarboxylate ferulic acid, all the *Saccharomyces diastaticus* strains initially studied produced 4-VG in the presence of ferulic acid. Assuming that the *POF* and *DEX* genes are independent characteristics, it could be possible to construct a strain containing the *DEX* but not the *POF* gene by means of hybridization. Thus, a haploid that was *DEX* positive and that carried the *POF* characteristic was mated with a dextrin negative phenolic-off- flavor negative haploid. The resultant diploid fermented dextrin and decarboxylated ferulic acid. When tetrad dissection was carried out, a 2:2 segregation for dextrin fermentation and a 2:2 segregation for phenolic-off-flavor was obtained. The *DEX* and *POF* genes segregated independently of each other, therefore it was possible to select haploids that were *DEX* positive and *POF* negative. Subsequently, a diploid with the genotype *DEX2/DEX2, pof/pof* was constructed and a fermentation of an 11.3°Plato wort was conducted (Figure 11-9). Although the initial wort attenuation rate was found to be slower than that of a polyploid ale yeast strain, the yeast was capable of superattenuating the wort, i.e., it was able to hydrolyze part of the dextrins into glucose which is readily fermentable, whereas the brewing strain was unable to utilize the dextrins. "Expert" taste panel assessment has deemed the beer produced from this dextrin positive diploid to be rather winey and to have a slightly sulfury character, however, the characteristic phenolic-off-flavor associated with the *POF* gene (4-VG) could not be detected.

The production of glucoamylase by strains of *Saccharomyces diastaticus* is subject to carbon catabolite repression by glucose and other sugars (17). Subjecting two such strains to the 2-DOG selection technique, stable spontaneous mutants derepressed for the formation of glucoamylase and mutants derepressed for maltose utilization have been isolated. In 12°P wort with a 2-DOG starch mutant of *Saccharomyces diastaticus* (strain 1393), the level of glucoamylase was increased five-fold when compared to the parental strain (Figure 11-10A). The overall fermentation rates of the starch and maltose mutants were increased when compared to the parental strain (Figure 11-10B).

A second strain of *Saccharomyces diastaticus* (strain 1400) gave similar results to strain 1393. In 12°P wort, the 2-DOG maltose mutants of this strain possessed increased maltose uptake rates together with an overall faster fermentation rate when compared to the parental strain (Figure 11-11A and 11-11B).

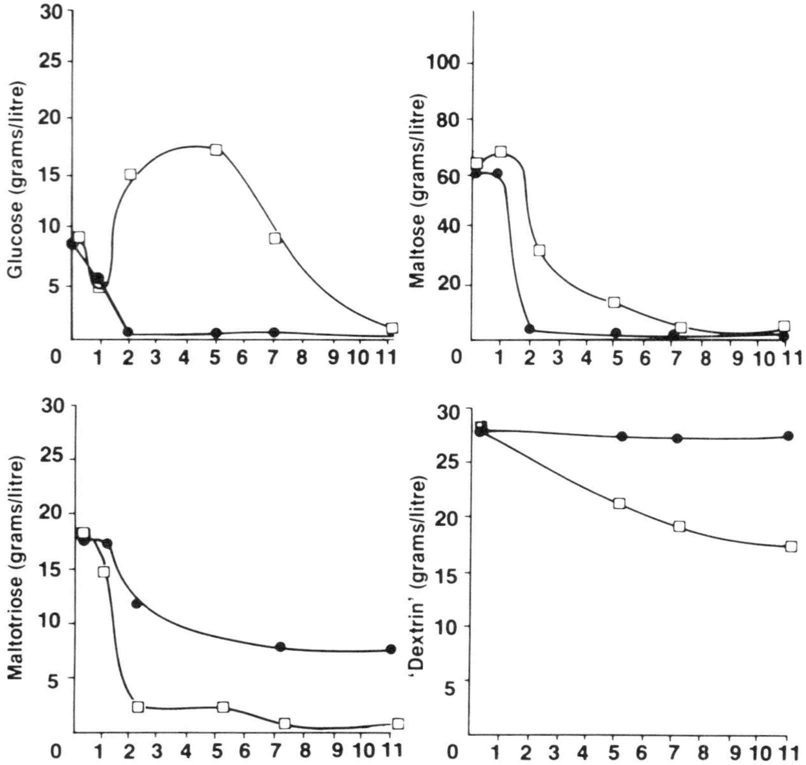

Figure 11-9. Fermentation Characteristics of a Polyploid *Saccharomyces cerevisiae* Strain (●-●) and a Diploid *DEX2/DEX2, pof/pof, Saccharomyces diastaticus* Strain (□-□) in a Brewer's Wort

A further problem with the use of strains of *Saccharomyces diastaticus* (*POF* or *pof*) for the production of beer is the fact that the extracellular glucoamylase is heat stable and remains active in the packaged beer after pasteurization. As a consequence, upon storage at 21°C an increasing concentration of glucose can be found in the beer (Figure 11-12).

This laboratory is committed to achieving the objective of obtaining a strain of *Saccharomyces* spp., that possesses the ability to hydrolyze starch entirely, i.e., able to synthesize and secrete α-amylase and glucoamylase with debranching ability (15). Although *Saccharomyces diastaticus* produces a thermostable glucoamylase, no traces of α-amylase or debranching ability could be detected (Table 11-1). As starch is a polysaccharide composed of two polymers, 20-25% in the form of amylose (linear chains of α-1,4 linked glucose residues) and 75-80% in the form of amylopectin (a highly branched polymer occurring by α-1,6 linkages) debranching activity is essential for complete hydrolysis of the polysaccharide. Two *Endomycopsis fibuligera* strains studied were found to possess α-amylase and glucomylase activity but no debranching activity (Table 11-1).

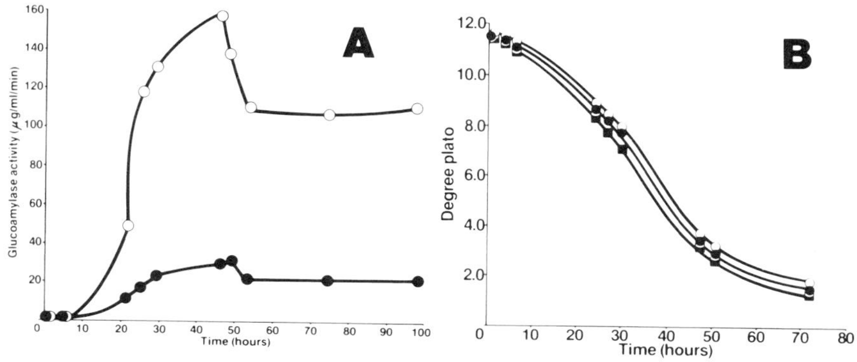

Figure 11-10. Fermentation Characteristics of a Polyploid Strain *Saccharomyces diastaticus* (*DEX1/DEX1, DEX2/DEX2, STA3/STA3, MAL6/mal6*) (strain 1393) and Its Derepressed Maltose and Starch 2 DOG Mutants in a 12°P Wort.
A) Production of Glucoamylase: Parent (●-●) and Derepressed Starch Mutant (○-○).
B) Fermentation Rate: Parent (○-○), Derepressed Maltose Mutant (●-●), and Derepressed Starch Mutant (■-■).

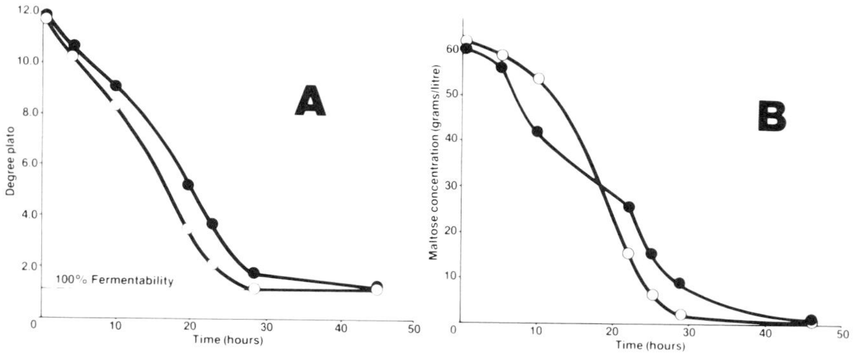

Figure 11-11. Fermentation Characteristics of a Strain of *Saccharomyces diastaticus* (strain 1400) and its Derepressed Maltose 2-DOG Mutants in a 12°P Wort.
A) Fermentation Rate: Parent (●-●) and Mutant (○-○).
B) Maltose Uptake: Parent (●-●) and Mutant (○-○).

A strain of *Pichia burtonii* produced very low levels of debranching activity, however, the yeasts *Schwanniomyces castellii* and *Schwanniomyces occidentalis* produced significant amounts of α-amylase, glucoamylase and debranching activity. These amylolytic systems have been isolated, purified and characterized and it has been found that *Schwanniomyces castellii* possesses a glucoamylase with debranching activity (15).

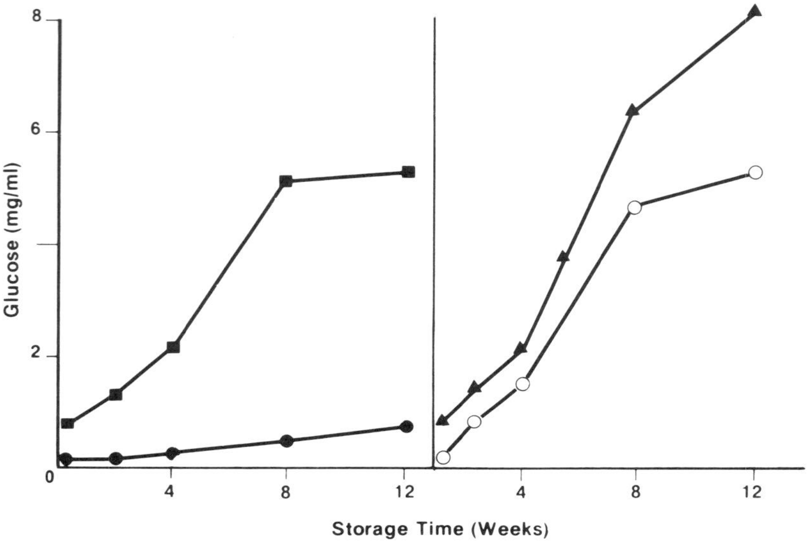

Figure 11-12. Production of Glucose in Pasteurized Beer During Storage at 21°C by a Polyploid *Saccharomyces diastaticus DEX* Strain (■-■), an Ale Brewing *dex* strain (●-●), a Haploid *DEX1, DEX2, STA3* Strain (▲-▲) and a Haploid *DEX2* Strain (O-O).

It has already been discussed in this chapter that one of the techniques for producing low carbohydrate (low calorie) "Lite" beers is to add fungal amylases to the wort during fermentation. Since 70-75% of the dextrins in wort are of the branched type, a debranching enzyme is essential for total hydrolysis of wort dextrins to fermentable sugars. The fungal glucoamylase used extensively in the production of light beer possesses debranching activity, therefore, it can hydrolyze the dextrins. However, normal pasteurization of the final product employing a conventional temperature/time cycle, does not completely inactivate this enzyme. This is a major impediment as the presence of active glucoamylase in the finished product means that the beer is likely to become sweeter due to the hydrolysis of residual dextrin to glucose, changing the beer's flavor characteristics and introducing a possible microbiological hazard. An important characteristic of amylases from *Schwanniomyces castellii* is their sensitivity to the normal pasteurization cycle employed in brewing. A representative curve of pasteurization temperature versus time

Table 11-1. Production of Amylolytic Enzymes by Several Yeast Species

	Enzymatic Activity		
Yeast Species	α-Amylase	Glucoamylase	Debranching
Saccharomyces diastaticus (1384)	−	+++	−
Endomycopsis fibuligera (240)	+	+++	−
Endomycopsis fibuligera (241)	++	+++	−
Pichia burtonii (222)	++	++	+/−
Schwanniomyces castellii (1402)	++	+++	+++
Schwanniomyces occidentalis (1401)	++	++	++
Aspergillus oryzae (245)	+++	+	−

+++ high activity ++ medium activity + low activity − no activity

indicates that there is an eight minute period during which the temperature is maintained a 60-62°C. It has been reported that 15 minutes at 60°C is the time required to inactivate the *Schwanniomyces castellii* glucoamylase, however, this study was conducted at pH 5.5 (the enzyme's optimal pH) and in the absence of ethanol (15). When a commercial fungal glucoamylase preparation derived from *Aspergillus niger* and the glucoamylase from *Schwanniomyces castellii* were compared for their sensitivity to pasteurization at pH 4.0 (normal beer pH) and pH 6.0 in the presence and absence of ethanol (Figure 11-13), the ethanol enhanced the inactivation effect of the pasteurization. In addition, at pH 4.0, the pasteurization cycle inactivated glucoamylase (as well as α-amylase—data not shown) from *Schwanniomyces castellii* with or without ethanol, but at pH 6.0 the presence of ethanol was necessary for enzyme inactivation.

A two stage fermentation system has been devised for the production of low carbohydrate beer. (16) Amylases (gluco- and α-amylase) from *Schwanniomyces castellii* are produced in a highly inducing medium containing maltose. Subsequently, the cells are removed, the culture filtrate is concentrated and added to wort previously inoculated with a genetically manipulated strain of *Saccharomyces diastaticus* or with a brewing production strain of *Saccharomyces uvarum* (*carlsbergensis*). The *Saccharomyces diastaticus* strain is a diploid containing both the *DEX1* and *DEX2* genes which code for glucoamylase production and, more importantly, this strain lacks the capability to decarboxylate ferulic acid to 4-vinyl guaiacol.

In an attempt to establish the optimal amount of enzyme culture filtrate to be added to the fermenting wort, several concentrations of enzymes were added (Table 11-2). As expected, increasing concentrations of enzymes had a direct correlation with apparent attenuation, i.e., with the maximum amounts of enzymes added, the degree of fermentation increased to 100% and up to 99.4% with 70% less added enzyme. Fermentations conducted with the diploid *Saccharomyces diastaticus* strain were more sluggish because this particular strain lacked *MAL* genes, therefore, maltose was only hydrolyzed by the extracellular glucoamylase and a longer time of fermentation was required to achieve 100% of apparent attenuation.

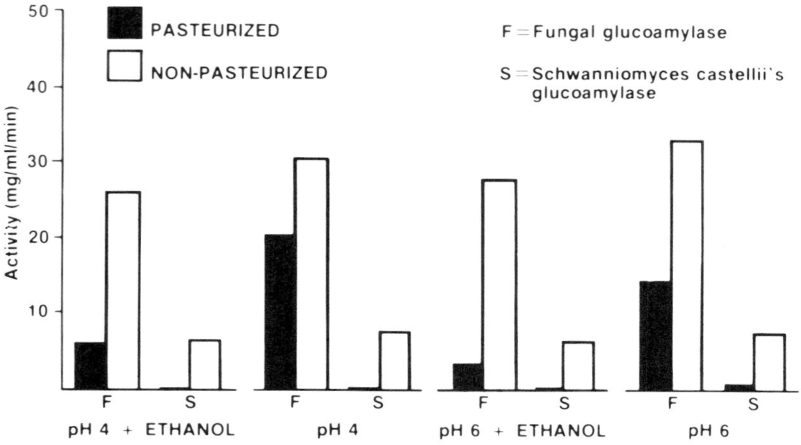

Figure 11-13. Effect of Pasteurization on the Enzymatic Activity of a Commercial Fungal Glucoamylase (F) and a Glucoamylase from *Schwanniomyces castellii* (S) with and without Pasteurization. Pasteurized (■) and Non-pasteurized (□).

A direct correlation has been found between ethanol production and enzyme concentration employed, i.e., the greater the enzyme concentration, the higher the level of ethanol obtained. Thus, with fermentations employing the lager yeast strain, a maximum increase of 19% ethanol (compared to the control) could be obtained in the final product, whereas with *Saccharomyces diastaticus* an improvement of 15.1% could be obtained with the maximal enzyme addition (Table 11-2).

DIASTATIC YEASTS FOR DISTILLED ETHANOL PRODUCTION

As previously discussed, the fermentation of starches to ethanol by yeasts requires pretreatment of the substrate in order to produce fermentable sugars. This pretreatment consists of three steps: gelatinization, liquefaction and saccharification. Gelatinization requires heat and free water and must precede liquefaction. Liquefaction, the dispersion of

Table 11-2. The Effect of Increasing Concentrations of Amylolytic Enzymes on Fermentation Characteristics of a 16°Plato Wort

	Sacch. uvarum (carlsbergensis)		*Sacch. diastaticus*	
Amount of Enzyme Added (mg/liter)	Final Ethanol (% w/v)	Apparent Attenuation (%)	Final Ethanol (%w/v)	Apparent Attenuation (%)
0	4.20	84.7	4.30	89.5
4	4.42	88.8	4.37	89.8
8	4.49	93.6	4.40	90.6
12	4.76	95.2	4.55	94.5
16	4.90	99.4	4.69	96.9
20	5.00	100.0	4.95	100.0

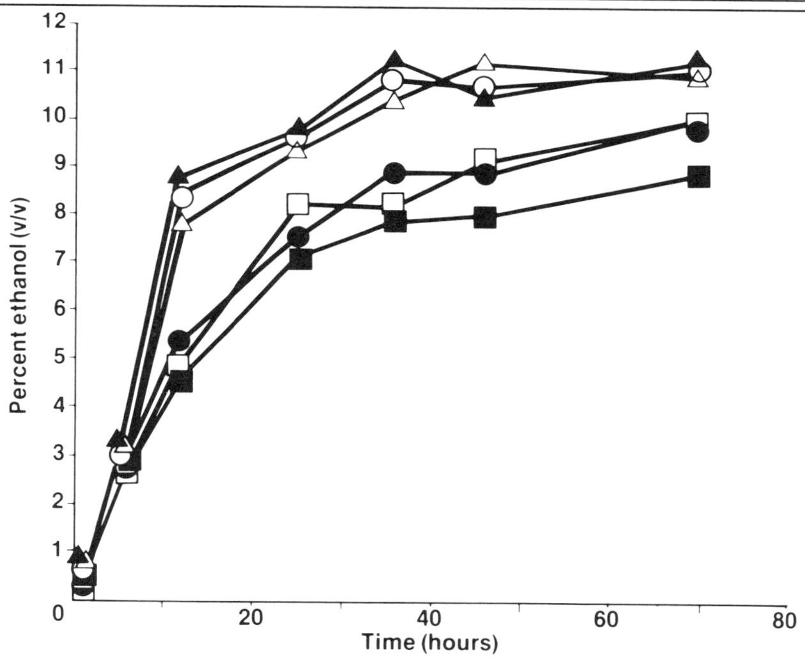

Figure 11-14. Ethanol Production by Hybrid Strain 1393.
0.0% v/w AMG (■-■)
0.025% v/w AMG (□-□)
0.04% v/w AMG (●-●)
0.05% v/w AMG (△-△)
0.075% v/w AMG (○-○)
0.10% v/w AMG (▲-▲)
AMG = glucoamylase

starch molecules into an aqueous solution, is accomplished by the use of heat and amylolytic enzymes. Heat stable bacterial α-amylases or malt enzymes may be employed. During liquefaction starch molecules are only partially hydrolyzed producing a form of carbohydrate which cannot be assimilated by ethanol-producing yeasts such as *Saccharomyces cerevisiae*. Therefore, the partially hydrolyzed starch molecules must be converted to lower molecular weight sugars such as glucose and maltose by a process known as saccharification. This may be accomplished enzymatically, usually by the addition of fungal glucoamylases to the fermentation vessel at the time of inoculation. The saccharifying glucoamylases represent a significant fraction of the total cost of producing ethanol. Reduction of the amount of added glucoamylase could significantly decrease the cost of the final product.

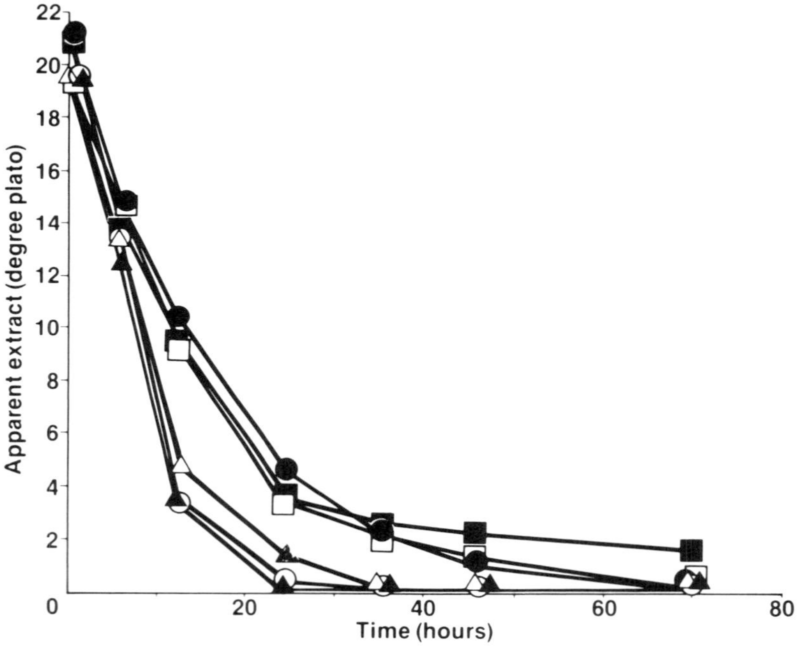

Figure 11-15. Apparent Extract Values for Hybrid Strain 1393.
0.0% v/w AMG (■-■)
0.025% v/w AMG (□-□)
0.04% v/w AMG (●-●)
0.05% v/w AMG (△-△)
0.075% v/w AMG (○-○)
0.10% v/w AMG (▲-▲)
AMG = glucoamylase

It has been found possible to decrease glucoamylase addition to starch mash fermentations by employing yeast which actively produce and secrete glucoamylase e.g., strains of *Saccharomyces diastaticus* (25). A genetically manipulated diploid strain of *Saccharomyces diastaticus* (strain 1393, *DEX1/DEX1, DEX2/DEX2, STA3/STA3, MAL6/mal6*) was studied as the glucoamylase producing strain and compared to a strain of *Saccharomyces cerevisiae* (strain 254) which was unable to produce and secrete glucoamylase. The fermentation performance of the two strains was compared in a corn mash with and without added glucoamylase. When the *Saccharomyces diastaticus* (strain 1393) culture was employed, added glucoamylase concentrations could be significantly decreased without reducing ethanol production or sugar uptake (Figures 11-14 and 11-15). Reduction of the added glucoamylase concentration from 0.1% by volume, based upon the substrate weight, to 0.05% resulted in no significant decrease in ethanol yield or sugar uptake; the reduction of added glucoamylase being possible because the hybrid yeast is able to produce and secrete its own glucoamylase.

Production of glucoamylase by the *Saccharomyces diastaticus* strain and subsequent reduction of added glucoamylase represents a potential financial saving for the ethanol producer. Since the cost of glucoamylase may exceed $3.75 (U.S.) per liter, a 50% reduction of this enzyme can result in a significant decrease in cost. For example, an ethanol manufacturer that is producing 100 million liters of 95% ethanol per year uses 190 million kg of starch (assuming that the fermentation of 20% (w/v) starch yields 10% (v/v) ethanol). For 190 million kg of starch, 0.19 million liters of glucoamylase are required (based upon an addition of 0.1 liter of glucoamylase per 100 kg of starch).

This results in an annual enzyme cost of approximately $700,000 U.S., assuming an enzyme cost of $3.75 U.S. per liter. Reduction of the added glucoamylase by 50% represents an annual saving of approximately $350,000 U.S.

CONCLUSIONS

Although yeast is the oldest of the biotechnology microorganisms, it is still a very important microorganism that is being employed in the modern biotechnology revolution. Current research is being directed into two areas, the first being to improve its capacity and efficiency to produce ethanol and carbon dioxide from a wide variety of substrates and the second is to employ yeast to synthesize and excrete a wide variety of proteins or peptides that possess therapeutic and commercial potential.

REFERENCES

1. Bailey. R.B. and Woodword, A. Isolation and characterization of a pleiotropic glucose repression resistant mutant of *Saccharomyces cerevisiae*. Mol. Gen. Genet. 193: 507-512 (1984)
2. Barnett, J.A., Payne, R.W., and Yarrow, D. In: "Yeasts - Characteristics and Identification". London, Cambridge University Press (1983)

3. Bevan, A., Herring, A.J., and Mitchell, D.J. Preliminary characterization of two species of dsRNA in yeast and their relationship to the "killer" character. Nature (London) 245: 81-86 (1973)
4. Bilinski, C.A., Sills, A.M. and Stewart, G.G. Morphology and genetic effects of benomyl on polyploid brewing yeast: induction of auxotrophic mutants. Appl. Environ. Microbiol. 48: 813-817 (1984)
5. Entian, K.D., Zimmerman. F.K., and Scheel, I. A partial defect in carbon catabolite repression in mutants of *Saccharomyces cerevisiae* with reduced hexose phosphorylation. Mol. Gen. Genet. 156: 99-105 (1977)
6. Erratt, J.A. and Stewart, G.G. Genetic and biochemical studies on glucoamylase from *Saccharomyces diastaticus*. In: "Current Developments in Yeast Research". (G.G. Stewart and I. Russell. eds.). pp 177-183, Toronto, Pergamon Press (1981)
7. Gunge, N. Yeast DNA plasmids. Annu. Rev. Microbiol. 37: 253-276 (1983)
8. Hammond, J.R.M. and Eckersley, K.W. Fermentation properties of brewing yeast with killer character. J. Inst. Brewing 90: 167-177 (1984)
9. Hitzman, R.A., Leung, D.W., Perry, L.J., Kohr, J.W., Levine, H.L., and Goeddel, D.W. Secretion of human interferons by yeast. Science 219: 620-626 (1983)
10. Kreger-van Rij, N.J.W. In: "The Yeasts: A Taxonomic Study". 3rd ed. Amsterdam, Elsevier Science Publishers B.V. (1984)
11. Lodder, J. In: "The Yeasts: A Taxonomic Study". 2nd ed. Amsterdam, North-Holland Publishing Company (1970)
12. Maule, A.P. and Thomas, P.D. Strains of yeast lethal to brewery yeasts. J. Inst. Brewing 79: 137-141 (1973)
13. Moir, D., Duncan, M., Kohno, T., Mao. J-I, and Smith, R. Production of cell chymosin by the yeast *Saccharomyces cerevisiae*. The World Biotechnology Report U.S.A. 2: 189-197 (1984)
14. Pfisterer, A., Garrison, I.F. and McKee, R.A. Brewing with syrups. MBAA Tech. Ouart. 15: 59-63 (1978)
15. Sills, A.M., Panchal, C.J., Russell, I., and Stewart, G.G. Genetic manipulation of amylolytic enzyme production by yeasts. In: "Proc. Ale Yeast Symposium". Vol. 1 (M. Korhola and G. Vaisanen, eds.). pp 209-228. Helsinki, Foundation for Biotechnical and Industrial Fermentation Research (1983)
16. Sills, A.M., Russell, I., and Stewart, G.G. The production and use of yeast amylases in the brewing of low carbohydrate beer. In: "Proc. 19th Int. Eur. Brew. Conv. Congr." pp 337-344. London, IRL (1983)
17. Sills, A.M., Sauder, M., and Stewart, G.G. Amylase activity in certain yeasts and fungal species. Dev. Ind. Microbiol. 24: 295-303 (1983)
18. Stewart, G.G. Fermentation—yesterday, today and tomorrow. MBAA Tech. Ouart. 14 (1): 1-15 (1977)
19. Stewart. G.G. The genetic manipulation of industrial yeast strains. Can. J. Microbiol. 27: 973-990 (1981)
20. Stewart, G.G., Bilinski, C.A., Panchal, C.J., Russell, I., and Sills. A.M. The genetic manipulation of brewer's yeast strains. In: "Microbiology 1985". Washington, D.C., ASM Publications (In Press)
21. Stewart, G.G., Russell, I. and Sills, A.M. Factors that control the utilization of wort carbohydrates by yeast. MBAA Tech. Quart. 20: 1-8 (1983)
22. Struhl, K. The new yeast genetics. Nature (London) 305: 391-397 (1983)
23. Tubb, R.S., Searle, B.A., Gooday, A.R., and Brown, A.J.P. Rare mating and transformation for construction of novel brewing yeasts. In: "Proc. 18th In. Eur. Brew. Conv. Congr." pp 487-496. London, IRL (1981)

24. Wackerbauer, K., Kossa, T. and Tressl, R. Die bildung von phenolen durch hefen. In: "Proc. 16th Int. Eur. Brew. Conv. Congr." pp 495-505. London, IRL (1977)
25. Whitney, G.K., Murray, C.R., Russell, I., and Stewart, G.G. Potential cost savings for fuel ethanol production by employing a novel hybrid yeast strain. Biotechnology Letters 7: 349-354 (1985)
26. Wickner, R.B. The killer double-stranded RNA plasmids of yeast. Plasmid 2: 303-322 (1979)
27. Young, T.W., Brewing yeast with anti-contaminant properties. In: "Proc. 19th Int. Eur. Brew. Conv. Congr." pp 129-136. London, IRL (1983)

Chapter 12

Lactobacilli in Food Fermentations

Bruce Chassy

INTRODUCTION

Lactobacilli are widely used in food and agricultural fermentations. Their uses and contributions to foods are described in this chapter. The status of methodology to do genetics, molecular genetics and recombinant DNA technology with members of the genus is reviewed. Possible future modifications of lactobacilli that would give rise to improved starter cultures are discussed in the final sections.

CURRENT APPLICATIONS OF LACTOBACILLI

Historically, fermented foods must have arisen by accidental or natural contamination of foods by microbes. Later, having observed the beneficial effects of some microbes and the deleterious effects of others, man learned to control the inoculation of foods (16,17,29). Today control has been enhanced by the widespread use of commercially prepared single and multiple strain starter cultures designed for specific products and markets (16,17,29). Among the numerous microbes that man has used in the preparation and preservation of foods, lactobacilli are common (16,29). Table 12-1 presents examples of the utility of members of the genus. A few non food-related applications are listed also.

The vast majority of bacterial species have no particular utility to man; they seem to coexist invisibly with us in our universe. A few bacteria have drawn our attention by virtue of their pathogenicity toward man or animals, or by their propensity to spoil or poison food. Fewer still have found their way into our food preparation processes by our intentional use of them. It is noteworthy that among food-related microorganisms, the majority fall into the group known as the lactic acid bacteria: the lactobacilli, the streptococci, (primarily the Group N or dairy streptococci), the pediococci and the leuconostocs (16). What are the properties that make these organisms desirable in food?

Table 12-1. Industrial Applications of *Lactobacillus* Strains

Product	Organism(s) used
yogurt	*L. bulgaricus* and *S. thermophilus*
fermented milks	*L. acidophilus, casei, bifidus,* bulgaricus
cheeses	*L. bulgaricus, helveticus*
soy sauce	*L. delbrueckii*
sour bread	*L. sanfrancisco*
crackers	*L. plantarum*
sauerkraut	*L. plantarum, brevis*
cucumbers (pickles)	*L. plantarum*
green olives	*Lactobacillus* spp.
cured ham	*L. casei, plantarum*
sausages, meats	*L. plantarum, reuterii,* spp, *Pediococcus* spp.
distillery mashes	*L. casei, fermentum, plantarum, delbrueckii*
feed additives	*L. acidophilus, bulgaricus, lactis*
silage starters	*L. plantarum, Pediococcus* spp.
lactic acid	*L. delbrueckii*

Table 12-2 lists the four major contributions of lactobacilli to foods. The modification of flavor caused by the *Lactobacillus* strains used in culturing buttermilk, *acidophilus* milk, or San Francisco sour dough bread are obvious examples of flavor being changed by a *Lactobacillus.* In these products, the appearance also changes as a result of bacterial action as it does with yogurt. The contribution that a fermentation may make to B-vitamin and free amino acid content is more subtle (17). The levels of some essential amino acids can be significantly raised by the fermentation process (17,28).

Table 12-2. Contributions of Lactobacilli to Foods

1. Alter flavor, texture and appearance
2. Enhance nutritional value
3. Retard spoilage and reduce contamination
4. Probiotic effects

Anaerobic glycolysis of carbohydrates by lactobacilli results in the production of lactic acid which lowers the pH and discourages growth of other contaminating microbes upon subsequent storage (16,17). Their use of carbohydrate substrates competitively contributes to their ability to retard spoilage (16,17). Some strains are known to produce bacteriocins and possibly other specific antimicrobials which may contribute to this effect (1,17).

Probiotic effects are far more difficult to document than those listed above. The controversy began in 1908 when Metchnikoff suggested that intestinal lactobacilli could play a role in the prolongation of life (25). What seems clear is that certain lactobacilli are commensal colonizers of the human and animal gastro-intestinal epithelium (17,30). Their exact role and function is hard to define. It is known that they can secrete enzymes such as lactase which might facilitate lactose digestion in less tolerant individuals (14,17). Lactobacilli have the potentia1 to metabolize dietary cholesterol (13). Strains of lactobacilli have been shown to detoxify potential carcinogens; while compounds with anti-tumor activity have been isolated from others (14,17). The tota1 effect on prolongation of life or enhancement of human health by lactobacilli is far from firmly established. It is tempting to believe that lactobacilli do us no harm, and that they may well have a positive effect on life.

The worldwide market value of *Lactobacillus* fermented products is hard to assess. From the applications listed in Table 12-1, it is clear that billions of dollars worth of commercial food products are affected by lactobacilli. Because fermentations start, in most cases, with highly specialized defined starter cultures, an industry directed at the production of large quantities of specific strains has emerged. These suppliers strive to provide metabolically active cells of high viability that can be added directly to fermentations. The strains are chosen for desirable physiological and growth characteristics, as well as for their flavor and end product effects. Other factors such as phage resistance and heat tolerance are often considered. Starters may be pure cultures or mixtures of strains of the same or differing species.

The economic value of preservation and retardation of spoilage is impossible to estimate, although in these days of improved refrigeration and transportation, it may no longer be of major importance. This is certainly not the case in less well developed countries where fermentation may still be a primary means of food preservation. The economic significance of enhancement of nutritional value and probiotic effects in foods are even more difficult contributions to evaluate. One can only conclude that *Lactobacillus* fermentations have a significant economic and social impact.

MOLECULAR GENETICS OF LACTOBACILLI

The opportunity to develop strains with new or enhanced properties by the techniques of biotechnology may significantly alter the economic and social impact of these organisms. It is likely that new applications may be found as well. The following sections give an overview of our current ability to accomplish genetic manipulation of lactobacilli. Table 12-3 outlines the four principle areas that will be reviewed.

Table 12-3. Molecular Genetics of Lactobacilli

1.	Classical mutagenesis and selection
2.	Transduction and phage
3.	Conjugation and cell fusion
4.	Transformation and recombinant DNA technology

Classical Mutagenesis and Selection

The classical techniques of bacterial genetics — mutagenesis followed by the selection of mutants with desired traits — can easily be applied to lactobacilli. For example, large-scale fermentations are sometimes ruined by bacteriophage contamination (10). Through the use of mutant selection, phage-resistant variants have been derived from a number of commonly used starters (10,11,31,41). These modified strains considerably reduce the losses associated with phage contamination.

As with other species, one may attempt to find mutants that accumulate a specific product because of a block in a metabolic pathway or an enhancement in the activity of a specific pathway. Mutagenesis and selection have been used to unblock cryptic amino acid biosynthetic pathways of lactobacilli giving rise to amino acid prototrophs with simplified nutritional requirements (27). In an extension of this approach, strains which enhance the nutritional content of corn by secreting lysine have been isolated (28). In a recent study, mutants of *Lactobacillus plantarum* that do not produce CO_2 from malate were described; these strains should be useful in reducing bloating in pickles that arises from the fermentation of malate (9).

Classical techniques are limited only by the gene pool available in lactobacilli. If a gene or operon is present it may be modified, blocked, or enhanced. Although many desirable new traits and properties may be obtained by classical genetics, totally new genes and gene products can not be introduced by these techniques.

Transduction and Phage

Phage-mediated transduction is a classical method to move genes between strains. Bacteriophage are common among lactobacilli and the majority of strains isolated from natural sources appear to be lysogenic for one or more phage (8,33,36,37). Transduction of the lactic streptococci has been reported; lactose plasmid DNA was packaged into defective phage particles and transduced into the chromosome of a lac$^-$ recipient (24). To date, however, no laboratory has reported successful transduction of a genetic trait between lactobacilli. Exploitation of this technique awaits investigation. When one considers the utility of the lambda and P phages to *Escherichia coli* genetics, the value of pursuing useful transductional systems and phage derived recombinant DNA systems becomes apparent: it offers a direct means for the introduction of new genes.

Conjugation and Cell-Fusion

Conjugal exchange of genetic information between bacteria presents another possible approach to strain development. An erythromycin resistance-encoding conjugal plasmid, pAMBl, originally isolated from *Streptococcus fecaelis*, has been transferred into and between several *Lactobacillus* species (12,39). In addition, a conjugal lactose plasmid has been isolated from *Lactobacillus casei* and shown to transfer the genes that determine the ability to ferment lactose into several strains of lactobacilli which had lost the ability to ferment lactose (5). Little more has been done to explore the potential of conjugation as a means of moving genes into and between lactobacilli. One might imagine, in the absence of a system to do molecular cloning in a specific strain, that a gene could be cloned in an organism such as *Escherichia coli*, moved into *Streptococcus sanguis* by transformation, and then transferred by conjugation into the target *Lactobacillus* strain.

Fusion of protoplasts of the same or dissimilar species also offers possibilities for the generation of hybrid strains with new characteristics. It would also serve as a means for introducing a transposon into lactobacilli on a dead-end or "suicide" vector. Once established, the transposon could be useful in tagging genes for molecular cloning, establishing genetic maps, and for the preparation of mutants. Transposons, at present, are unavailable as a genetic tool for the genus *Lactobacillus*. However, the first steps necessary to carry out such experiments have been reported; that is, there exists a system for the production and regeneration of protoplasts (4,21,40).

Transformation and Recombinant DNA Technology

What then is needed to be able to directly move genes into a given strain? Table 12-4 provides a minimal list of the necessary factors. The status of each is discussed in the following paragraphs.

Table 12-4. Factors Required to Do Recombinant DNA Technology

1. Plasmids that will replicate in the target organism
2. Information about plasmid maintenance and incompatibility
3. Stable vectors with selectable markers
4. Specific target genes isolated by molecular cloning
5. A knowledge of factors controlling gene expression
6. A means to escape host nucleases and restriction systems
7. A system of transformation to introduce DNA into organism

Plasmids isolated directly from the organism for which a system of molecular cloning is sought are the most suitable starting point, since one has at least the confidence that they will replicate in the target organism. Chassy et al. (2,4) were initially successful some years ago in isolating plasmids from a number of strains of *Lactobacillus casei*. Some of the plasmids isolated determined the ability of the host organism to ferment lactose (3). Small cryptic plasmids representing several homolgy groups were also present in many of the strains (22). Other workers have isolated plasmids from various lactobacilli (15,18,19,23,26,35,38). They appear widely distributed. Antibiotic resistance genes are encoded by some of these plasmids (15,26,38), which may be important in the development of cloning vectors. They are likely candidates for the direct transformation of the strains from which they were isolated. Little is known about plasmid maintenance and stability in lactobacilli; some of the plasmids appear quite stable, while others can be eliminated by curing agents or are lost spontaneously (3,40).

Vectors developed for molecular cloning in other gram-positive bacteria might be useful for the transformation of lactobacilli, however, it is difficult to develop a transformation system without the certain knowledge that the vectors will function in the target organism. Therefore, bifunctional "shuttle vectors" containing *E. coli* or *S. sanguis* and *Lactobacillus* plasmid replicons along with selectable markers known to function in lactobacilli (eg. the erythromycin resistance gene from pAMBl) were constructed (22). Also under study is the use of naturally occurring lactose plasmids that can be isolated directly from lactobacilli and then used to transform an isogenic lactose negative mutant previously cured of its lactose plasmid.

Ultimately the objective of such studies is to introduce a novel gene into a new host, thus specific genes must be isolated. At present it appears less difficult to isolate and characterize genes in *E. coli* because a wide diversity of tools are available to work with the better understood *E. coli* host system. Once characterized and understood in *E. coli*, genes could be transferred into a chosen *Lactobacillus* strain. Chassy et al. (20) transferred the gene encoding phospho-β-galactosidase from *L. casei* into *E. coli*. The gene expressed in *E. coli* from a *Lactobacillus*-derived promoter (20). Recently, this group succeeded in cloning β-galactosidase from *Lactobacillus*. The sequence of the phospho-β-galactosidase gene will soon be available. Of particular interest will be the nucleotide sequence of its promoter, since it is hoped to compare sequences of promoters derived from constitutive and inducible variants of the same gene in order to help understand gene regulation and expression in lactobacilli (6,7). In cases where *E. coli* is not a suitable host, *S. sanguis* or *Bacillus subtilis* might be employed.

The major barrier to genetic technology in lactobacilli has been the lack of a transformation system to introduce DNA into the cells. The introduction of foreign DNA into protoplasts by the action of fusants such as polyethylene glycol (PEG) has been successful for a wide variety of bacterial, plant and animal cell types. Therefore, methods were developed to generate protoplasts of lactobacilli and formulate a medium on which the protoplasts could be regenerated (21). The successful transfection of *L. casei* protoplasts with phage DNA encapsulated in liposomes formed from lecithin was encouraging (32). The transfection has been optimized to deliver more than 10^6 transfectants/μg phage DNA; a frequency suitable to insure efficient introduction of DNA into the target bacterium.

Success for the first time has occurred in introducing lactose plasmid DNA into lactobacilli and selecting transformants that have acquired the ability to metabolize lactose by genetic transformation. Southern transfers of *Eco*RI digests of chromosomal DNA isolated from the transformants demonstrated that a single copy of phospho-β-galactosidase, and presumably the genes encoding the proteins of the lactose PEP:PTS, were integrated into the *L. casei* chromosome during the transformation process. The frequency of the transformation process was approximately 10^{-4}/protoplast regenerated; the efficiency was 10^4 transformants/μg lactose plasmid DNA. Thus, it appears that a usable system of transformation will be developed in the near future. The capacity to do recombinant DNA manipulations depends principally on the availability of a transformation system.

FUTURE OBJECTIVES FOR BIOTECHNOLOGY WITH LACTOBACILLI

The list presented in Table 12-5 suggests some of the general directions that strain development by rDNA technology might take. These are but a few examples, the number of other possibilities is limited only by ones imagination and the level of investment that is made. Modifications of strains used in industry today are easily conjectured. The distant future will see more new and novel applications. The directions suggested here are very general; no doubt many have already set their sights on specific proprietary objectives. If the pattern of biotechnology research with other organisms is followed by experiments with lactobacilli, many exciting and innovative new strains will be developed.

It should be noted that lactobacilli have a number of traits that make them particularly attractive as producers of end-products or secondary metabolites. As such, they might represent a useful alternative to the host-vector systems in use today. Some of their more desirable properties are given in Table 12-6. Not to be overlooked is the propensity of *L. acidophilus* strains to colonize the gastro-intestinal tract. This property opens the way for their use as engineered probiotic organisms, once appropriate

Table 12-5. Objectives of Biotechnology in Modifying Starter Cultures

Objective	Example(s)
1. Stabilize trait	Prevent essential plasmid loss by integration
2. Combine traits	Eliminate the need for binary starters
3. Add new traits	Proteases to eliminate rennin
	Phage resistance genes
	Bacteriocin and antibiotic secretion
	Faster growth
	Ability to manipulate physiology
	Enhanced flavor or end product formation
	Altered or improved physical properties
	Added nutritional value
	Enhanced probiotic effect
4. New applications	Secondary metabolite production
	New fermented foods
	Vaccines and pharmaceuticals

genes are introduced by rDNA technology. Their tissue-specific colonization might even be extended into the area of antigen secretion for the purpose of vaccination. The National Institute for Dental Research is particularly intrigued by this possibility and its implications for vaccination against dental caries and periodontal diseases.

Table 12-6. Advantages of *Lactobacillus* Strains for Industrial Processes

1. Methods already exist for large-scale cultivation
2. Non-pathogenic
3. No toxin or toxic product formation
4. Aerotolerant
5. No aeration required
6. Moderately thermophillic (51-54°C)
7. Withstand low pH
8. Natural products discourage contamination and spoilage
9. Ferment a diversity of carbohydrate feedstocks (whey, silage, plant juices and wastes, hydrolyzed starch)
10. Relatively rapid and abundant growth
11. Cultures are stable and viable
12. Non-sporeforming

How quickly the biotechnology community is able to capitalize on these possibilities is directly related to the level of investment in research and the numbers of investigators focusing their attention on the area. At the moment, there is a scarcity of trained investigators with experience in the genetics of lactic acid bacteria. This may change, since it appears that numerous university laboratories, government-associated or sponsored health and biotechnology research institutes, and industrial and biotechnology companies have become interested in these organisms. While secrecy and confidentiality make it difficult to assess how much research is being done, it is clear that the field is becoming increasingly popular. The possible rewards merit the new attention that is being received by lactobacilli and other lactic acid bacteria. Recently, a patent for the process of introducing DNA into lactobacilli has been filed (34).

CONCLUSIONS

It should be noted that legal and regulatory decisions will have to be made before new strains are fully utilized. The regulations are clear with respect to use of these organisms as producers of end products or metabolites; no particular problems should arise with their use. However, when advocating dissemination of engineered organisms in the human food supply, it is clear that new ground may have to be broken. To the extent that the efforts are restricted to the mixing of genes between strains that are already used in food and industry, perhaps it will not be difficult to demonstrate their safety and efficacy. Doubtless, the antibiotic resistance markers used in most vector constructions can not be used in engineered food organisms except in development steps. Integration of specific genes into the host chromosome may become the primary objectives. When genes from new sources are introduced into starter cultures, or are used to develop new foods using engineered organisms, approval will depend on satisfying more demanding criteria. The most difficult approval to obtain would be that for *Lactobacillus* strains designed to colonize the human host. History tells us, that one must proceed cautiously and satisfy the most stringent criteria of health and safety. The next decade of research with lactobacilli should be exciting and the economic and societal returns great.

REFERENCES

1. Barefoot, S.F., and Klaenhammer, T.R. Detection and activity of lactacin B, a bacteriocin produced by *Lactobacillus acidophilus*. Appl. Environ. Microbiol. 45:1808-1815 (1983)
2. Chassy, B.M., Gibson, E.M., and Giuffrida, A. Evidence for extrachromosomal elements in *Lactobacillus*. J. Bacteriol. 127:1576-1578 (1976)
3. Chassy, B.M., Gibson, E.M., and Giuffrida, A. Evidence for plasmid-associated lactose metabolism in *Lactobacillus casei* subsp. *casei*. Current Microbiol. 1: 141-144 (1978)
4. Chassy, B.M., and Giuffrida, A. A method for the lysis of gram-positive apsorogenous bacteria with lysozyme. Appl. Environ. Microbiol. 39: 153-158 (1980)

5. Chassy, B.M., and Rokaw, E. Conjugal transfer of lactose plasmids in *Lactobacillus casei.* In: "Molecular Biology, Pathogenesis and Ecology of Bacterial Plasmids". (S. Levy, R. Clowes, and E. Koenig, eds) Plenum Press. New York, p. 590 (1981)
6. Chassy, B.M., and Thompson, J. Regulation and characterization of the galactose-phosphoenol pyruvate-dependent phosphotransferase system in *Lactobacillus casei.* J. Bacteriol. 154: 1204-1214 (1983)
7. Chassy, B.M., and Thompson, J. Regulation of lactose-phosphoenol pyruvate-dependent phosphotransferase system and β-D-phosphogalactoside galactohydrolase activities in *Lactobacillus casei.* J. Bacteriol. 154: 1195-1203 (1983)
8. Coetzee, J.N., and De Klerk, H.C. Lysogeny in the genus *Lactobacillus.* Nature (London) 194:505 (1962)
9. Daeschel, M. A., McFeeters, R.F., Fleming, H.P., Klaenhammer, T.R., and Sanozky, R.B. Mutation and selection of *Lactobacillus plantarum* strains that do not produce carbon dioxide from malate. Appl. Environ. Microbiol. 47: 419-420 (1984)
10. Davies, F.L., and Gasson, M.J. Reviews of the progress of dairy science: genetics of lactic acid bacteria. J. Dairy Res. 48: 363- 376 (1981)
11. Gasson, M.J., and Davies, F.L. Prophage cured derivatives of *Streptococcus lactis* and *Streptococcus cremoris.* Appl. Environ. Microbiol. 40: 964-966 (1980)
12. Gibson, E.M, Chace, N.M., London, S.B., and London, J. Transfer of plasmid-mediated antibiotic resistance from streptococci to lactobacilli. J. Bacteriol. 137:614-619 (1979)
13. Gilliland, S.E., Nelson, C.R., and Maxwell, C. Assimilation of cholesterol by *Lactobacillus acidophilus.* Appl. Environ. Microbiol. 49: 377-381 (1985)
14. Goldin, B.R., and Gorbach, S.L. The effect of oral administration on *Lactobacillus* and antibiotics on intestinal bacterial activity and chemical induction of large bowel tumors. Devel. Indust. Microbiol. 25:139-150 (1984)
15. Ishiwa, H., and Iwata, S. Drug resistance plasmids in *Lactobacillus fermentum.* J. Gen Applied Microbiol. 26: 71-74 (1980)
16. Kandler, O. Current taxonomy of lactobacilli. Devel. Indust. Microbiol. 25:109-123 (1984)
17. Kilara, A., and Treki, N. Uses of lactobacilli in foods—unique benefits. Devel. Indust. Microbiol. 25:125-138 (1984)
18. Klaenhammer, T.R., and Sutherland, S.M. Detection of plasmid deoxyribonucleic acid in an isolate of *Lactobacillus acidophilus.* Appl. Environ. Microbiol. 35: 592-600 (1980)
19. Klaenhammer, T.R. A general method of plasmid isolation in lactobacilli. Curr. Microbiol. 10:23-28 (1984)
20. Lee, L.-J., Hansen, J.B., Jagusztyn-Krynicka, E.K., and Chassy, B.M. Cloning and expression of the β-D-phosphogalactoside galactohydrolase gene of *Lactobacillus casei* in *Escherichia coli* K- 12. J. Bacteriol. 152: 1138-1146 (1982)
21. Lee-Wickner, L.-J., and Chassy, B.M. The production and regeneration of protoplasts of *Lactobacillus casei.* Appl. Environ. Microbiol. 48: 994-1000 (1984)
22. Lee-Wickner, L.-J., and Chassy, B.M. Molecular cloning and characterization of cryptic plasmids isolated from *Lactobacillus casei.* Appl. Environ. Microbiol. 49: 1154-1161 (1985)
23. Lin, J.H.-C., and Savage, D. Cryptic plasmids in *Lactobacillus* strains isolated from the murine gastrointestinal tracts. Appl. Environ. Microbiol. 49: 1004-1006 (1985)

24. McKay, L.L., Cords, B.R., and Baldwin, K.A. Transduction of lactose metabolism in *Streptococcus lactis* C2. J. Bacteriol. 115: 810-815 (1973)
25. Metchnikoff, E. "The Prolongation of Life". G.P. Putnam's Sons, New York (1908)
26. Morelli, L., Vescovo, M., and Bottazzi, V. Plasmids and antibiotic resistances in *Lactobacillus helviticus* and *Lactobacillus bulgaricus* isolated from natural whey culture. Microbiologica. 6: 145-154 (1983)
27. Morishita, T., Fukada, T., Shirota, M., and Yura, T. Genetic basis of nutritional requirements in *Lactobacillus casei*. J. Bacteriol. 120: 1078-1084 (1974)
28. Newman, R.K., and Sands, D.C. Nutritive value of corn fermented with lysine excreting lactobacilli. Nutrit. Reps. Internatl. 30: 1287-1293 (1984)
29. Rose, A.H. The microbial production of food and drink. Scientific Amer. 245: 127-138 (1981)
30. Savage, D.C. Adherence of normal flora to mucosal surfaces. In: "Bacterial Adherence". (E.H. Beachey ed.). Chapman and Hall, London, pp. 33-59 (1980)
31. Shimizu-Kadota, M., Kiwaki, M., Hirokawa, H., and Tsuchida, N. A temperate *Lactobacillus* phage converts into virulent phage possibly by transposition of the host sequence. Devel. Indust. Microbiol. 25: 151-159 (1984)
32. Shimizu-Kadota, M., and Kudo, S. Liposome-mediated transfection of *Lactobacillus casei* spheroplasts. Agric. Biol. Chem. 48: 1105-1107 (1984)
33. Shimizu-Kadota, M., and Tsuchida, N. Physical mapping of the virion and the prophage DNAs of a temperate *Lactobacillus* phage oFSW. J. Gen. Microbiol. 130: 423-430 (1984)
34. Shimizu-Kadota, M., Kudo, S., and Masahiko, M. European Patent Application # 84305178.0. (1984)
35. Smiley, M.B., and Fryder, V. Plasmids, lactic acid production, and N-acetyl-D-glucosamine fermentation in *Lactobacillus helveticus* subsp. *juqurti*. Appl. Environ. Microbiol. 35: 777-781 (1978)
36. Stetter, K.O. Evidence for frequent lysogeny in lactobacilli: temperate bacteriophages within the subgenus *Streptobacterium*. J. Virology. 24: 685-689 (1977)
37. Stetter, K.O., Preiss, H., and Delius, H. *Lactobacillus casei* phage PL-1: Molecular properties and first transcription studies *in vivo* and *in vitro*. Virology 87: 1-12 (1978)
38. Vescovo, M., Morelli, L., and Bottazzi, V. Drug resistance plasmids in *Lactobacillus acidophilus* and *Lactobacillus reuteri*. Appl. Environ. Microbiol. 43: 50-56 (1982)
39. Vescovo, M., Morelli, L., Bottazzi, V., and Gasson, M.J. Conjugal transfer of broad-host-range plasmid pAMBl into enteric species of lactic acid bacteria. Appl. Environ. Microbiol. 46: 753-755 (1983)
40. Vescovo, M., Morelli, L., Cocconcelli, P.S., and Bottazzi, V. Protoplast formation, regeneration and plasmid curing in *Lactobacillus reuteri*. FEMS Microbiol. Letts. 23: 333-334 (1984)
41. Watanabe, K., Ishibashi, K., Nakashima, Y., and Sakurai, T. A phage resistant mutant of *Lactobacillus casei* which permits phage adsorption but not genome injection. J. Gen. Virol. 65: 981-986 (1984)
42. Yokokura, T., Kodaira, S., Ishiwa, H., and Sakurai, T. Lysogeny in lactobacilli. J. Gen. Microbiol. 84: 277-284 (1974)

Chapter 13

Food Fermentation with Molds

Robert Buchanan

INTRODUCTION

The past several years can be characterized as a period when our basic understanding of the principles underlying the behavior of biological systems has increased dramatically. As a result of this new knowledge, there is an increasing appreciation of the potential benefits that can be derived from the successful application of biological approaches to the needs of both industry and society in general. Of course, prime examples of new biological techniques that have piqued the interest of several industrial sectors, including the food industry, are the recent advances in recombinant DNA technology, and the accompanying advances in fermentation technology. A variety of bacterial and (more recently) yeast recombinant DNA systems have been developed for potential biotechnological applications for the needs of the food industry, and the work that is ongoing in these areas is impressive. An area that has lagged behind, even though it may have an even greater impact in terms of potential applications, has been the use of filamentous fungi in recombinant DNA-related biotechnology.

The filamentous fungi potentially offer several distinct advantages in regard to their use as biotechnological tools. Of particular importance is the extensive experience that can be drawn upon in regard to the use of fungal species in large scale fermentations. For example, the pharmaceutical industry is well versed in the use of various molds for the large scale production of antibiotics. Likewise, the food industry has extensive experience in the use of filamentous fungi for the production of both various fermented foods and fermentation derived food ingredients.

Molds as a general class of microorganisms have several characteristics that make them excellent candidates for industrial fermentations. One such characteristic is that they are generally capable of reasonably rapid growth coupled with the ability to achieve high levels of biomass. Further, recent work has demonstrated that molds lend themselves nicely to use with immobilized support systems. For example, it is a relatively easy matter to entrap, grow, and use fungal species in conjunction with support systems such as carrageenan beads.

Another advantage of fungal systems is that they typically have a wider range of biochemical capabilities than bacteria or yeast, and accordingly have the potential for offering greater flexibility. For example, fungal species characteristically have the ability to utilize a wide range of carbon sources, which should offer distinct advantages in terms of potential substrates that could be used to carry out fermentation processes. Further, fungal species are well known for their ability to carry out extensive secondary metabolic processes that can generate an extremely wide range of biochemical products, many of these having unique and often exotic chemical structures.

A third advantage to the use of filamentous fungi as systems for recombinant DNA applications is that they can be readily used to integrate recombinant and classic genetics. For example, the molds *Aspergillus nidulans* and *Neurospora crassa* have been employed for classic genetic studies. Accordingly, their genes have been mapped extensively, and a large number of useful mutant strains with well defined genetic markers is available. This is extremely useful in regard to genetic manipulations and strain improvement once desired genetic characteristics have been established.

The same advantages that fungal species offer as tools for recombinant DNA technology are also the reasons why work with them has lagged behind that with bacteria and yeast. Their complexity, coupled with the fact that there are relatively few individuals in the field of fungal genetics, has delayed their use as recombinant DNA vectors. However, since 1982 there has been increased activity in this field, including several important break-throughs, such that the use of fungi for recombinant DNA research is now feasible. The purpose of this chapter is to highlight some of the important advances that have been made in the use of fungal systems, as well as discuss potential food applications and limitations.

TRANSFORMATION SYSTEMS

Since 1982, reports of successful protocols for the transformation of fungi have been published. Currently, available fungal transformation systems are limited to *Neurospora crassa* (5,6,19,20,23,24) and *Aspergillus* species, particularly *Aspergillus nidulans* (3,10,11,12,16,25,27). However, it appears likely that the techniques that have been developed for these species can be readily modified for use with a variety of other fungi. This review will focus on work done with *Aspergillus* due to the importance of this genus in regard to commercial fermentations.

The transformation systems for *Aspergillus* species have been very similar both in terms of approaches and results. The various investigators have employed recipient strains of *A. nidulans* that are either auxotrophic mutants or have some other related nutritional marker. Transformation is then detected by transferring the missing genetic element from a wild type strain of *A. nidulans*, and selecting for colonies that are capable of growing vigorously in the absence of the required nutrient. Using an *A. nidulans* pyrimidine auxotroph, Ballance et al. (3) was able to obtain prototrophic transformants by replacing the missing gene with one derived from *N. crassa*, indicating that the transfer of genetic information among fungal genera is feasible. Different genetic markers that have been used in *A. nidulans* transformation systems include uridine auxotrophy/orotidine-5'-phosphate decarboxylase (3), acetamide utilization/acetamidase (25), tryptophan utilization/*trp*C gene (27), and arginine auxotrophy/*arg*B gene (10,11,27).

The specific techniques that have been successfully employed to produce transformants in *A. nidulans* are very similar regardless of the specific gene to be transferred. The protocols involved cloning the desired gene into a plasmid from *E. coli* (eg. pBR322) or a plasmid constructed from genetic material from *E. coli* and *Saccharomyces cerevisiae*. Protoplasts of the recipient strain were then formed and mixed with the plasmid. After incubating for a short period to permit uptake of the plasmid DNA, the protoplasts were allowed to undergo regeneration, and the cells screened for the presence of the specific nutritional marker.

The results obtained with the different transformation systems employing *A. nidulans* have been amazingly similar, and indicate that a low frequency of transformation (10 - 20 transformants/μg DNA) is readily obtainable. Johnstone et al. (11) reported that they were able to improve transformation rates by up to 500 transformants/μg DNA through modification of the bacterial plasmid employed as the cloning vector. This suggests that other improvements in transformation rates are likely in the future. Typically, isolated transformants had the transferred gene integrated into the chromosome of the recipient strain, often at the site of the original genetic lesion. Most investigators concluded that at least some transformants had incorporated multiple copies of the plasmid into their chromosome. Generally these integrated transformations were stable to mitosis, but did have a relatively high degree of instability in conjunction with meiosis. Most investigators also observed a relatively high number of "abortive" transformants which most likely represent isolates where the plasmid was taken up but not integrated into the chromosome. These isolates were unstable upon subculturing. To date, there have been no reports of the successful development of autonomously replicating plasmids in either *Aspergillus* or *Neurospora*.

Using the *amd*S gene from *A. nidulans*, Kelly and Hynes (12) were able to transform commercial, non-mutant strains of *Aspergillus niger*. This species normally grows poorly on acetamide as the sole nitrogen source and does not contain an *amd*S gene. The extent to which the transformed

A. niger isolates grew on acetamide medium was proportional to the number of copies of the plasmid integrated into the chromosome. They were able to isolate a transformant that was estimated to have incorporated > 100 copies of the *amdS* gene into its chromosome. Kelly and Hynes (12) also reported that this system had a high rate of co-transformation for other plasmids, which greatly enhances its potential use as a vector for recombinant DNA research. They concluded that the *amd* S gene of *A. nidulans* has great potential as a means for introducing genes into commercially important fungal species provided that they are normally unable to grow well on acetamide in a defined medium.

Due to the newness of the techniques, specific applications of the various *Aspergillus* transformation systems to biotechnology and recombinant DNA research largely await the results of future research efforts. However, Johnstone et al. (11), and Lockington et al. (15) have already successfully employed these techniques to clone a developmental gene and the ethanol utilization regulon of *A. nidulans*. As previously mentioned, Ballance et al. (3) were able to transform *A. nidulans* using a gene from *N. crassa*, suggesting that *Aspergillus* species have the potential for expressing genetic information from other eukaryotes.

Overall, the results of these early transformation studies indicate that filamentous fungi are very promising for use in recombinant DNA research. This, coupled with their extensive use in fermentations, suggests that *Aspergillus* species are likely to have an important role in emerging biotechnologies.

In addition to recombinant DNA techniques discussed above, there are a number of other methods that can be employed for transferring or modifying the genetic information of a microorganism. Protoplast fusion appears to offer significant potential with the filamentous fungi, particularly in regard to improving the productivity of strains employed for fermentation. This technique involves removing the cell walls of two strains or species, fusing the cells, and then allowing intact cells to regenerate. During this process, recombination events between the chromosomes of the two fused cell types can occur, and as the hybrid microorganism reverts back to normal state, cells that combine the desirable characteristics of the two original cell lines can often be isolated. With the commercial availability of lytic enzyme preparations such as Novozyme 234 or snail gut enzyme preparations, protoplasts of a variety of fungal species can be readily obtained (14,21,26). It is expected that this will allow increased research into the use of protoplast fusion as a means of producing fungal isolates with desirable fermentation characteristics. For example, Furuya et al. (9) have demonstrated that protoplast fusion techniques can be successfully employed to develop improved strains of *Aspergillus oryzae* for the production of soy sauce.

CURRENT AND POTENTIAL APPLICATIONS

In the United States there is a tendency to think of food fermentation as primarily involving the use of bacterial or yeast species; however, on a worldwide basis, the use filamentous fungi is equally important in terms of the direct use of fermentation techniques to produce food products (4). This difference between our impressions and the actual use of filamentous fungi largely reflects the fact that the consumption of foods produced via a fungal fermentation process are predominately restricted to specific geographical locations in Asia and Africa. In those regions, fungal fermented foods are both traditional and often a staple part of the diet, while to us these products seem exotic. Some of the better known examples of Oriental foods that employ either a fungal fermentation or a mixed fungal/yeast/bacterial fermentation include soy sauce, miso, natto, tempeh, and oncom. Likewise, some of the better known examples of fungal food fermentations indigenous to Africa (particularly West Africa) include gari and ogi. While these products are certainly not a staple part of the U.S. diet, it should be noted that over the past decade the American consumer has tended to partake of a wider range of foods, and there has been a sharp increase in interest in Oriental food products, especially tofu and tempeh.

An area which more directly impacts the U.S. food processing industry is the use of fungal fermentations for the production of food ingredients. The extent to which fungal fermentation products are employed is not fully appreciated by professionals within the food industry, even though a number of important food ingredients or processing aids are derived in this manner. Some of the classes of food ingredients that are produced commercially by fungal fermentation techniques include acidulants, amino acids, enzymes, polyols, and flavors. Further, there is an even wider range of food ingredients that could be competitively produced by fungi if there were even marginal improvements in available strains and fermentation techniques. This is the area where recent advances in biotechnology could have its most immediate impact on the food processing industry.

Recent advances in biotechnology offer the opportunity for employing a whole new set of tools for manipulating and altering fermentation processes. This, in turn, allows individuals to stretch their imagination in regard to potential applications presently unattainable. However, enthusiasm for this type of research and development must be tempered and guided by sound assessment of underlying economic considerations. For example it makes little sense to devote a significant portion of a R & D budget to develop a modified microorganism capable of utilizing new or exotic substrates if the desired fermentation is already being successfully achieved using an easily obtained, inexpensive substrate such as corn. On the other hand, significant economic gains may be realized by using new biotechnological approaches to make existing fermentations more cost effective, thereby generating a significant market advantage. Accordingly, decisions to become involved in biotechnology research and the subsequent selection of specific target applications must be backed by a strong assessment of business opportunities.

Based on this type of economic assessment, it appears unlikely that the U.S. food processing industry will develop major opportunities applying new biotechnological techniques to improve the use of fungal fermentations for the direct production of foods. Currently the U.S. population consumes little in the way of fermented foods employing fungal species, and it is unlikely that any advance in biotechnology will significantly change consumers' dietary preferences and patterns. This is not to imply that biotechnology cannot impact this area, but simply indicates that this type of work is probably more appropriate in those geographical regions where this type of fermented food is an accepted part of the diet. Instead, the area that seems to hold the greatest potential for the U.S. food processing industry is the use of fungal species for the production of food ingredients and food additives.

In considering how the various new techniques in genetics could impact the production of food ingredients via fungal fermentations, it is convenient and appropriate to subdivide potential applications into two broad areas: 1) improvement of existing fungal fermentations, and 2) development of new products and processes. Of these two areas, the former is most likely to yield successful applications in a short period of time, while the latter involves greater risk-taking, but may ultimately have a greater long-term impact. Each of these areas will be discussed individually.

Strain Improvement

As indicated previously, fungal species are currently employed for the production of a number of important food ingredients, particularly acidulants and enzymes. As with any fermentation process, there are ongoing efforts to improve the fungal strains employed in order to maximize both the rate and extent of production. This is the area where recombinant DNA techniques are most likely to have an immediate impact. This appears particularly true in regard to the use of filamentous fungi for the production of commercially important enzymes such as amylases or proteases. This supposition is based on the observations of the earlier cited transformation studies with *Aspergillus* species. Those investigations consistently indicated that a portion of the transformed fungi had integrated multiple copies of the transferred gene. For example, Kelly and Hynes (12) estimated that *A. niger* incorporated as many as 100 copies of the *amd* S gene from *A. nidulans*, with the resulting production of acetadimase by the transformed *A. niger* isolates being directly proportional to the number of gene copies integrated into the fungal chromosome. This suggests that by isolating the gene for a target enzyme, and using it in conjunction with an appropriate plasmid vector, it should be possible to produce strains with an increased number of gene copies. This strain would be expected to have an amplified level of enzyme production.

Another area where the manipulation of the genetic material of filamentous fungi may have application for producing improved strains is in generating hybrid isolates that incorporate the best characteristics of two or more parent strains. An example of this approach is the recent report by Furuya et al. (9) on their efforts to develop improved strains of *A. oryzae* for the production of soy sauce. They employed a protoplast fusion protocol in conjunction with two *A. oryzae* strains, one that grew rapidly but produced low levels of protease, and a second that produced high levels of protease but grew slowly. Furuya et al. (9) demonstrated that protoplast fusion could be used to generate stable heterozygous strains that grew rapidly and produced elevated levels of protease.

Again based on the results of available transformation studies with filamentous fungi, it appears reasonable to expect that it should be possible to transfer genetic information between different species and genera. This capability is likely to be important in regard to improving fungal strains such that they can employ a greater range of potential substrates or produce a wider range of fermentation end products. As a hypothetical example, one can consider the commercially important filamentous fungus, *A. niger*. This species is known to produce relatively low levels of cellulase (4), and accordingly will slowly utilize and grow on cellulose. However, other fungal species are known to produce significantly greater amounts of cellulase. Examples include closely related species such as *Aspergillus terreus* and more taxonomically distant genera such as *Trichoderma*. An area that warrants investigation would be determining if the cellulolytic activity of *A. niger* could be amplified by transferring the cellulase gene from another species. In this manner, fermentations employing *A. niger* could potentially be improved by allowing the mold to more effectively use cellulose-containing substrates for its initial growth, or possibly for the actual production of useful products such as organic acids.

This application of recombinant DNA technology was selected because it also points out one of the problems that has to be considered when developing fermentation techniques. Successfully modifying a fungal species to better carry out a fermentation process is often more involved than simply transferring a desirable gene. The synthesis of enzymes is typically under tight metabolic control in fungal species, including carbon catabolite repression of a variety of enzymes (1,2,7,8). Therefore, to effectively express amplified cellulase activity on complex substrates, it is likely that an appropriate regulatory mutant would have to be developed in order to achieve constitutive enzyme synthesis (2). However, it is possible that by transferring the gene for cellulase from another species or genera, the regulatory regions or genes controlling synthesis could be deleted or modified such that enzyme synthesis is derepressed regardless of other available substrates.

Overall, the outlook appears good for utilizing gene manipulation techniques such as plasmid vectors or protoplast fusion to improve fungi used for the production of food ingredients. By integrating these approaches with more classical approaches such as generation and selection of mutants, sexual and parasexual crosses, and identification of nutritional and bioregulatory parameters, it should be possible to significantly impact this class of commercial fermentations.

New Fermentation Products

The aspect of recombinant DNA research that has most stimulated the imagination of the scientific and business communities is the possibility of using a relatively simple fermentation process to cheaply produce previously expensive natural products. In the area of medical applications, this type of research has been actively pursued, and is currently reaching the point of commercialization. Agriculture is a second area where there has been a relatively high level of activity directed toward the development of new classes of products and processes, including some fermented food applications. Biotechnology has largely involved the use of bacteria. Relatively little research has employed filamentous fungi. Lately, however, there has been increased interest in identifying the products of fungal metabolism. One key issue would be to identify new classes of fungal metabolites with potential application as food ingredients that can be produced efficiently by modern fermentation techniques. Flavoring compounds, food pigments, and new enzymes would most likely yield a rate of return that would justify the necessary research and development costs. The first two areas will be discussed further.

Flavors: Each vegetable, fruit, spice and herb has its own distinctive flavor and fragrance, reflecting the highly individualized synthesis of specific volatile compounds by plant species and varieties. The synthesis of these distinctive compounds in turn reflects specific genetic differences among the various plant species such that they have highly individualized anabolic biochemical pathways and reactions. Filamentous fungi share many of the same general classes of enzymatic reactions that produce volatile flavor compounds in higher plants, and a variety of fungal species produce flavors or fragrances similar, if not identical, to those occurring in foods. Schindler and Schmid (22) recently reviewed some of the fragrances associated with specific fungal species. Examples of potentially important fragrances detected in fungal species include the aromas of apples, anise, pepper, lemon, banana, and vanilla.

Currently, flavors are obtained via agricultural production or through the chemical synthesis of specific flavor compounds. Fungal fermentation would eliminate the seasonal and geographical limitations associated with agricultural production and may surpass the cost-effectiveness of chemical synthesis (particularly for more complex compounds). This is not without precedent in that fungal fermentations have been employed commercially to produce blue cheese and mushroom flavors for the food industry.

The first step would be to identify fungal species that produce desirable flavor notes. Once prospective species have been identified, the various biotechnology techniques discussed in regard to strain improvement could then be brought to bear to maximize production of the desired volatile products.

An alternate biotechnological approach to the production of flavor compounds is the use of fungi or other appropriate microorganisms to transform chemically synthesized flavors or flavor precursors (22). This approach entails presenting the fungus with a preformed flavor compound or precursor, and using the inherent biochemical capabilities of the fungus to modify the starting material to a more desirable end product. This approach has been employed successfully by the pharmaceutical industry for the production of various antibiotics. Further, these investigators have noted that polyurethane-entrapped yeast cells can be successfully employed for the commercial scale conversion of DL-menthol mixtures to L-menthol of 100% optical purity (22).

An even more ambitious approach to the production of flavor compounds would be the transferring to fungal species of those genes responsible for the synthesis of flavor compounds in plants. This is likely to be a long-term project in that few of the pathways responsible for the production of flavor compounds or their precursors have been elucidated. Further, it is likely that these pathways involve multiple gene loci. Before specifically engineered transformations could be achieved, the individual genes responsible for the synthesis would need to be identified and isolated. It might be possible to short cut this process by using protoplast fusion techniques to achieve the transfer of genetic information from plant cells to a recipient fungal species. Alternatively, if the pathway responsible for the synthesis of a flavor compound involves a precursor that can be readily synthesized by chemical means, it may be possible to transfer to appropriate fungal species the plant genes encoding the enzymes responsible for the conversion of that precursor to the final flavor compound. The resultant fungal strain could then be employed to biochemically transform the chemically synthesized precursor. Using this approach, it may be feasible to greatly reduce the number of genetic loci that would have to be transferred from the plant.

Pigments: The use of synthetic (artificial) colors in the food processing industry has had a long history of controversy, and these compounds are continually being reviewed in regard to their safe use. For this reason there has been an ongoing interest in identifying naturally occurring pigments suitable for use in foods. It should be noted that legally, a natural pigment made by biotechnology and extracted may still be considered an artificial color and might require rigorous safety testing. This may place some drawbacks in the pursuit of colors through the use of fungi unless one gets approval for using the fungi directly as a food ingredient.

Filamentous fungi are noted for producing high levels of pigments, particularly various carotenoids and anthraquinones. If useful and safe pigments could be identified and produced cost effectively by a fermentation process, this would eliminate the problem of seasonal availability that is often associated with agriculturally produced natural products.

A traditional fermentation that has received renewed interest in regard to its potential use in foods is the Oriental product, ang-khak (red rice) (4). It is produced by fermenting rice with *Monascus purpureus* which produces a red pigment, monascorubrin. Traditionally, the fermented rice was dried, ground and used as a colorant. During the past several years there has been substantial research to determine if this pigment has wider applications in the food industry and if it can be produced by modern fermentation techniques. One major question would be whether the rice is a food ingredient or a color additive.

Another class of pigments that has received substantial attention is the carotenoids, particularly beta-carotene (13,18). Some carotenoids do not require safety testing as they are already approved for food use. In fact, this is a good example of how biotechnological approaches can be employed to develop a successful fungal fermentation. A number of the molds of the order *Mucorales* are capable of producing beta-carotene; however, the normal production levels were considered too low for commercial applications. In a series of basic research studies with *Phycomyces blakesleeanus*, *Choanephora cucurbitarum*, and *Blakeslea trispora* it was demonstrated that synthesis is controlled by several environmental and bioregulatory factors that can be manipulated to greatly elevate production. Murillo and Cerda-Olmedo (17) demonstrated that it was possible to obtain deregulated mutants of *P. blakesleeanus* capable of producing beta-carotene at levels up to 6000 μg/g. Further, Murillo et al. (18) demonstrated that the mold could be genetically manipulated such that strains produced a variety of other carotenoid pigments such as lycopene, alpha-carotene, gamma-carotene, phytofluene, and neurosporene. Lampila et al (13) recently reviewed the potential use of filamentous fungi in conjunction with food processing byproducts for the production of beta-carotene. They concluded that while the biotechnological approach to the production of food colorants is not currently being used, the process is economically feasible.

Genetically modified filamentous fungi can also be used to produce pigments that occur in higher plants. For example, a major class of pigments existing in nature is the xanthophylls which are synthesized by the biochemical modification (e.g., hydroxylation, carboxylation, etc.) of carotenoid compounds. It should be feasible to transfer the genes encoding the enzymes that catalyze these reactions in higher plant species to a strain of fungi capable of synthesizing high levels of carotenoid precursor. In this manner, it would be possible to develop a group of fungal strains that could produce a spectrum of pigments for potential food use applications. Alternatively, the genes for the modifying enzymes could be transferred to a fungal strain that doesn't produce carotenes, but that could transform exogenously supplied carotenoids to the appropriate xanthophyll.

CONCLUSIONS

The use of filamentous fungi to produce useful products has a long history both in terms of pharmaceuticals and food materials. This class of microorganisms has a number of distinct advantages with regard to fermentation technology, including vigorous and extensive growth, ability to utilize a range of substrates, and a highly complex anabolic biochemistry. Research into the use of filamentous fungi for recombinant DNA applications has lagged behind that of other microorganisms such as bacteria and yeast; however, recent breakthroughs in the development of fungal transformation systems indicate that the genetics of commercially important fungal strains can be manipulated and modified effectively. It is likely that this class of microorganisms will become an important tool for biotechnological applications, and is most likely to impact the U.S. food processing industry in the area of food ingredient/additive production.

REFERENCES

1. Arst, H.N., Jr. Regulation of gene expression in *Aspergillus nidulans*. Microbiol. Sci. 1:137-141 (1984)
2. Bailey, C. and Arst, H.N., Jr. Carbon catabolite repression in *Aspergillus nidulans*. Eur. J. Biochem. 51:573-577 (1975)
3. Ballance, D.J., Buxton, F.P. and Turner, G. Transformation of *Aspergillus nidulans* by the orotidine-5'-phosphate decarboxylase gene of *Neurospora crassa*. Biochem. Biophys. Res. Commun. 112: 284-289 (1983)
4. Beuchat, L.R. (ed.). "Food and Beverage Mycology". AVI Pub. Co., Inc. Westport, Conn. (1978)
5. Buxton, R.P. and Radford, A. The transformation of mycelial spheroplasts of *Neurospora crassa* and the attempted isolation of an autonomous replicator. Molec. Gen. Genet. 196:339-344 (1984)

6. Case, M.E., Schweizer, M., Kushner, S.R. and Giles, N.H. Efficient transformation of *Neurospora crassa* by utilizing hybrid plasmid DNA. Proc. Natl. Acad. Sci. USA 76:5259-5263 (1979)
7. Cohen, B.L. Regulation of protease production in *Aspergillus*. Trans. Brit. Mycol. Soc. 76:447-450 (1981)
8. Erratt, J.A., Douglas, P.E., Moranelli, F. and Seligy, V.L. The induction of alpha-amylase by starch in *Aspergillus oryzae*: evidence for controlled mRNA expression. Can. J. Biochem. Cell. Biol. 62:678-690 (1984)
9. Furuya, T., Ishige, M., Uchida, K. and Yoshino, H. Koji-mold breeding by protoplast fusion for soy sauce production. Nippon Nogeikagaku Kaishi 57:1-8 (1983)
10. John, M.A. and Peberdy, J.F. Transformation of *Aspergillus nidulans* using the *arg*B gene. Enz. Microb. Technol. 6:386-389 (1984)
11. Johnstone, I.L., Hughes, S.G. and Clutterbuck, A.J. Cloning an *Aspergillus nidulans* developmental gene by transformation. EMBO J. 4:1307-1311 (1985)
12. Kelly, J.M. and Hynes, M.J. Transformation of *Aspergillus niger* by the *amd*S gene of *Aspergillus nidulans*. EMBO J. 4:475-479 (1985)
13. Lampila, L.E., Wallen, S.E. and Bullerman, L.B. A review of factors affecting biosynthesis of carotenoids by the order *Mucorales*. Mycopathol. 90:65-80 (1985)
14. Laurila, H.O., Nevalainen, H. and Makinen, V. Production of protoplasts from the fungi *Curvularia inaequalis* and *Trichoderma reesi*. Appl. Microbiol. Biotechnol. 21:210-212 (1985)
15. Lockington, R.A., Sealy-Lewis, H.M., Scazzocchio, C. and Davies, R.W. Cloning and characterization of the ethanol regulon in *Aspergillus nidulans*. Gene 33:137-149 (1985)
16. Miller, B.L., Miller, K.Y. and Timberlake, W.E. Direct and indirect gene replacement in *Aspergillus nidulans*. Molec. Cell. Biol. 5:1714-1721 (1985)
17. Murillo, F.J. and Cerda-Olmedo, E. Regulation of carotene synthesis in *Phycomyces*. Molec. Gen. Genet. 148:19-24 (1976)
18. Murillo, F.J., Calderon, I.L., Lopez-Diaz, I. and Cerda-Olmedo, E. Carotene-superproducing strains of *Phycomyces*. Appl. Environ. Microbiol. 36:639-642 (1978)
19. Paietta, J. and Marzluf, G.A. Plasmid recovery from transformants and the isolation of chromosomal DNA segments improving plasmid replication in *Neurospora crassa*. Current Genet. 9:383-388 (1985)
20. Paietta, J.V. and Marzluf, G.A. Gene disruption by transformation in *Neurospora crassa*. Molec. Cell. Biol. 5:1554-1559 (1985)
21. Peberdy, J.F. and Ferenczy, L. (eds.). "Fungal Protoplasts". Marcel Dekker Inc. New York
22. Schindler, J. and Schmid, R.D. Fragrance or aroma chemicals - microbial synthesis and enzymatic transformation - a review. Process. Biochem. 17:2-6 (1982)
23. Schweizer, M., Case, M.E., Dykstra, C.C., Giles, N.H. and Kushner, S.R. Identification and characterization of recombinant plasmids carrying the complete q^a *gene cluster from Neurospora crassa including the* q^a*-1*$^+$ *regulatory gene. Proc. Natl. Acad. Sci. USA 78:5086-5090 (1981)*
24. Stohl, L.L. and Lambowitz, A.M. Construction of a shuttle vector for the filamentous fungus *Neurospora crassa*. Proc. Natl. Acad. Sci. USA 80:1058-1062 (1983)
25. Tilburn, J., Scazzacchio, C., Taylor, G.C., Zabicky-Zissman, J.H., Lockington, R.A. and Davies, R.W. 1983. Transformation by integration in *Aspergillus nidulans*. Gene 26:205-221 (1983)

26. Yabuki, M., Kasai, Y., Ando, A. and Fujii,T. Rapid method for converting fungal cells into protoplasts with a high regeneration frequency. Exp. Mycol. 8:386-390 (1984)
27. Yelton, M.M., Hamer, J.E. and Timberlake, W.E. Transformation of *Aspergillus nidulans* by using a *trp*C plasmid. Proc. Natl. Acad. Sci. USA 81:1470-1474 (1984)

Chapter 14

Unitization of Fermented Foods: An Application of Fermentation Technology

Michael Sfat

INTRODUCTION AND CONCEPT

This chapter describes a potential for the use of fermentation in food processing. Because of past activity in commercial biotechnology with emphasis in malting, brewing, and aseptic aerobic fermentations, there is a special confidence and enthusiasm for the idea that new fermented foods can be created from existing fermented foods through application of available biotechnology, especially fermentation. This attitude is buttressed by encouraging comments from Dr. Jean Mauron, of Nestle's, who wrote "...we are at the eve of a new biological revolution. We might rightly ask ourselves whether this time it will lead to the development of basically new fermented foods and beverages or whether tradition will continue to impose its exclusive rule...." (10). Two distinguished American food scientists who concur say "Foods prepared by fermentation, aside from those well known in the West, will increase in amount and use and will spread to other parts of the world, including the Western developed countries..." (C.W. Hesseltine) (8); "...it is only relatively recently that the Western world has taken a closer look at the indigenous fermented foods and discovered that they are literally a gold mine of food science and technology waiting to be tapped...." (K.H. Steinkraus) (21).

By unitizing the traditional fermented food process into fermentation elements which control analytical and functional properties such as flavor, texture, alcohol, and nutrition, the food technologist is provided with building blocks which can be recombined into improved versions of existing fermented foods, new foods, fermented flavors, and ingredients. Although exploitation of Western fermented foods could yield a rich harvest of candidate fermentation elements, the traditional oriental fermented products provide a wider spectrum of organisms, substrates, and conditions for the matrix. Because these organisms and substrates are well established in the food system, there should be a high probability for success. Table 14-1 lists several examples of fermented food unitization.

Table 14-1. Examples of Fermented Foods Unitization

Traditional Fermented Food Process	Application of Unitization	Reference
Production of wine vinegar	Oxidation of ethanol to acetic acid (white vinegar) by *Acetobacter*	—
Natural fermentation of sausage	Development of starter culture (*Pediococcus*)	—
Traditional yogurt process	Development of starter culture (*Lactobacillus*)	—
Aspergillus oryzae koji (used in soy sauce manufacture)	Koji mixed with fish meal	20
Brewing process	Concentrated beer flavor	11
Brewing process	Optimization of separate flavor and alcohol fermentations for blending into a wider range of products	16,18
Tempeh	*Rhizopus oligosporus* isolated from tempeh and urea added to cassava to upgrade protein content	19
Tempeh	New tempeh-like products substituting wheat, oats, barley, rice, or mixtures of cereals with soybean	23

MARKET POTENTIAL

Before spending money and time on exploiting this concept, it would be helpful to know the market potential for the projected products. Unfortunately, the author had no direct method for doing this; but by analogy and deduction, a market picture can be formed.

In the United States, traditional fermented foods and beverages such as beer, wine, cheese, bread, sausage, sauerkraut, and pickles, are well established and continue to grow. Since the early 1980's, a heretofore oddity, yogurt, has established a solid position on grocery shelves and cafeteria counters, illustrating how a combination of imagination, food technology, and superb marketing can convert a Balkan traditional fermented food from a variable, often bitter product into a winner. The case for oriental fermented foods is abetted by several factors: 1) Proliferation of oriental restaurants, widening the taste spectrum beyond the entrenched Americanized Chinese establishments. Japanese, Korean, and Southeast Asian restaurants, originally catering to the recent influx of oriental businessmen and immigrants, have been discovered by the American public who have transferred this new cuisine to their homes; 2) Increased travel by Americans to the Orient for pleasure and business is bound to stimulate interest in fermented foods; 3) The rapid upward mobility of oriental immigrants accelerates the spread of their culture, especially food; and, 4) Concern about health and appearance has increased interest in natural, low-calorie, low-cholesterol, high-fiber foods, which translates into more cereal grains and vegetables in the diet, an established feature of oriental diets for centuries.

Other evidence of a trend towards oriental fermented foods is the building of a Japanese soy sauce factory in Wisconsin to accommodate a growing demand (15% per year). Recently, the meat substitute, tempeh, a fermented soy bean product, has joined tofu, a nonfermented soy curd, in the fast growing (25 to 30% per year) soy food industry. Tempeh is being incorporated into the National Hot Lunch Program for schools and the Armed Forces Food Program.

APPLICABLE FERMENTATION TECHNOLOGY

Unitization

Candidate fermented foods are broken down (unitized) into main substrates, microorganisms, and conditions such as pH, moisture or solids content, aerobiosis, temperature, lesser nutrients, and osmotic control. A substantial literature has evolved which serves as a starting point (4,5,6,7,8,9,12,14,15,20). The foods should also be evaluated for desirable properties such as stability, flavor, texture, alcohol content, nutrition, and color. Then comes the task of matching fermentation elements with desirable properties, much of which will require a substantial laboratory effort. Simplified examples of unitization for cheese and beer are illustrated in Tables 14-2 and 14-3 (16). Examples of known fermentation

Table 14-2. Simplified Elements of Cheese Fermentation (16)

Property	Substrate	Microorganism
Texture (curd)	Carbohydrate	Lactic acid bacteria
Texture (Swiss cheese)	Carbohydrate	*Propionibacterium*
Flavor	Fatty acids	*C. lipolytica*
	Carbohydrates	*L. casei*
	Proteins	Other lactic acid bacteria
Flavor (surface-ripened)	—	*Brevibacterium linens*
Flavor (mold-ripened)	—	*Penicillium*

Table 14-3. Simplified Elements of Beer Fermentation (16)

Property	Substrate	Microorganism
Alcohol	Carbohydrates	Yeast
Flavor	Nitrogenous compounds	Yeast
	Carbohydrates	Yeast
	Lipids	Yeast
	Above + hops	—
CO_2	Carbohydrates	Yeast
Color	Carbohydrates and nitrogenous compounds	—
Foam	Protein	—
Colloidal instability	Anthocyanogens and proteins	—

elements for cheese are: 1) the action of lactic acid bacteria on carbohydrates, producing lactic acid which causes curd formation from protein, thus providing a desirable texture; and, 2) modification of fatty acids by *Candida lipolytica* yielding flavors characteristic of Italian-type cheese.

In the case of beer, the primary elements of alcohol and flavor formation are known and, as illustrated in Figure 14-1, the two processes are different; beer flavor formation substantially lags behind the main fermentation (11). Table 14-4 identifies the role of traditional raw materials in the fermentation element matrix (16). It is interesting to note that malt is essentially the sole source of protein for yeast growth and, more importantly, for flavor derived from protein. In spite of this, barley, the ungerminated precursor of malt, is usually neglected in the marketplace for protein content. An example of the effect of fermentation conditions on a desirable property, beer flavor, is illustrated in Figure 14-2 (16); lowering the temperature and raising the solids content both improve beer flavor.

Table 14-4. Main Functions of Materials Used in Beer Fermentation (16)

Material	Key Component	Main Beer Characteristic Affected
Malt	Enzymes	Alcohol, CO_2
	Carbohydrates	Alcohol, flavor, CO_2
	Protein	Flavor, foam, color
Adjunct (corn, rice)	Carbohydrate	Alcohol, CO_2
Hops	Alpha acids	Flavor

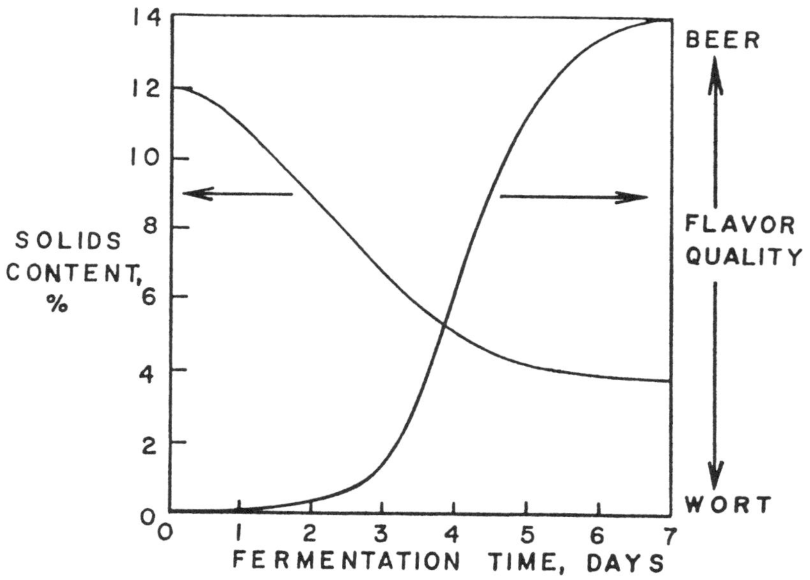

Figure 14-1. Beer Fermentation Flavor Development (11)

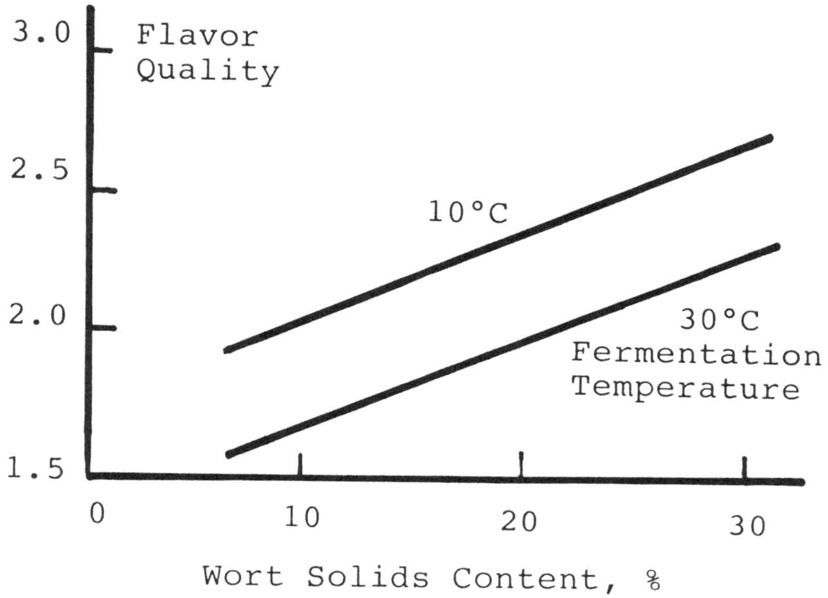

Figure 14-2. Effect of Wort Solids Content on Beer Flavor Quality (16)

Optimization

With a unit identified as to desirable property (e.g., flavor), microorganism (e.g., yeast), substrate (e.g., protein), and other conditions (e.g., temperature, pH, solids content), it now is possible to optimize the unit by measuring changes in the desirable property as fermentation elements are varied. Genetic manipulation of cultures and preprocessing of substrates should not be excluded. The end product could be a concentrated form of the desirable property such as flavor, which could be used as a food ingredient. On the other hand, the knowledge gained could be used to improve the original fermented food. Past experience shows that optimization is more difficult than unitization.

Recombination

Having established a library of flavors, textures, nutritive properties, etc., one can now create new fermented foods by combining the units in novel and imaginative ways. For example, would a cheese-tempeh recombination work?

Fermentation Technology and Economics

Much has been written about the fermentation process (1, 13, 22) and a substantial experience in process design, engineering, and economics is available through consultants, engineering firms, and equipment manufacturers.

In general, fermentation equipment can be classified as aerobic or anaerobic, aseptic or sanitary, liquid or solid. The most probable configurations for food ingredient or food production are: 1) aerobic-aseptic-liquid, used to produce vitamins and monosodium glutamate; 2) aerobic-sanitary-liquid, used to produce bakers yeast; 3) aerobic-sanitary-solid, employed for koji; 4) anaerobic-sanitary-liquid, used for beer and pickles; and, 5) anaerobic-sanitary-solid, used for cheese and bread. These are not hard and fast classifications, but serve to illustrate the range of conditions that can be employed.

Three fermentation configurations including recovery which can be compared for capital investment costs and operating costs based on 1979 data are listed in Table 14-5. A schematic view of the fermentation portion of an aerobic-aseptic-liquid plant is illustrated in Figure 14-3. The anaerobic-sanitary-liquid scheme is essentially the same, except that aeration, mixing and antifoam equipment are not needed. Aspetic tank and valve construction are substantially more sophisticated to accommodate sterilization, aseptic transfer, and fermenter mass and heat transfer needs.

Table 14-5. List of Fermentation Configurations Used in Economic Comparisons

Configuration	Source of Data
Case 1: Anaerobic-sanitary-liquid	5,000,000-gal/yr (2) ethanol plant using corn. Fermenter charged at 12% solids for total substrate usage of 84,000,000 lbs/yr. Three-day fermenter cycle.
Case 2: Aerobic-aseptic-liquid	Flexible stainless steel plant (2) of type used to produce glutamic acid charged at 12% substrate solids for annual consumption of 84,000,000 lb. Three day fermenter cycle.
Case 3: Aerobic-sanitary-solid	Malt plant converting 84,000,000 lbs. of barley to malt per year (2, 17) employing three-day germination.

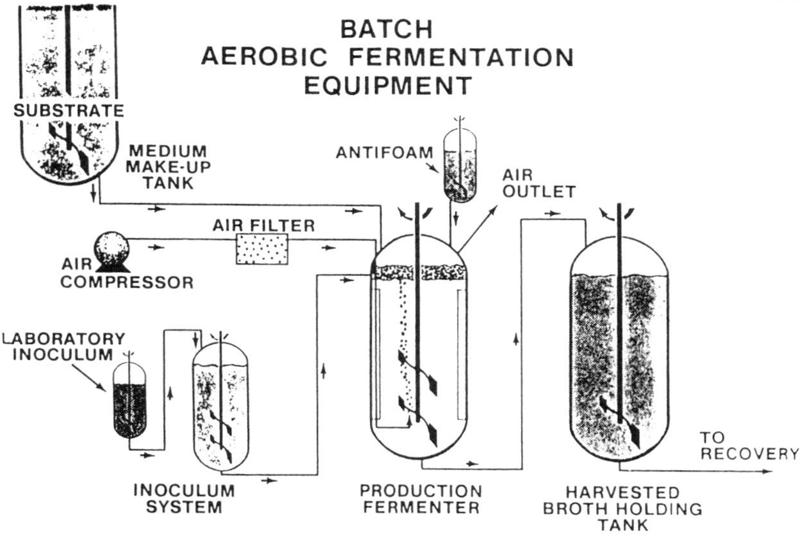

Figure 14-3. Batch Aerobic Fermentation Equipment

Figure 14-4 shows a schematic arrangement of equipment for the solid state fermentation of koji, a rice-*Aspergillus oryzae* system (3). This is similar to the layout of the germination portion of a malting plant illustrated in Figure 14-5. The strong similarity suggests that the engineering and economics developed for many years in the international malting industry provides an excellent starting point for solid-state fermenter design and economic planning. Another piece of malting equipment, the rotating drum, also has excellent applicability to food fermentation. The scale, annual substrate usage, and bioconversion time are based on a five million gallon per year ethanol plant (Case 1).

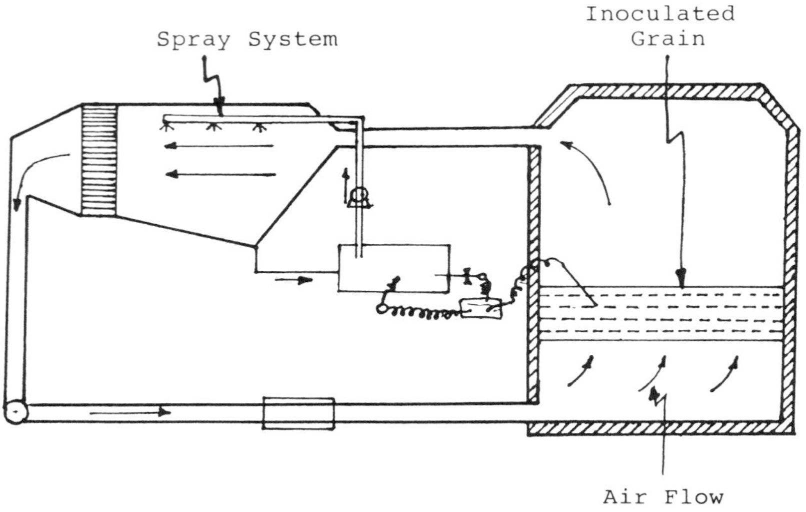

Figure 14-4. Koji-Making Apparatus

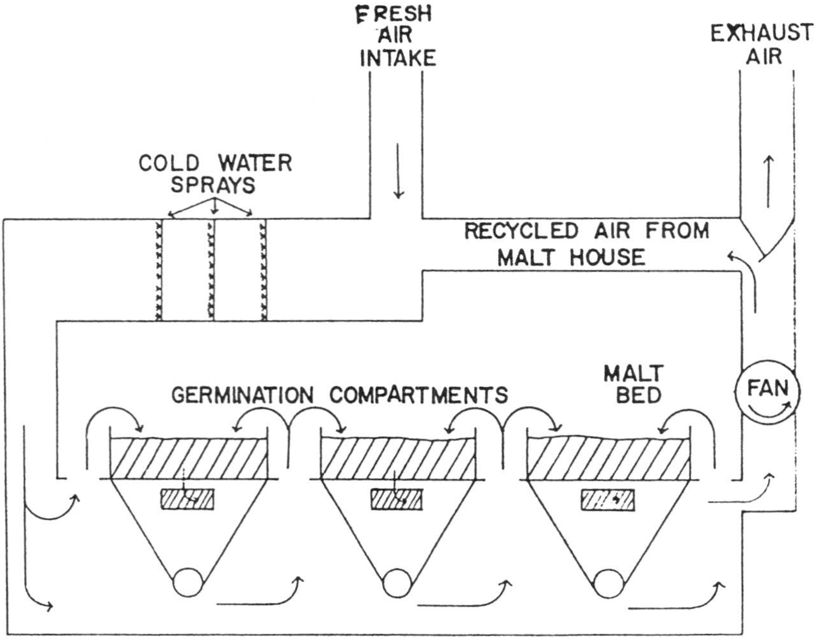

Figure 14-5. Compartment Malt House

The best fermentation cost data available are for the aerobic-aseptic-liquid configuration but are, nevertheless, characterized by a ±40 percent error. The Case 3 (aerobic-sanitary-solid) data source, the malting process, although not a fermentation, employs equipment similar to that used in koji processing at the Kikkoman soy sauce plant in Wisconsin (24). If correction is made for the substantial elevator capacity required for barley and malt storage (50% of yearly production), a reasonable preliminary approximation of investment and cost for a solid-state food fermentation can be made.

Table 14-6 summarizes capital investment comparisons and shows that the aerobic-aseptic-liquid plant costs about four times the anaerobic-sanitary-liquid (Case 1) and aerobic-sanitary-solid (without storage) (Case 3) configurations. Investments for Case 1 and Case 3 would be somewhat higher if stainless steel or other corrosion-resistant alloys were employed. The Case 2 data are based on type 304 stainless steel construction. It is interesting to note that the unit investment, based on installed capacity, of the aerobic-sanitary-solid system and the capital intensive aerobic-aseptic-liquid system are alike, but that based on substrate, the solid system is cheaper. The trade-off is due to the vast air-conditioning system needed for cooling the actively-respiring germinating barley. Undoubtedly, cooling requirements for a koji-type fermentation will be higher. This is only one of several bioengineering problems that must be considered when planning a solid-state fermented food plant (3).

Table 14-6. Comparison of Fixed Assets Investment for Three Fermentation Configurations

1979 Data

Three-Day Fermentation Cycle
84 Million Pounds Substrate Processed Per Year

			Fixed Assets Investment		
Configuration	Substrate Solids Concentration, %	Installed Capacity	Total, Millions of Dollars	$/ft³ Installed Capacity	$/lb Substrate in Process
Case 1: Anaerobic-Sanitary-Liquid	12	770,000 gal	11	103	14
Case 2: Aerobic-Aseptic-Liquid	12	770,000 gal	48	462	62
Case 3: Aerobic-Sanitary-Solid	50	20,000 ft³	11 (without storage)	500	14

Critical operating cost elements i.e., depreciation, labor, and utilities, expressed in cents per pound of substrate processed, are compared in Table 14-7. Assuming that raw material costs are identical for all three cases and that depreciation, labor, and utilities are operating costs most sensitive to technology, the three technological combinations listed in descending operating costs order are: aerobic-aseptic-liquid, anaerobic-sanitary-liquid, and aerobic-sanitary-solid. It is interesting to note that the order of process and product flexibility for these fermentation systems is identical. Obviously, for the same profitability, the higher the investment and high operating cost, the higher the selling price for products, an economic fact that must be considered when planning unitization strategy.

Table 14-7. Comparison of Selected Operating Costs for Three Fermentation Configurations

1979 Data

84 Million Pounds Substrate Processed Per Year
Three-Day Fermentation Cycle

	Elements of Operating Cost, ¢/lb substrate processed			
Configuration	**Depreciation**	**Labor, Supervision**	**Fuel, Power, Water**	**Total**
Case 1: Anaerobic-Sanitary-Liquid	0.9	1.2	3.4 2.7 (without distillation)	4.8-5.5
Case 2: Aerobic-Aseptic-Liquid	3.8	1.0	5 (without recovery) 6 (est. including recovery)	9.8-10.8
Case 3: Aerobic-Sanitary-Solid	1.6 (without storage)	1.0	0.8 0.2 (without drying)	2.8-3.4

A helpful first look at the possibility of using aerobic-aseptic-liquid technology can be gleaned by inspecting a 1980 profitability map for fermentation products in Figure 14-6.

The economics and technical objectives of producing a concentrated Romano cheese flavor by aerobic fermentation can be used as an example. The results, although dated (1974 data), are presented to show how technical objectives can be defined by an economic analysis. The basic assumptions and variables tested are listed in Table 14-8.

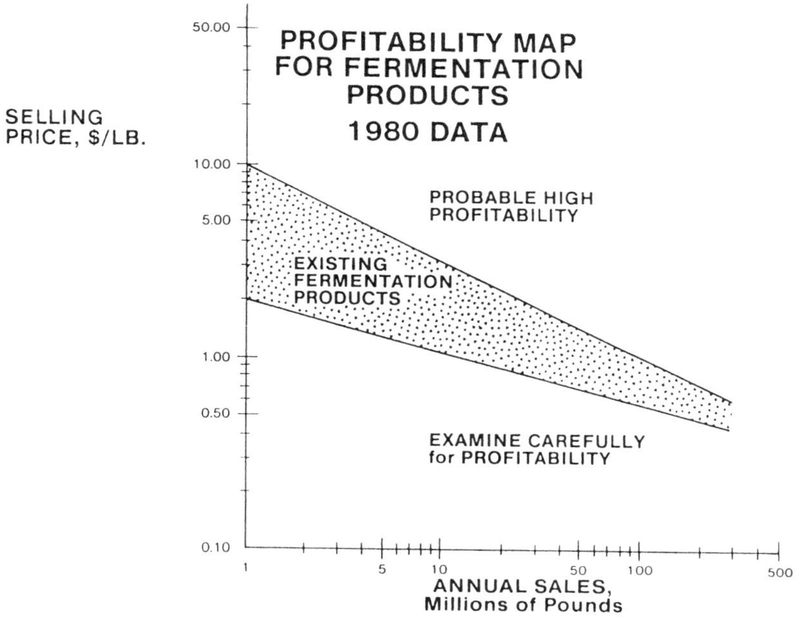

Figure 14-6. Profitability Map for Fermentation Products (1980 Data)

Figure 14-7 shows the effect of scale on elements of fermentation cost. As scale increases, all other variables being constant, labor-related costs decrease substantially; raw material and utility costs are constant, and capital-dependent costs decrease slightly. As the plant becomes larger, raw material (milk) accounts for about half of the cost of fermentation.

The relative effects of flavor yield, fermentation cycle, whole milk costs, substrate solids, and scale are pictured in Figure 14-8. Profitability calculations are based on Bio-Technical Resources' fermentation information, the assumptions in Table 14-8, and the following assumptions: 1) selling price is half that of Romano cheese which is at $1.20 per pound. For example, at 1 X flavor yield, selling price would be sixty cents per pound, as is, and at 10 X it would be six dollars per pound, as is; and, 2) the income tax rate is fifty percent. The base case is 10 X flavor yield, 13 percent substrate solids, 120-hour fermentation cycle, milk at $8.30 per one hundred pounds, as is, and a scale of five million pounds of cheese equivalent per year. For the base case using 1974 costs, installed fermenter capacity needed is 8,500 gallons, which would produce 500,000 pounds of cheese flavor at ninety percent solids, selling at six dollars per pound, and yielding a return on investment after taxes of sixty percent. After unitizing the Romano cheese fermentation into organism and substrate, the key optimization objective is clearly flavor yield improvement.

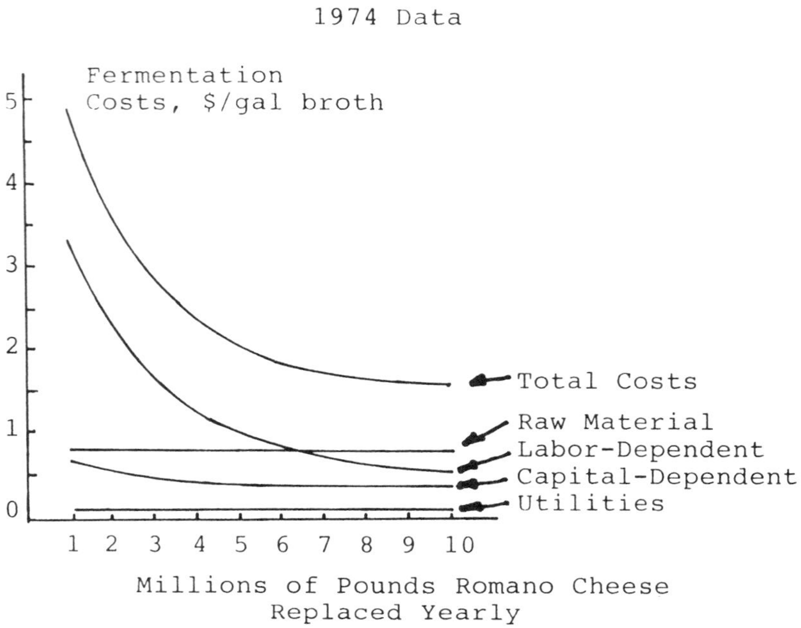

Figure 14-7. Effect of Scale on Cheese Flavor Fermentation Costs

Table 14-8. Assumptions and Variables for Concentrated Cheese Flavor Project

1974 Data

Assumptions:
1. A highly flexible aerobic aseptic fermentation plant
2. Recovery costs equal to fermentation costs without raw materials
3. Total plant investment is twice that for fermentation

Variables:
1. Market
   From 1,000,000 to 10,000,000 lbs per year of Romano cheese replaced with cheese flavor valued at 50 percent of $1.20 per pound Romano cheese.
2. Cheese Flavor Yield
   Expressed as multiplier of natural product, e.g., 2 X means that 1 lb of concentrated product at 10 percent moisture replaces 2 lbs of the 40 percent moisture cheese.
   From 1 X to 50 X
3. Fermentation Cycle
   72 to 168 hours
4. Raw Material Costs
   $4.15, $8.30, and $16.60/cwt of whole milk (13% solids)
5. Substrate Solids
   13% and 26%

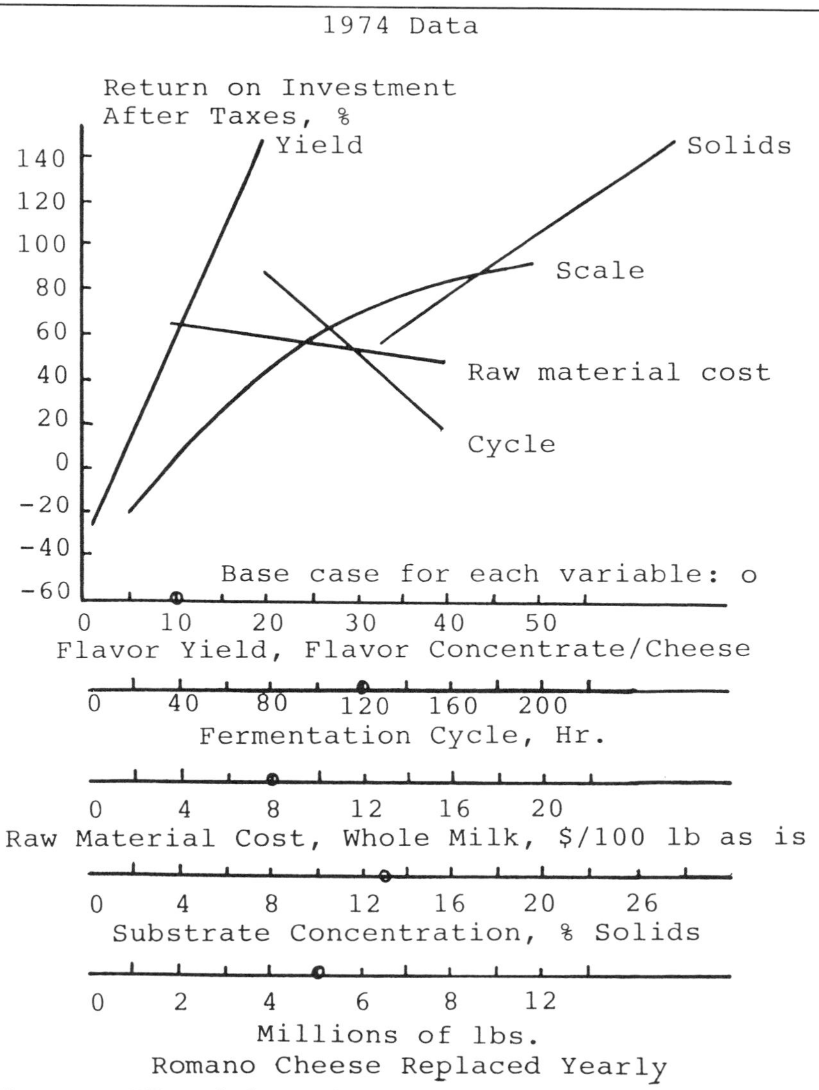

Figure 14-8. Effect of Cheese Flavor Fermentation Variables on Profitability

CONCLUSION

Market conditions and fermentation technology currently available provide a feasible basis for creating new fermented foods, new fermented flavors, food ingredients, and better versions of extant fermented foods employing unitization of traditional fermented products, especially those from the Orient, followed by optimization and recombination of the fermentation and functional elements.

REFERENCES

1. Aiba, S., Humphrey, A.E., and Millis, N.F. "Biochemical Engineering". Academic Press, New York (1973)
2. Bio-Technical Resources, Inc. files
3. Cannel, E. and Moo-Young, M. Solid state fermentation systems. Proc. Biochem. p. 2, June/July; p. 24, Aug/Sept (1980)
4. Desrosier, N.W. "The Technology of Food Preservation". AVI Publishing Co., Westport, CT (1963)
5. Findlay, W.P.K. (ed.) "Modern Brewing Technology". The Macmillan Press, London, England (1971)
6. Hesseltine, C.W. and Wang, H.L. Fermented soybean food products. In: "Soybeans: Chemistry and Technology". (Smith, A.K. and Circle, S.J., eds.) pp. 389-419 AVI Publishing Co., Westport, CT (1972)
7. Hesseltine, C.W. and Wang, H.L. The importance of traditional fermented foods. Bioscience. 30:402 (1980)
8. Hesseltine, C.W. Future of fermented foods. Proc. Biochem. p. 2, April/May (1981)
9. Kosikowski, F. "Cheese and Fermented Milk Foods". Published by Author, Distributed by Edwards Brothers, Inc. Ann Arbor, MI (1970)
10. Mauron, J. Nestle Research News 1980/81. Nestle Products Technical Assistance Co. Ltd., Technical Documentation Centre, 1814, La Tour-de-Peilz, Switzerland
11. Morton, B.J. and Sfat, M.R. Developments in Industrial Microbiology. 20:217 (1979)
12. Pederson, C.S. "Microbiology of Food Fermentations". AVI Publishing Co., Westport, CT (1971)
13. Peppler, H.J. and Perlman, D. (eds.) "Microbial Technology". Academic Press, New York (1979)
14. Pollock, J.R.A. (ed.) "Brewing Science". Vol. 1. Academic Press, London, England (1979)
15. Rose, A.H. (ed.) "Economic Microbiology: Alcoholic Beverages". Vol 1. Academic Press, London, England (1977)
16. Sfat, M.R., Skatrud, T.J., and Doncheck, J.A. Design of food fermentations based on analyses of existing processes. MBT Div. of ACS, Annual Mtg., Las Vegas, NV. August (1980)
17. Sfat, M.R. and Doncheck, J.A. Malts and Malting. In: "Encyclopedia of Chemical Technology", 3rd ed. Kirk-Othmer 14:810-823 (1981)
18. Skatrud, T.J. and Sfat, M.R. Designing new fermented foods. Food Eng. p. 7, July (1981)
19. Stanton, W.R. and Wallbridge, A. Fermented food processes. Proc. Biochem. 4:45-51 (1969)
20. Steinkraus, K.H. Traditional food fermentations as industrial resources. 8th Conf. of Assoc. for Sci. Coop. in Asia. Medan, Indonesia, February (1981)
21. Steinkraus, K.H. Nestle Research News 1980/81. p. 23 (See Reference 10 for publisher)
22. Vogel, H.C. (ed.) "Fermentation and Biochemical Engineering Handbook". Noyes Publications, Park Ridge, NJ (1983)
23. Wang, H.L. Oriental soy bean foods: Simple techniques produce many varieties. Food Dev. p. 29, May (1981)
24. Ziemba, J.V. Authentic soy sauce brewed in Wisconsin. Food Eng. p. 59, February (1974)

Chapter 15

Separation Technology for Bioprocesses

Roy Grabner

INTRODUCTION

The biotechnology revolution has demonstrated the feasibility of producing many unique products by biological systems. Although potential products may be identified, commercialization depends on reliable separation techniques to isolate saleable products from the complex biological mixture. These multi-step purification processes are highly product dependent reflecting both the product characteristics and specifications. Therefore the commercial processes may necessitate a unique combination of conventional purification technology with the development of novel techniques or scale-up of laboratory biochemical techniques.

BACKGROUND

The field of biotechnology has generated considerable interest as a result of the highly publicized achievements in genetic engineering in the past two decades. Actually, biotechnology is a new name for processes that have roots in antiquity. However, the new biotechnology has expanded these horizons through the union of classical biochemistry and traditional food and pharmaceutical technology with advances in genetic engineering and related developments in process technology. Although the biological techniques utilized for the discovery that foreign DNA fragments could be inserted into plasmids which could then be transformed into bacteria have become routine, commercial development of important products and processes remains a formidable task requiring the cooperation of a number of diverse disciplines. The transfer of this new biotechnology from the laboratory to commercialization will depend on sequential developments first in basic biological sciences, (i.e., genetic engineering); then in biological systems (i.e., fermentation) and final downstream processing and product characterization which are unique to each application (1,2,3).

The enthusiasm based on almost daily reports of new basic biological discoveries must however be tempered by the substantial technical problems, development time, and cost required to bridge the gap between scientific feasibility and an economically viable industrial scale process. The transition from a feasible laboratory procedure to commercial process often introduces time lags of years.

The burden and challenge of developing these economically viable products is thus gradually shifting downstream towards selection and definition of purification processes to isolate the desired product with acceptable purity and activity. Additional interest in downstream purification has been generated by recent estimates that these multi-step operations may actually contribute 80-90% of the total process costs. Therefore, it is appropriate to review both the expectation for downstream purification as well as the available separation technology.

SEPARATION TECHNIQUES

The isolation and recovery of a desired biologically derived product with adequate purity and activity usually depends on the combination of a series of recovery and purification operations into a facile, economical and reliable manufacturing process. It is important to consider the conditions that the product characteristics will impose on the final purification process since a broad range of products can be produced by biotechnology. Some of the critical product parameters are summarized in Table 15-1. The production of a complex, fragile, high purity product for a health care application (i.e., vaccines) with high unit value and low volume may be scaled up directly from the batch laboratory process to minimize the risk of changing the product. However, biologically produced chemicals (i.e., ethanol, citric acid) are frequently high volume, low unit value products which demand highly efficient, large scale processes with minimum environmental impact. This may necessitate a substantial purification process development effort.

Table 15-1. Product Characterization

Product Volume
Product Unit Value
Product Complexity
Product Stability
Product Purity

A list of potential separation techniques available for bioprocesses is shown in Table 15-2. However these techniques are not equally effective or applicable to the broad range of biotechnology applications which can be categorized into three general areas (4).

Table 15-2. Representative Separation Techniques

Cell Rupture
Solid-Liquid Separations
Membrane Processes
Chromatographic Separations
Precipitation
Multiphase Equilibrium Separations
Electrical Separations
Surface Phenomena
Enzyme Reactions
Inactivation
Structural Reactions

The first area can be categorized as biochemical laboratory separation methods. The phenomenal progress in biological science has produced some sophisticated and powerful tools for biological production as well as purification and characterization. These unique purification techniques based on immunological and electrophoretic concepts, however, are predominantly suited for laboratory and small scale preparative operations. Indeed these techniques are frequently used to isolate the initial quantities of material and establish feasibility of the basic science. Unfortunately these techniques are not readily scaleable since low productivity, high cost, inefficiency, and delicate control do not meet industrial requirements. Nevertheless these techniques are frequently used to develop sensitive and sophisticated analytical techniques to monitor product quality.

The second group of methods involves the conventional large scale industrial separation methods. A number of large scale separation processes such as liquid-liquid extraction, precipitation, adsorption, chromatography, membrane processing, and electrodialysis have been commercialized by the food and pharmaceutical industries. These techniques have already demonstrated the scaleability, efficiency, economics, and control required for consistent routine operation. Operating conditions, however, must be optimized to maximize selectivity and efficiency as well as preservation of the activity of the unique biological products, especially if handling crude raw materials with low product concentrations.

The third area involves use of novel separation methods. The unique characteristics of each new biotechnology product may necessitate the development or adaptation of a novel purification process for the evolution of a safe and economical product. Although these requirements are highly product dependent, many are based on biochemical concepts such as "selective enzymatic" transformation and digestion, selective oxidation and re-folding or physical operations such as electrophoresis in space and two phase aqueous extraction (5,6). The limits and demands on these processes will vary with the complexity of the product, raw materials, and degree of purification required. The biological origin of the products may also impose some unique process requirements such as inactivation of recombinant culture and removal of trace quantities of poorly defined impurities.

UNIQUE ASPECTS OF BIOPROCESSES

Although biotechnology may facilitate the production of products already available by chemical synthesis, it may also make it possible to manufacture unique biological products on a commercial scale. There is some precedent in the chemical and pharmaceutical industries for processing the chemically derived products which are stable and isolatable as relatively pure, well defined chemical entities (7). There is, however, limited experience to date for processing products with unique biological characteristics (8,9,10,11). The downstream processing must handle relatively large volumes of a complex poorly defined multi-component mixture containing the desired product usually in low concentration and low purity. The biological products are frequently labile materials demanding processing conditions of a specific temperature and pH for maximum stability. The actual biological activity of some important proteins (i.e., insulin) depends on secondary structures. These secondary structures depend on the formation of disulfide bonds which may require special processing (12,13,14). In addition, the analytical support essential for process development may be more qualitative than quantitative to further complicate monitoring process performance. Characterization of the final product quality must evaluate the purity, activity, and safety through a series of complementary assays. Regulatory concerns also have imposed some demanding specifications for unique impurities such as pyrogens, nucleic acids, and extraneous proteins as well as containment conditions for processing (15,16,17).

It is important to consider the purification of biologicals as a series of dependent cascade reactions. Therefore, any change upstream, including fermentation, will eventually impact downstream operations in addition to product quality. In many instances the process itself becomes a ritual to insure consistent product quality.

DOWNSTREAM PROCESSING

Downstream processing begins with the fermentation broth and ends with the production of a saleable product. Although the operations are highly product dependent, there are some common characteristics in the initial broth handling operations. The initial recovery phase of the processing is generally used for the gross separation of components such as cells and broth. Subsequent operations may then be introduced depending on whether the desired product is intracellular or extracellular. The recovery phase has three basic objectives; namely, clarification, concentration, and partial purification yielding an intermediate suitable for sale or further purification.

The options for the recovery phase of the process are shown schematically in Figure 15-1. The solid liquid separation is based on relatively well defined techniques such as centrifugation, filtration, or ultrafiltration (18,19,20).

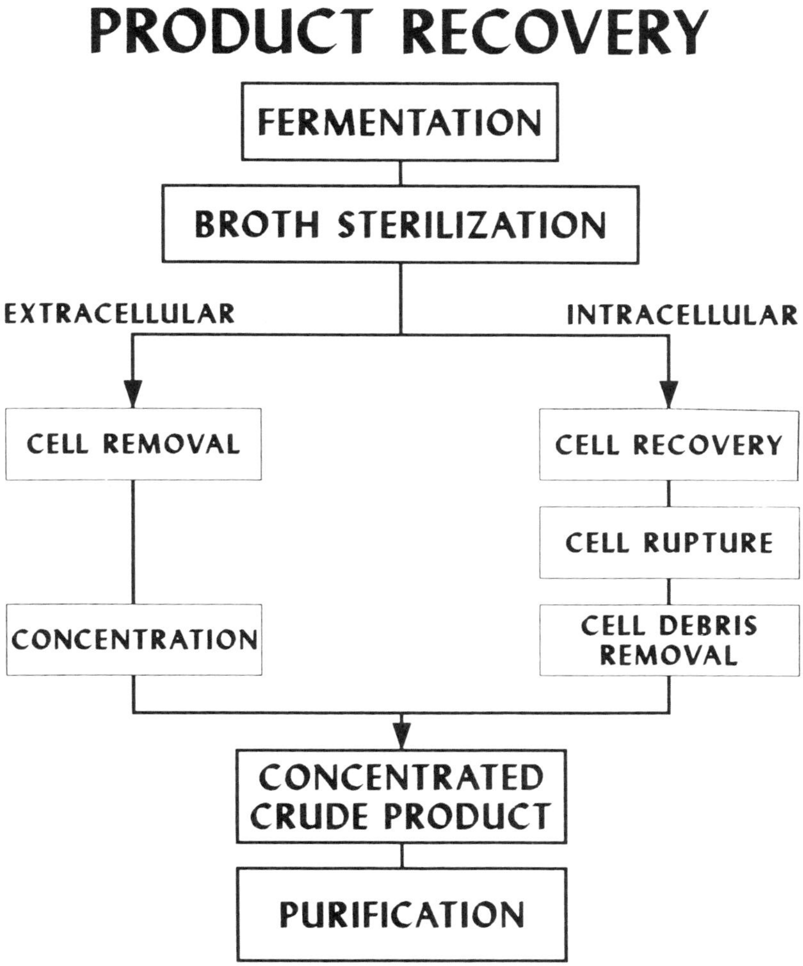

Figure 15-1. Product Recovery

Recovery of Extracellular Products

Extracellular products are secreted into the fermentation broth. These may be small molecules or macromolecules which are transported through the cell membrane. Although usually readily separable from the cells, these products are often present in low concentrations; therefore, the initial recovery steps consist of clarification and concentration. Product concentrations usually range from 0.1 to 100 grams per liter of broth.

Primary impurities include a number of nucleic acids and proteins excreted through the membrane plus the broad range of soluble nutrients added during fermentation. Effective and economical water removal is a practical consideration for these process streams. Membrane technology has been effectively introduced for this relatively high volume, low resolution separation based on molecular size.

A variety of membranes suitable for large scale separation operations are available for this application with a wide selection of porosity, configuration, and composition to maximize purification and product recovery. Although membranes can be operated in reverse osmosis, ultrafiltration, or microfiltration modes, these already low resolution techniques are susceptible to membrane fouling which may modify membrane characteristics during processing. These techniques are, however, excellent for water and salt removal. Removal of water is an integral part of these recovery operations since water is a major impurity and may be as much as 90% of the total mass. Some operating parameters are summarized in Table 15-3.

Table 15-3. Membrane Processes

Separation by Molecular Size	
	Reverse Osmosis
	Dialysis
	Ultrafiltration
	Micro-filtration
Low Resolution	
Operating Parameters	
	Pressure
	Flow Rates
	Concentration
	Membrane Porosity
	Membrane Composition
	Membrane Configuration

Recovery of Intracellular Products

Recovery of intracellular products is a multi-step operation based first on concentrating of cells to a 50% v/v solution followed by a cell rupture step to break the cells and release the product (21,22,23,24). Typical cell rupture techniques are summarized in Table 15-4. Industrial applications generally use mechanical methods which are commercially available, simple, effective, and readily scaleable. Operating conditions must be selected based on the specific application, considering product stability and cell resistance to rupture. The cell rupture technique can adversely affect subsequent purification steps if the solution is too viscous or cell disintegration is extreme. Practical operating conditions are balanced between capacity, operating pressure, number of passes, temperature, and the requirements of downstream processing.

Table 15-4. Cell Disruption

- Methods
  - Mechanical
    - High Pressure Homogenization
    - Bead Mills
  - Ultrasonics
  - Enzymes
  - Chemicals
    - Acid
    - Base
    - Detergents
- Operating Parameters
  - Cell Concentration
  - Temperature
  - Residence Times
  - Energy Input

Intracellular products are usually macromolecules retained inside the cell. Many recombinant products are in this category since the product is not native to the organism and no mechanism exists for transport through the cell membrane. Products retained within the cell may exist as a soluble form readily released when the cell is ruptured or an insoluble inclusion body necessitating another step for solubilization (25). These solutions are relatively concentrated but contain high levels of nucleic acids, complex proteins, and cellular debris.

The recovery phase of these downstream processes is generally based on conventional physical separation techniques to produce a partially purified concentrate. A wide range of products need further purification; however, the degree of purification necessary is highly dependent on the product. Since the cost of purification can be expensive (product losses and operating costs), it is vital to define the requirements.

The subsequent purification steps are selected based on specific product characteristics to achieve a high degree of selectivity with high product recovery. The purification techniques are based on one or more of the product characteristics in Table 15-5 coupled with process requirements shown in Table 15-6.

Table 15-5. Product Characteristics Important in Purification

Electrical Charge
Molecular Size
Hydrophobicity
Sterochemical Specificity
Solubility
Chemical Stability
Density

Table 15-6. Process Consideration

Stability of Activity
Resolution
Throughput
Cost
Reliability and Simplicity
Scalability

Chromatographic Techniques

Assuming a high purity product is both desirable and necessary it is essential to incorporate some high resolution purification steps into these later phases of the downstream processing. Chromatographic purification techniques have demonstrated excellent resolution, capacity, and flexibility on both the laboratory and industrial scale (26,27). The primary chromatography methods of ion exchange, size exclusion, affinity, and reverse phase are directly associated with the respective product characteristics of electrical charge, molecular size, sterochemical specificity, and hydrophobicity.

The chromatographic systems are based on operating a packed column in a batch mode with the appropriate selection of packing and operating conditions for optimum resolution. This simple statement ignores the considerable effort required to balance practical factors such as cost, throughput, capacity, activity, scalability, product stability, and resolution while making the appropriate selections.

High performance liquid chromatography (HPLC) is another technique that has received considerable attention recently (28,29,30,31). This is not actually a purification technique but rather a special method of operating chromatography systems to achieve excellent resolution in relatively short time. The key element of HPLC is the unique packing materials. These packings have average particle diameters as low as 5 microns, thereby significantly reducing diffusion distance in and out of the particle with a corresponding decrease in equilibration time resulting in more rapid separation. The small particles, however, significantly increase the pressure drop through the bed. Therefore, high pressure (500 psi) operation is necessary to obtain high flow rates to realize the benefits of rapid separation.

HPLC has been successfully applied to a wide variety of laboratory and preparative scale operations that demand high resolution in a single step. This is particularly attractive for analytical applications and the rapid production of gram quantities of materials.

Some characteristics of conventional low pressure liquid chromatography and high pressure liquid chromatography are listed in Table 15-7. The obvious differences are reduction in packing particle size, increase in pressure and smaller columns for a smaller loading but more rapid turnaround. The selection between high pressure and more conventional low pressure liquid chromatography is a balance between resolution, time cycles, and capacity. Although HPLC is an essential analytical tool and offers direct scale-up for preparative studies, it is not necessarily the process for larger scale processing.

Table 15-7. Comparison of Chromatography Systems

	Conventional	LPLC	HPLC
Beads			
Size (μ)	250-800	50-150	5
Pore Size (Å)	$\leq$ 100		$\leq$ 1000
Surface Area (m^2/g)	500		300
Porosity	50%	90%	70%
m_{eq}/ml	1-2	1-5	1-5
Materials			
	Polymer Polystyrene Divinyl Benzene	Dextrans Cellulose Agarose	Silica Alumina Glass Polymeric
Nominal Operation			
Bed Diameter (cm)	100	75	10
Bed Height (cm)	300	100	30
Bed Volume (liters)	10,000	1,000	5
Flow Rates ($ml/hr \cdot cm^2$)	100	10-100	100
Pressure (psig)	10-100	10	500

Affinity chromatography (32,33) is another recent development for biopurification separation based on a highly specific sterochemical interaction such as antigen/antibody reactions. Ideally a pure product should be obtainable in a single step. This assumes that non-specific binding to the support does not occur and there is no leakage of the attached function group. In addition, the availability and cost of the custom designed functional group is a major consideration for this technology.

Less specific but highly effective separations have also been realized with group specific adsorbents. These functional groups include metal chelates (34,35), Protein A, Concanavalin A, lysine, dyes, and lectins. These highly specific and therefore high resolution chromatography adsorbents offer the potential for a single step purification as well as a technique for removing trace impurities from partially purified products.

CONCLUSIONS

Current conventional purification technology has evolved in response to specific applications, opportunities, and problems in the past. Many of these techniques are directly applicable to the biopurification processes, assuming adequate process conditions can be developed in a timely and economical manner. New separation techniques will also evolve to meet unique product requirements and specifications. The basis for these techniques will probably evolve from laboratory biochemical or analytical methods introduced to demonstrate the feasibility of biologically produced products.

REFERENCES

1. Michaels, A.S. The impact of genetic engineering. Chem. Eng. Prog. 80:9-15 (1984)
2. Michaels, A.S. Adapting modern biology to industrial practice. Chem. Eng. Prog. 80:19-25 (1984)
3. Brunt, J. van. Scaleup: the next hurdle. Biotechnology 3:419-424 (1985)
4. Humphrey, A.E. Commercializing biotechnology: Challenge to the chemical engineer. Chem. Eng. Prog. 80:7-12 (1984)
5. Kula, M.R., Kroner, K.H. and Hustedt, H. Purification of enzymes by liquid-liquid extraction. In: "Advances in Biochemical Engineering, v. 24. Reaction Engineering" (A. Fechter, ed.) Springer Verlag, New York (1982)
6. Goklen, K.E. and Hatton, T.A. Protein extraction using reverse micelles. Biotechnol. Prog. 1:69-75 (1985)
7. Atkinson, B. and Mauitone, F. "Biochemical Engineering and Biotechnology Handbook". The Nature Press, New York (1983).
8. Grinnan, E. L. L-Asparaginase: A case study of an *E. coli* fermentation product. In: "Insulin, Growth Hormone and Recombinant DNA Technology". (J.L. Gueriguian, ed.) Raven Press, New York (1981)
9. Brewer, S.J. and Sassanfield, H.M. The purification of recombinant proteins using C-terminal polyarginine fusion. Trends in Biotechnology 3:119-122 (1985)
10. Street, G. Large scale industrial enzyme production. Critical Rev. Biotechnol. 1:59-85 (1984)
11. McGregor, W.C. Large scale isolation & purification of proteins from recombinant *E. coli.* Annals New York Academy of Sciences. Biochemical Eng. III 413:231-237 (1983)
12. Thannhauser, T.W., Konishi, Y. and Scheraga, H.A. Sensitive quantitative analysis of disulfide bonds in polypeptides and proteins. Anal. Biochem. 138:181-188 (1984)
13. Konishi, Y., Ooi, T. and Scheraga, H.A. Regeneration of ribonuclease A from the reduced protein. Biochemistry 20:3945-3955 (1981)

14. Anfinsen, C.B. Principles that govern the folding of protein chains. Science 181:223-230 (1973)
15. Sofer, G. Chromatographic removal of pyrogens. Biotechnol. 2:1035-1038 (1984)
16. Points to consider in the production and testing of new drugs and biologicals produced by recombinant DNA technology. Food and Drug Administration, National Center for Drugs and Biologicals, Bethesda, MD (1985)
17. Guidelines for research involving recombinant DNA molecules. U.S. Department of Health and Human Services, National Institute of Health, Federal Register 49 FR 46266-46291 (1984)
18. van Hement, P. Strictly aspetic techniques for continuous centrifugation. J. Microbiology 46:501 (1980)
19. Higgins, J.J., Lewis, D.J., Daly, W.H., Mosqueira, F.G., Dunnill, P. and Lilly, M.D. Investigation of the unit operations involved in the continuous flow isolation of β-galactosidase for *Escherichia coli.* Biotechnol. Bioeng. 20:159-182 (1978)
20. Ricketts, R.T., Lebrerz, W.B., Klein, F., Gustafson, M.E., and Flickinger, M.C. Application, sterilization and decontamination of ultrafiltration systems for large scale production of biologicals. In: "ACS Symposium Series, Purification of Fermentation Products". (D. LeRoith, J. Shiloach, and T.J. Leahg, eds.) ACS, Washington, D.C. 271:21-49 (1985)
21. Kroner, K. H., Shutte, H., Stach, W., and Kula, M. R. Scale-up of dehydrogenase by partition. J. Chem. Tech. Biotechn. 32:130-137 (1982)
22. Darbyshire, J. Large scale enzyme extraction and recovery. In: "Topics in Enzyme and Fermentation Biotechnology", Vol. 5 (A. Wiseman, ed.) John Wiley & Sons, New York (1981)
23. Mosqueira, F.G., Higgins, J.J., Dunhill P. and Lilly, M.D. Characteristics of mechanically disrupted bakers yeast in relation to its separation in industrial centrifuges. Biotechnol. Bioeng. 23:335-343 (1981)
24. Schutte, H., Kroner, K.H., Hustedt, H. and Kula, M.R. Experience with a 20 liter industrial ball mill for the disruption of microorganisms. Enzyme Microb. Tech. 5:143-148 (1983)
25. Marston, F.A.O., Lowe, P.A., Doel, M.T., Schoemaker, J.M., White, S. and Angal, S. Purification of calf prochymosen (prorennin) synthesized in *Escherichia coli.* Biotechnology 9:800-806 (1984)
26. Cooney, J. M. Chromatographic gel media for large scale protein purification. Biotechnology 2:41-55 (1984)
27. Bjurstrom, E. Biotechnology, fermentation & downstream processing. Chem. Eng., 126-158 (1985)
28. Regnier, F.E. and Gooding, K.M. High performance liquid chromatography of proteins. Anal. Biochem. 103:1-25 (1980)
29. Regnier, F.E. High performance ion exchange chromatography. In: "Methods in Enzymology". (W.G. Jakoby, ed.) Academic Press, New York 104:170-189 (1984)
30. Unger, K. High performance size exclusion chromatography. In: "Methods in Enzymology". (W.G. Jakoby, ed.) Academic Press, New York 104:154-169 (1984)
31. Sitrin, R.D., Chan, G., DePhillips, P., Dingerdissen, I., Valenta, J., and Snader, K. Preparative reversed phase high performance liquid chromatography. In: "ACS Symposium Series, Purification of Fermentation Products". (D. LeRoith, J. Shiloach, and T. J. Leahg, eds.) ACS, Washington, D.C. 271:71-90 (1985)
32. Practical Guide for use in Affinity Chromatography and Related Techniques, Reactifs IBF-Societe Chimique Pointet-Girard, France (1983)

33. "Affinity Chromatography - Principles & Methods". Pharmacia Fine Chemicals, Uppsala, Sweden (1983)
34. Porath, J., Carlsson, J., Olsson, I., and Becfrage, G. Metal chelate affinity chromatography. Nature 258:598-599 (1975)
35. Sulkowski, E. Purification of protein by IMAC. Trends in Biotechnol. 3:1-7 (1985)

Chapter 16

Scale Up of a Fermentation Process

Peter Senior

INTRODUCTION

A scientific paper begins with following the "Instructions to Authors" so as to arrange the information with the usual sections for Materials and Methods, Results, Discussion, etc. In describing the design of the Imperial Chemical Industries (ICI) process for production of single cell protein (SCP), the Materials and Methods began as follows: 1) take $200 million and the idea of making 50,000 metric tons/year of single cell protein for animal feeds; 2) employ a small army of biochemists, microbiologists, fermentation technologists, economists, chemical engineers, designers, mathematicians, toxicologists, metallurgists, market researchers, and managers; 3) build the world's largest continuous culture vessel containing 1200 m^3 of culture consuming 14 metric tons/hour of methanol; and, 4) coordinate the effort effectively and as efficiently as possible. This chapter will describe the scale up exercise, the well disguised man traps ICI fell into, and the methods used to resolve the problems. The purpose is to demonstrate how to proceed from a plate culture of an organism to the world's largest continuous microbial culture process.

ECONOMIC JUSTIFICATION

It is pertinent to first review the pre-project economic justification for SCP as it existed at ICI Agricultural Division in 1968 when the project was initiated. In the early 1960s the United Kingdom discovered massive natural gas reserves in the southern basin of the North Sea. In 1966, ICI invented a low pressure, high efficiency process for chemical production of methanol; today approximately 70% of the world's methanol capacity uses this technology. Plants were built at Billingham in Northern England to manufacture methanol and ICI negotiated a long-term gas contract with the supplier.

The question which arose then was how else to add value to relatively cheap methane gas? Methanol was commercially attractive but what else was there? Being the United Kingdom's major manufacturer of fertilizers and agrochemicals, the ICI contact with the animal feed industry was extensive and the price trends of soy bean and fish meal, two of the staples of that business, were such that a competitive product looked possible in the long term.

At the same time, various discoveries in the academic world were coming to light. The first isolation and growth on a small scale of methane utilizing microorganisms had been published. ICI had good contacts with the academic centers and thus the idea was born. Methane, ICI's agriculture business, microorganisms and economics came together and the SCP challenge was born in 1968.

PRODUCT VOLUME DECISION

The first thing to decide when scaling up a process is how much of "X" do you want to make? These were purely economic arguments based on market projections and substrate availability. ICI decided on a process to produce 50,000 metric tons per year. It is not exactly certain where that figure came from but the fact that at the right price ICI could sell the entire output of the plant within a forty mile radius, may have influenced the decision.

THE BATCH VERSUS CONTINUOUS QUESTION

At roughly the same time, the batch versus continuous question arose. This was the late 1960s and nobody anywhere had any experience of large-scale continuous culturing of microorganisms. With a dedicated plant, for maximum return on capital, the plant should be designed and built around the organism and the product. One simply cannot build an off-the-shelf fermentation process with ancillary equipment and expect a good return upon start up. One must look at the organism and the product. For example, an industrial enyzme producer might want maximum flexibility of capital assets, as they may be growing 20 different organisms for 60 different products. What is therefore most attractive to the enzyme producer: a custom designed continuous fermentation plant capable of optimum cost production of one product or a battery of relatively flexible units capable of multiple product manufacture?

SCALE

Scale is the next most important consideration. For example, for a production of 50,000 metric tons per year of dry biomass; taking 250 m^3 batch fermenters with a turn-around time of four days and a three hundred day year, one needs (assuming batching to 30 g/L dry cell weight) 90 vessels, a total fermentation capacity of 22.5 million liters, a large value. Furthermore, large power input is needed in terms of agitation to get the fermentation activity. In the ICI development it was suggested that the newly developed Rolls Royce Olympus engine in the Concorde was adequate. However, 90 of those at full throttle complete with an after-burner was unacceptable. Thus continuous culture was chosen. At the time of that decision, ICI was still scraping the colonies off agar plates. ICI was not daunted by continuous processes or the fact that the one under consideration would be the world's first large scale fermenter.

In a review of available processes, it appeared that most fermentations were mechanically stirred batch vessels in stainless steel. One simply had to use stainless steel for growing microorganisms. It was then pointed out that for the size of the fermenter under contemplation, the annual production of United Kingdom stainless steel would have been barely adequate and furthermore, rather expensive. In addition, using methane with air in the presence of water vapor would require explosion proofing and an armoured vessel was not out of the question based on safety considerations.

By 1968 ICI had laboratory one liter continuous culture systems in operation with the organism. However, research and development was rather difficult, yields were poor, efficiency of gas utilization was poor because of poor CH_4 solubility, oxygen yield was poor because of the reduced nature of the carbon source, and the organisms were hard to grow as a monoculture. ICI persevered from behind armoured glass screens. A fermenter was even blown up to make sure the safety controls would work.

Once it worked reasonably well at the laboratory scale the question was whether the product was any good. It was decided to proceed to the next scale in order to get enough material to test in animal feeding and toxicology models and the scale selected reflected that need. It was also the minimum needed to provide design data on mass balances and heat balances for the construction of a proper pilot plant embodying most of the concepts that the main plant would possess. Two continuous methane fermenters (75 L and 1 m^3) were built. These were housed in concrete bunkers (the walls were two feet thick) with a large port in the top so that in the event of an explosion the plant would conveniently exit via the roof falling into the scrap equipment yard adjacent. It should be noted that the pilot plant fermenters never exploded.

METHANE VERSUS METHANOL

The next decision was whether or not to abandon methane as the carbon and energy source and operate with methanol alone. The problems with methane were related to gas solubility, overall efficiency, heat removal, oxygen efficiency and intensity. At the time there was no vision as to which of these problems could be overcome. The answer became apparent later when ICI fell into the trap set by the biochemistry of methane utilization by microorganisms. The oxidation of methane to methanol in microorganisms is a net energy consuming process which is opposite to an energy yielding step as found in the chemical reaction. Energy here is defined as net useful chemical energy, heat is useless to a microorganism in terms of biosynthesis. The methane mono-oxygenases required oxygen and a reducing agent in order to activate oxygen to attack methane and produce biologically available carbon. The reducing agent consumed in this initial attack on methane had to be generated by the further oxidation of methanol or formaldehyde thus reducing the overall carbon efficiency of methane carbon conversion to cell carbon.

In the late 1970s, ICI therefore changed the feedstock for the process to methanol and began selecting methylotrophic organisms. This had a dramatic effect on the scale up plans. Methylotrophs were much easier to grow in terms of process control. The substrate was soluble and easily sterilized, yields in laboratory fermenters were excellent at 68% carbon conversion (68% of the methanol carbon was incorporated into cellular material). In addition, the oxygen demand by the fermenter was much reduced with a subsequent lowering of the heat load. All these factors combined to facilitate process intensification.

FERMENTER CONFIGURATION

Despite the ease of laboratory fermentation in achieving about 20 g/L dry cell weight at high efficiency, the target for the full-scale plant was 30 g/L. Thus ICI was still presented with the formidable problem of selecting a reactor configuration for the pilot plant that would improve yield. It was decided to move to a production unit of intermediate scale, manufacturing 1000 metric tons per year of dried cells. This figure was decided on as an overestimate. Given the initial troubles of a new technology, it was expected that a few hundred tons of a material that had a specification suitable for animal feeding trials and toxicology studies could be achieved. The concepts for the full-scale plant were now hardening along with the problems they presented: 1) stirred reactors were impractical - a new type of fermenter of mild steel manufacture had to be invented; 2) separating the cells from the exit culture was a problem - it appeared that dozens of the largest commercially available centrifuges would be needed; 3) drying the concentrated cell cream would have to be done - calculations showed that spray drying would give the correct product in terms of a physical and nutritional specification, but many driers would be required; and, 4) nominally setting the dilution rate at 0.18 hr^{-1} with a culture working volume of one million liters produced in the region of 180 m^3 of culture/hr. At 30 g/L this equated to 5.4 metric tons of product/hr. If that was dried as a 20% (w/v) solids cream this meant initially rejecting

153 metric tons/hour of culture supernatant as plant effluent. This would have high BOD and COD loading and the economics were undesirable if not disastrous for waste treatment. It was decided to separate the cells from the culture and recycle the spent medium aseptically.

AIR LIFT PRESSURE CYCLE FERMENTER (PCF)

Based on all these concepts worked through simultaneously, the ICI pressure cycle fermenter was developed in the early 1970s. This vessel has no moving parts and is simply a massive air lift reactor with a separate downcomer for fluid from which bubbles have disengaged.

Everything involving the fermenter was designed to maintain monosepsis. New valve assemblies had to be invented. Process controls had to be of the highest integrity with multiple sensors at key points. A pilot plant was built on the PCF design and after about two years ICI achieved a reliable fermentation operation. Initially the culture was poured down the drain, not having yet invented a harvesting system.

CELL HARVESTING

The harvesting system, when developed, included simple heat treatment to 85°C to permeabilize the cells, followed by acidification to pH 4.2 with a mixture of sulfuric and phosphoric acids to flocculate and entrain bubbles in the flocs. A flotation process was developed which was difficult to control.

It is difficult to describe the complex web of interacting problems that ensued. The heat/acid separation stage required the culture to be physiologically correct to work. This yielded a concentrated cream of disrupted cells and acid precipitated proteins. The supernatant from this step contained much acid soluble protein. To satisfy the economic target it had to be recovered. Going back to the needs of recycle, it rapidly became obvious that re-sterilizing this so-called underflow was essential. However, when attempting re-sterilization by either shell and tube or plate exchanger, salts and protein fouled all the surfaces and heat transfer declined to a point where sterility could not be assured. The only alternative was to heat shock and acidify the culture aseptically, float the cream off likewise and then recycle the underflow back to the fermenter having first cooled it. This extended the plant's sterility boundary, making for greater complexity and risk of ingress by unwanted organisms. However, in solving that major problem it revealed another problem.

CORROSION

The old adage of "for want of a nail etc., the Empire was lost" was true of a very complex set of problems which emanated from an early decision to exclude chloride ions from the process. In heat stressed stainless steel, stress corrosion cracking enhanced by chloride is relatively common, thus it was specified that chlorides were to be eliminated. That meant that hydrochloric acid could not be used for acidification. Nitric acid was also excluded, leaving only sulfuric and phosphoric acids as cost effective alternatives.

The microbial nutritional requirement for P and S on an hourly basis could be just satisfied by the amount of phosphoric and sulfuric acids going into the culture separation acidification downstream process. This was all right for the first cycle but recycling steadily built up the buffering capacity of the entire system such that after a few recycles the product went out of specification on phosphorus and sulfur content. In addition, pH control in the fermenter required more ammonia gas to be put into the system and everything ultimately ground to a halt.

UNDESIRABLE PROCESS EFFECTS

At high sulfate concentrations the organism began to manufacture an extracellular polysaccharide. This had two effects: 1) it diluted the product in terms of its valuable protein content; and, 2) on drying it took part in the Maillard browning reaction with the epsilon-amino groups of lysine, causing an undesirable darkening of the product and loss of available lysine.

BUBBLE PROBLEMS

A much more serious effect of protein solution recycle was related to bubble coalescence and disengagement. Rather rudimentary fundamental knowledge of multiphase interactions in fermentation existed at that time. To exemplify the "bubble problems," it is useful to describe a typical timescale of events.

A transient inefficient operation of flotation led to periods when fluid with a high soluble protein content returned to the fermenter and stabilized the surfaces of bubbles such that the rate of coalescence and thus disengagement fell. These small bubbles carried around the PCF became substantially enriched in CO_2 and depleted in oxygen, the solution ions of CO_2, and HCO_3^- increased in concentration leading to two major effects. The greater buffering capacity of the ions lead to a greater acid requirement in flotation with the effects on the microorganism as described previously. Another problem was a physiological effect such that at high pCO_2 the organism made more extracellular carbohydrate and dropped its yield coefficient leading to the generation of more CO_2 from the unincorporated methanol carbon source. These spiralling, self amplifying effects, often rapidly progressed beyond the grasp of process control and the result was the need to dilute and start over again. Most of these considerations of process control were not addressed, nor did they manifest themselves until progressing to the large-scale operation of the "PRUTEEN" plant.

TOXICOLOGICAL CONSIDERATIONS

As the large fermenter was being built, the toxicological results showed that 25% "PRUTEEN" in the diets of the target species (poultry broilers) caused half of them to die of extensive visceral gout. Although that was fivefold the recommended commercial feeding rate, it caused extreme concern to ICI. There was a massive effort to find the culprit molecule. It turned out that because of the chloride ion decision referred to earlier on, the diets were deficient in NaCl. This affected renal transport of uric acid and by rebalancing the diets the problem went away. The problem occurred only in broilers, not in rats, mice, calves, ducks or pigs.

IRON PROBLEM

The major market for powdered "PRUTEEN" is as a veal calf milk replacer. Due to the European taste for pale veal, a key specification for veal feeding was low iron content in the feed. In the United Kingdom veal calves are penned in wood and drink out of plastic buckets to minimize iron ingestion. Thus, the growth medium specification required iron sufficiency for microbial growth needs and deficiency in the final product. A deviation from the target level meant either an iron deficient culture (bright green) or too much iron in the product yielding the eventual undesirable red veal meat. The control window was very narrow and with all the other sources of process control deviation, control was difficult.

YIELD

In the process of scale up from a laboratory one liter fermenter, through the pilot plant 75 L, 1 m^3 and 40 m^3 operation, yet another and much more serious trend was apparent in terms of yield loss. In the laboratory fermenter, yields of cells were at roughly 68% carbon conversion. In the pilot plant this fell to 40%. The main plant had to operate at 62% yield which presented a major problem. Everything was simulated in the laboratory, including, for example, oscillations in pH, temperature, dissolved oxygen tension, CO_2 partial pressure, individual nutrient sterilization regimes, etc. This included all perturbations imagined possible as one scaled up the fermentation process. Obviously as one increased the scale, power input per unit volume declined with subsequent repercussions on mixing times and efficiencies. Those were simulated and it was during this program that another interesting difference came to light. Certain of the laboratory fermenters always gave better yields than others and these were segregated. There were two classes of fermenters, "drippers" and "squirters" depending on whether the methanol inlet port was above or below the culture fluid respectively. In order to get accurate flow measurements at the small scale the methanol was invariably diluted. Depending on the steady state dry weight desired in a particular experiment, one could dilute or use concentrated methanol. The yield decreased as the frequency of drips decreased and as the concentration of methanol in the drip increased. On a purely empirical basis, it was determined that growth yield was dependent on the steady state level of methanol in the culture even when that culture displayed all the characteristics of carbon limitation. One had to match exactly the assimilation rate of carbon coming through to formaldehyde in the organism and its subsequent incorporation, with the ability of the methanol dehydrogenase to oxidize methanol.

In the metabolic pathway of methanol utilization, formaldehyde could be oxidized by at least three known routes. One route is via the hexulose phosphate cycle, one via the NAD-linked formaldehyde dehydrogenases and the last via methanol dehydrogenases (MDH). The efficiency in energy yield of the first two routes were high, the $K_m$of the third route was high but of low efficiency in ATP yield. Furthermore, there was no respiratory control on the electron transport chain from MDH. Thus, when MDH was exposed to excess methanol it simply oxidized all of it to HCHO and the high transient concentration of formaldehyde was again "burnt" off inefficiently by the MDH. Once at the level of formate, carbon incorporation into cells was very inefficient. Summarizing, the cells responded to a transient high concentration of methanol by oxidizing the excess to CO_2 with little or no energy conservation and yet still gave all the appearances of continued carbon limitation of growth. The problem came down to the cell's growth acceleration potential. Starting from a baseline growth rate of 0.2 hr^{-1} with a very low steady state methanol concentration (a few parts per million) if the culture was given a sharp transient pulse of methanol to 20 ppm, the cell, now presented with a higher flux of formaldehyde attempted to accelerate its growth rate. Depending on where one started on the baseline, this capability differed. At 0.2 hr^{-1} the organism could instantaneously change its growth rate up to a certain value.

If one pulsed the culture with methanol such that the steady state methanol concentration dictating this maximum instantaneous high growth rate exceeded that value, then the organism could not grow at the higher rate but could oxidize the methanol inefficiently. The growth rate would only accelerate to higher levels over a longer period of time. As the baseline moved up to 0.4 hr^{-1} a culture would take a pulse of methanol in its stride, it could accelerate its growth rate very rapidly and oxidize and assimilate efficiently.

The events described have a time base of seconds and it was required to calculate a series of curves relating growth rate, frequency of methanol additions, calculated profiles of growth rate acceleration and deceleration, methanol concentration and oxygen/CO_2 use and evolution. The concept of a critical concentration of methanol above which yield penalty was incurred gave a great deal of trouble mainly because it was so low. How could one almost instantaneously disperse something like 14 metric tons per hour of methanol into the big fermenter? According to calculations and experiments, this had to be done on a timescale of seconds to avoid yield losses.

The solution was a compromise between what was possible in engineering terms and the yield penalty affordable. In effect, it was required to turn the one million liter PCF into about 30,000 fermenters of 33.3 L each with an inlet port for methanol. The banks of nozzles were distributed through the fermenter in order that independent control was possible to match the various oxygen supply profiles that existed around the PCF loop. In this way the main plant achieved its original design efficiency.

CONCLUSIONS

This chapter has attempted to highlight the way in which the process of scale up was both multidisciplinary and highly interactive at ICI. The design had to meet the economics and the product's specifications. The product dictated the process alongside the available technology, but the most important message learned was that during the process of scale up the biologists and engineers had to interact and talk the same language. Major problems had a way of masking minor problems which in time became major problems themselves. What appeared on the surface to be a chemical engineering problem was sometimes rooted in a fundamental biochemical feature of the organism. The converse was also true.

How one anticipates all these things before embarking on the scale up trail with a new product and process is impossible to answer. Certainly the experience proves that hindsight is an exact science and that foresight is an elusive commodity. Scale up is not easy because if it were, one would learn very little. By its very nature the process is challenging, exciting, wearying beyond belief and when it all finally works, immensely satisfying.

Chapter 17

The Use of Enzymes for Waste Management in the Food Industry

Sharon Shoemaker

INTRODUCTION

Future approaches to food production and processing will have to be more concerned with waste management and energy conservation. Environmental concerns and economic issues necessitate better utilization of raw materials and further reduction of food processing waste. The term waste will refer to byproducts and residuals from food processes that are relatively low value. Indeed interest in this issue was expressed by several committees participating in the IFT Workshop on Research Needs, held in November, 1984 (1,2,3,4).

The committee addressing environmental issues noted the need for an overall "systems" approach, in contrast to focusing only on unit operations or the by-product streams "at the end of the pipe" (1). This committee provided several recommendations, including the following needs:

> "To devise processing methods that incorporate more of the raw product into the finished product and to develop new products that utilize liquid and solid residues and/or upgrade existing by-products to value added products" (1).

This chapter will address these needs from the standpoint of using cell-free enzymes. Selected examples that illustrate the potential and the utility of enzymes in treating food processing waste will be presented. This will be followed by a discussion of the potential impact of recombinant DNA technology on enzyme cost and utility.

To date most approaches to the problems of food waste involve physical or chemical processing. Biological approaches to this problem have largely employed intact microorganisms either to produce a single cell protein (SCP) product from food residues or to aid in the treatment of wastewater streams by anaerobic digestion (6). However, an enzyme - based process offers, in theory, a highly selective, controlled and efficient process for use in food processing. Enzymes are quite specific and thus it is often possible to avoid undesirable side reactions that are associated with the use of less specific catalysts. The ability of enzymes to exert their affect at relatively low temperatures can also help to avoid undesirable side reactions. Since free enzymes can usually be used at low concentrations, their subsequent removal from the reaction system is often unnecessary or less difficult than is the case with less efficient catalysts which must be used at higher concentrations (5). Furthermore, immobilization of enzymes has been shown to improve enzyme efficiency under relevant process conditions (7).

The use of cell-free enzymes as a viable option to more conventional approaches is only recently being pursued. In part, the reason for this current interest rests in the availability of cheaper and better commercial enzyme preparations. Figure 17-1 illustrates the reduction in cost over an 8-year period for a representative commercial hydrolase. It is likely that this trend will continue and enzymes produced and/or modified by recombinant DNA technology will offer new opportunities for use in food processing and waste management.

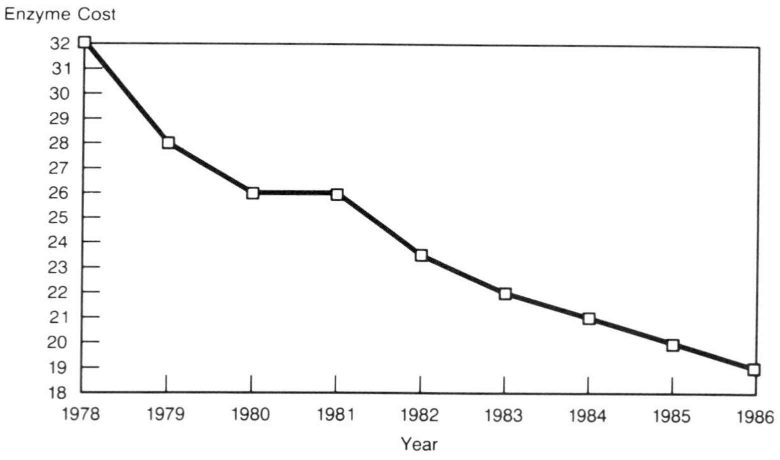

Figure 17-1. Reduction in Cost of Commerically Available Hydrolase

ENZYMES AND THEIR APPLICATIONS TO FOOD WASTE

The application of enzymes to the treatment of food waste has been discussed (5,8,9,10). For the purpose of this chapter selected examples of enzymes which have been used, or should be considered, in better utilization of the raw materials or in processing of food residuals to new value added products (Table 17-1) will be discussed. Clearly, the single most important enzyme class is the hydrolases.

Table 17-1. Application of Enzymes to Food Processing Waste

Enzyme	Application
Hydrolase	
Pectinase	Fruit, Vegetable
Amylase	Grain, Rice, Waste Water
Cellulase	Grain, Fruit, Vegetable
Mannanase	Coffee
Hemicellulase	Coffee, Grain, Fruit, Vegetable
Chitinase	Shellfish
Protease	Meat, Fish
Lactase	Dairy (Whey)
Oxido-Reductase	
Peroxidase (Ligninase)	Waste Water

Polysaccharidases

Polysaccharide hydrolases represent a diverse group including pectinase, amylase, cellulase, mannanase, hemicellulase and chitinase. These enzyme systems share a number of common characteristics; namely: 1) most active enzymes are derived from fungi; 2) extracellular, glycosylated proteins; 3) relatively stable (pH, T); 4) relatively simple (monomeric, no cofactors); and, 5) multicomponent enzyme systems, including exo- and endo-acting enzymes.

These enzyme systems consist of exoglycanases, endoglycanases and beta-glycosidases. The combined action of the enzyme components results in the breakdown of polymer substrate to monomer. With substrates that contain mixed linkages and/or points of branching along a linear polymer chain, the conversion process is more complicated. Furthermore, the multi-component nature of these enzyme systems contributes to the high degree of variability between enzyme preparations, which is not necessarily apparent by the activities reported. In some cases we do not have the ability to distinguish separate components by activity.

The potential for using single isolated enzymes should not be overlooked. Purified enzymes have more limited specificity and reactivity and, in some cases, can impart specific structural changes without causing significant depolymerization. This can result in changes in the functional properites of food residues which might prove useful in certain applications. The usefulness of polysaccharidases for particular applications can only be assessed properly with a thorough understanding of the enzyme preparation and testing on relevant substrates.

Food residues from fruit and vegetables (3), grain (11) and shellfish (12) processing are rich in carbohydrate. The potential of using residues from the U.S. fruit and vegetable industry was examined in detail by Cooper (8). A summary of this study is shown in Table 17-2. It is clear that although the total figures are large, the amount of residue generated at any single geographic location is rather small. In addition, most of the residue is sold as animal feed and this does not therefore serve as an available raw material (8).

Table 17-2. US Fruit and Vegetable Residuals: (1970-1974 Data)

Amount (Million Tons)

	Total Harvested for Processing	Residual		Dry Weight Residual Carbohydrate		
	(Wet Weight)	(Wet Wt.)	(Dry Wt.)	Total	Crude Fiber	Extractives
Fruit	14.4	4.9	1.9	1.5	0.3	1.2
Vegetable	19.7	5.7	1.2	1.0	0.2	0.8

Use of Residuals (%)

	Animal Feed	By-Products	Land and Water Disposal
Fruit	84	5	11
Vegetable	76	0	24

However, if one re-examines the total process and considers the use of enzymes to help minimize residue, then an enzyme step, separate or in conjunction with another processing step, might prove useful. For example, it has been shown that treatment of fruits with enzymes (cellulase, SPSase, pectinase) increases juice yield and dissolved solids and improves the separation of residual solids from the juice product (13,14). In another example, chitinases have been used to depolymerize the shells of shellfish (12). The product of this reaction, N-acetyl glucosamine was found to be a good substrate for yeast SCP production.

Amylases have been used in the treatment of starch-containing food wastewaters (15,16,17). In one example, McCarty, Bales and Ramalingam (15) demonstrated the economic feasibility, at pilot plant scale, of recovering starch and process quality water from rice wastewater (15). The starch from the rice cookers (0.5-1.5%) was first concentrated using ultrafiltration to 14% solids and then brought to 20-30% solid by the addition of other waste starch solids. The resultant slurry was reacted with alpha-amylase and glucoamylase for a total residence time of six hours producing a glucose syrup. This syrup was either concentrated (80%) or used as a raw material for yeast alcohol production (15).

In other examples, amylases have been used in wastewater treatment but without attempting to remove the carbohydrate (16,17). Iwanowski and Linowska (16) found that adding a combination of purified amylases and proteases to activated sludge significantly reduced the treatment time. The application of amylases to activated sludge of food processing wastewater appears to be a promising area for further development.

Mannanases of varying specificity have been used the processing of coffee beans (18). It might be desirable, therefore, to have an enzyme to direct the conversion of linear 1,4-β-D mannan polymer of coffee beans (19) to soluble oligomers, but not to mannose, as mannose is known to give rise to bitterness. Work at Cetus Corporation showed that a component of the cellulase system reacts to a great extent on mannan and produces predominately mannobiose as product. This example illustrates the broad specificity of certain hydrolases. Thus, the availability of reasonably priced, specific (not crude) hydrolases should provide new applications for enzymes in food processing.

Lactases

Lactases have been used in the processing of dairy wastes (20,21,22,23,24). There are currently two commercial plants using immobilized lactase: Nutrisearch (Winchester, Kentucky) and Specialist Dairy Ingredients, SDI (Maelor, Whales, UK). Both processes use whey as the raw material (Figures 17-2 and 17-3). Whey is the lactose-rich effluent from the manufacture of cheese, casein, cottage cheese and whey proteins (23). There was some 37 billion pounds of whey produced in the U.S. in 1979, of which 1.8 billion pounds is lactose. The key step in both processes is the enzymatic hydrolysis of lactose to glucose and galactose.

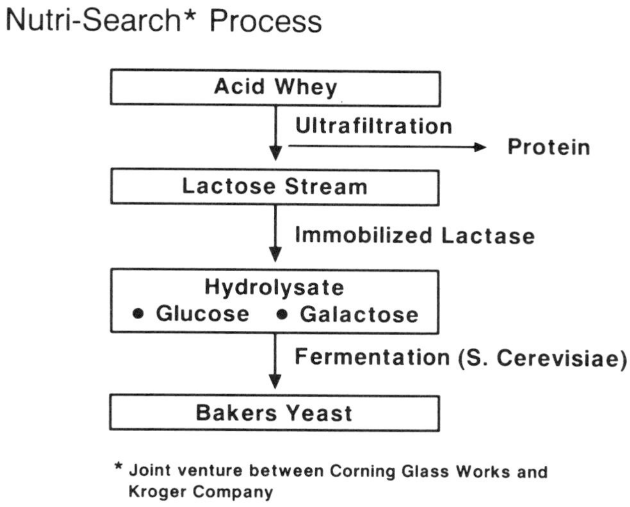

Figure 17-2. Schematic for Nutri-Search Process

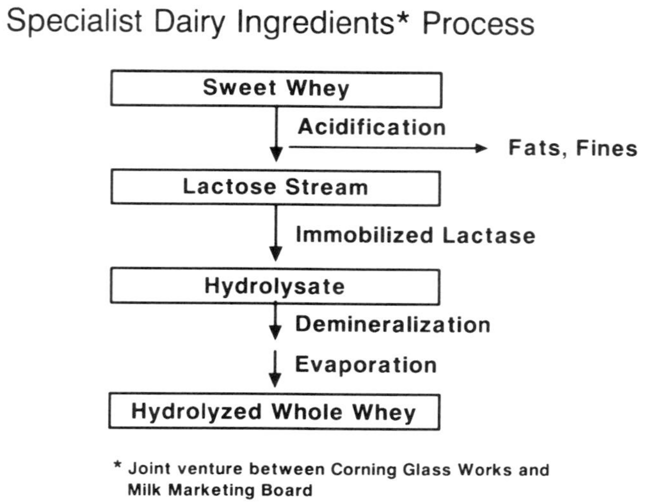

Figure 17-3. Schematic for Specialist Dairy Ingredients Process

Both commercial processes use the Corning immobilized lactase technology for this purpose.

The immobilized lactase system developed at Corning has been described in the literature (20,25). Lactase is derived from *Aspergillus niger*. The enzyme is covalently bound to a controlled-pore silica carrier using the silane gluteraldehyde technique of Weetall and Havewala (26). The performance of the immobilized lactase at commercial scale has been good. The same column has been in place since the SDI plant began opertion in May, 1985. This plant operates 16-20 hours/day at 360 liters/hour and acheives 80% hydrolysis of lactose (20). While the system is cleaned by standard dairy cleaners, the lactase bed is fluidized every day for 30 minutes with dilute acetic acid.

Proteases

A third group of hydrolases which have been used widely in the food industry are the proteases. In addition to the use of proteases discussed previously for wastewater treatment, they have been used successfully in fish waste processing and processing of meat byproducts.

Fish waste in the form of trash fish (26) and stickwater (27,28) has been "recycled" and upgraded for use as fish meal. A commercial protease preparation, available from Genencor, has been used to prepare fish pellets (Figure 17-4). The enzyme solubilizes the proteins of comminuted whole fish. The solubilized proteins can be recovered in reasonable yield to a liquid concentrate or dry solids of nutritional value for use as fish food.

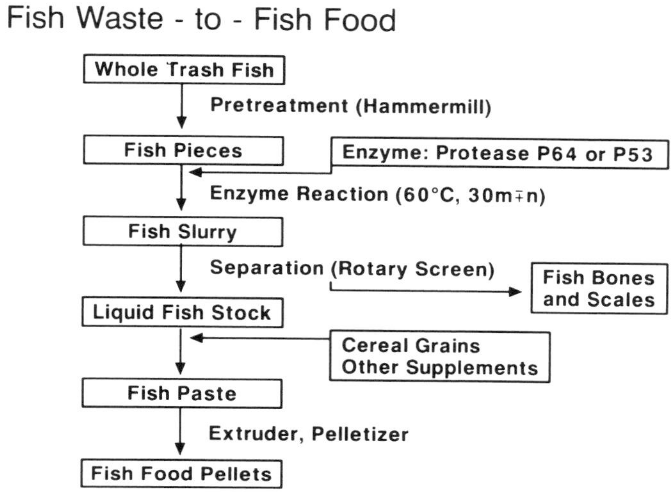

Figure 17-4. Schematic Process for Treating Trash Fish

In a similar manner, proteases have been added to the liquid phase in fish meal production, called stickwater, in order to reduce viscosity and, thereby, increase the final solids content (27). The recovery of fish oils, improves the efficiency of evaporator tubes and decreases the deposition of sludges in centrifuges. For this application the protease, Rhozyme 64, has been used at concentrations of 0.5% based on fish solids and a holding time of 30-60 minutes at 50-60°C (27).

Proteases have also been used in processing meat wastes (29-30). Indeed, there is a need for recovering proteins and for improving conservation and reuse of water from the standpoint of minimizing waste volumes and conserving energy. Blood is one of the most valuable and under-utilized protein sources generated by the meat industry. The proteins are already in solution and its limitations in use are only its dilute concentrations and intense red color (30).

The resulting blood cell hydrolysate can be used in a similar manner as other protein hydrolysates. It also has been found to be good temperature and freeze/thaw characteristics with respect to flavor.

Oxido-reductases

Oxido-reductases are enzymes which catalyze oxidation and reduction reactions. Among the oxido-reductases, peroxidases could be useful in the treatment of food processing wastewater. Klibanov (31, 32) has shown that horseradish peroxidase and hydrogen peroxide can be used to purify wastewater streams. The enzyme causes the polymerization and the precipitation of phenolics from wastewater streams.

Moreover, ligninases of peroxidase-type are being developed at Repligen (Cambridge, MA). Ligninases have been shown to effectively decolorize pulp mill effluents. This application may have a significant impact on the treatment of food processing waste.

FUTURE APPROACHES TO WASTE MANAGEMENT

Continued development in enzyme production and in the ability to manipulate enzyme structure should provide cheaper and more functional enzymes. The use of molecular biololgy in conjunction with protein biochemistry and X-ray crystallography will likely give rise to a new generation of enzymes better tailored to specific food applications.

For example cellulases could find wide use in food processing, but because of low specific activity and high cost they have not been tested. Recent studies in fermentation development by Dr. Bill Dean and Mr. Henry Heinsohn of Genencor, Inc., have resulted in the production of cellulases at commercial scale (33) at productivities in far excess of the best reported in the literature (34,35). The use of recombinant DNA techniques for the molecular cloning, gene characterization and expression of *Trichoderma* cellulases (36,37,38) has revealed some interesting patterns among the cellulase gene family. The application of protein engineering techniques is expected to produce second generation cellulases of higher activity, better stability and lower cost.

The molecular biology approach to protein engineering can help to determine where structural changes should be made. Examples of structural changes which may lead to changes in function are given in Table 17-3. At the current level of understanding, the functional consequence of a structural change cannot be predicted with any degree of certainty. However, there are examples like those given by Dr. Ronald Wetzel of Genentech where structural modifications of the gene at specific codons have resulted in the desired functional change at the protein level.

In addition, the development of fungal transformation systems which will allow the expression of "modified" genes at high yield will help to make a wide variety of "tailored" enzymes available at reasonable cost. Genencor indeed has already developed a secretion system for heterologous enzymes from an *Aspergillus* species (39).

Molecular cloning, characterization and expression in fungal hosts of genes encoding enzymes will result in lower cost and more stable enzymes under relevant process conditions. This will encourage the testing and use of natural and recombinant enzymes in the following areas: 1) decrease food waste through more complete utilization of raw materials; 2) decrease food waste through enzymatic processing of waste streams to give higher value by-products; 3) aid in processing of raw materials; and, 4) aid in clean-up of food waste by-product streams.

Table 17-3. Examples of Potential Structure Changes in Protein Engineering

Targets for Structural Change	Possible Affect on Function
Regional mutagenesis (alteration or deleteion of amino acid)	$>V_m$ $<K_m$ $\triangle$stability $\triangle$pH
Changing specific amino acids	$>V_m$ $<K_m$ $\triangle$stability $\triangle$pH
Changing type and extent of glycosylation	$\triangle$stability $\triangle$solubility
Expression in filamentous fungi	>amount of desired enzyme(s)

ACKNOWLEDGEMENT

The author wishes to gratefully acknowledge the following in the preparation of the manuscript for technical assistance , W. Pitcher, S. Silver, B. Dean, D. Munnucke and J. Zelko of Genencor, Inc., A. Ramalingam of General Foods, R. Farrell of Repligen, Inc., G. A. Levine of Nutrisearch and R.J. Price of U.C. Davis; for secretarial assistance, M. Pasternik and R. Whiteley.

REFERENCES

1. Cooper, J.L., Dostol, D. and Carroad, P.A. Report of the workshop session on environmental issues. Food Technology 39:33R-35R (1985)
2. Hopper, P.F. Implementing food research recommendations: Industrial viewpoint. Food Technology 39:39R-40R (1985)
3. Karel M., Wellbourn, J.L. and Singh, R.P. Report of the workshop session on innovative methods of food preservation/processing. Food Technology 39:17R-20R (1985)
4. Sinskey, A.J., Knorr, D.W. and Stevens, A. Report of the workshop session on biotechnology. Food Technology 39:23R-24R (1985)
5. Reed, G. (ed.) "Enzymes in Food Processing," New York, Academic Press (1980)
6. Herzka, A. and Booth, R.G. (eds) "Food Industry Wastes: Disposal and Recovery," New Jersey, Applied Science Publishers (1981)
7. Pitcher, Jr., W.H. (ed) "Immobilized Enzymes for Food Processing," Florida, CRC Press, Inc. (1980)
8. Cooper, J.L. The potential of food processing solid wastes as a source of cellulose for enzymatic conversion. Paper presented at Symposium on Enzymatic Conversion of Cellulosic Materials: Technology and Applications, Newton, MA. (1975)
9. Tsao, G.T. In: "Enzymes: Use and Control in Foods," pp 12-1 - 12-12, IFT, Chicago (1976)

10. Chemical and Engineering News. Mammalian protein: Secretion gained in filamentous fungi. Chemical & Engineering News 63:4 (1985)
11. Ladisch, M.R., Lin, K.W., Voloch, M. and Tsao, G.T. Process considerations in the enzymatic hydrolysis of biomass. Enzyme Microbial Technology 5:82-102 (1983)
12. Cosio, I.G., Fisher, R.A. and Carroad, P.A. Bioconversion of shellfish chitin waste: Waste pretreatment, enzyme production, process design, and economic analysis. Journal of Food Science 47:901-905 (1982)
13. Kilara, R. Enzymes and their uses in the processed food industry: A review. Process Biochemistry 17:36-41 (1982)
14. Dorreich; U.S. Patent 4,483,875; Nov. 20, 1984; assigned to Novo Industri A/S, Denmark
15. McCarty, L., Bales, A. and Ramalingam, A. Recovery of starch and process quality water from rice processing effluent. Paper presented at AICHE Summer Meeting Abstracts, Colorado (1983)
16. Iwanowski, H. and Linowska, M. An attempt of application of enzymes to wastewater treatment. Environment Protection Engineering 10:39-46 (1981)
17. R.M. El-Sayed. U.S. Patent 4,237,003. Dec. 2, 1980. assigned to R.M. El-Sayed
18. Ehlers, G.M. Possible applications of enzymes in coffee processing. Paper presented at ASIC 9 Colloque, Lourdes (1980)
19. Clifford, M.N. The composition of green and roasted coffee besns. Process Biochemistry 11:13-19 (1976)
20. Weetall, H.H. and Royer, G.P. (eds) In: "Enzyme Engineering," Vol. 5, pp 279-291, Plenum Press, New York (1980)
21. Sprossler, B. and Plainer, H. Immobilized lactase for processing whey. Food Technology 37:93-95 (1983)
22. Jelen, P. Reprocessing of whey and other dairy wastes for use as food ingredients. Food Technology 37:81-84 (1983)
23. Muller, L.L. Physico-chemical separation processes for food wastes. CSIRO Fd Res. Q. 41:37-43 (1981)
24. Varna, N. and Chawla, R.C. In: "Industrial Waste: Proceedings of the Thirteenth Mid-Atlantic Conference," pp 554, U. Delaware, Delaware (1981)
25. Pitcher, W.H. Immobilized Lactase Systems in Proceedings Whey Products Conference. USDA ERRC Publ #996:104-115 (1974)
26. Weetall, H.H. and Havewala, N.B. In: "Enzyme Engineering" (L.B. Wingard, ed.), Vol. 3, p 249, Wiley, New York (1972)
27. Genencor, Inc. Technical Bulletin-Rhozyme Proteases. Genencor, Inc., 180 Kimball Way, South San Francisco, CA 94080 (1985)
28. Jacobsen, F. and Rasmussen, O. Energy-savings through enzymatic treatment of stickwater in the fish meal industry. Process Biochemistry 19:165-169 (1984)
29. Leward, D.A. and Lawrie, R.A. Recovery and utilisation of by-product proteins of the meat industry. Journal of Chemical Technology and Biotechnology 34B:223-228 (1984)
30. Hald-Christensen et al.. U.S. Patent 4.262,022. April 14, 1981. assigned to Hald-Christensen et al.
31. Alberti, B.N. and Klibanov, A., Enzymatic Removal of Dissolved Aromatics from Industrial Aqueous Effluents, Biotechnology and Bioengineering Symposium 11:373-379 (1981)
32. Klibanov, A., Tu, T.M. and Scott K. P., Reports: Peroxidase-Catalyzed Removal of Phenols from Coal-Conversion Waste Waters, Science 221:259-261 (1983)
33. Genencor, Inc. Genencor Cellulase 150L, Genencor Cellulase 250P. Genencor, Inc., 180 Kimball Way, South San Francisco, CA 94080 (1985)

34. Hendy, N.A., Wilke, C.R. and Blanch, H.W. Enhanced cellulase production in fedbatch culture of *Trichoderma reesei* C30. Enzyme Microb. Technol. 6:73-77 (1984)
35. Allen, A.L. and Andreotti, R.E. Cellulase production in continuous and fed-batch culture by *Trichoderma reesei* MCG80. Biotechnology and Bioengineering Symposium 12:451-459 (1982)
36. Shoemaker, S., Schweichart, V. Ladner, M., Gelfand, B., Kwok, S., Mayambr, K. and Innis, M. Molecular cloning of exocellobiohydrolase I derived from *Trichoderma reesei.* Biotechnology 1:691-696 (1983)
37. Shoemaker, S.P. In: "Biotech 84 USA," pp 593-600, Online Publications, UK (1984)
38. Teeri, T., Salovouri, I. and Knowles, J. The molecular cloning of the major cellulase gene from *Trichoderma reesei.* Bio/Technology 1:696-699 (1983)
39. Chemical and Engineering News. Mammalian protein: Secretion gained in filamentous fungi. Chemical & Engineering News 63:4 (1985)

Chapter 18

Biosensors for Biological Monitoring

Renee Fitts

INTRODUCTION

The ability to monitor foods for microbial contamination, the presence of toxins, or the levels of chemical components has dramatically changed in the last ten years. Classical microbiological assays were and are still designed with two goals in mind: quantitation of bacteria, yeast, molds, or viruses present in a food sample, and the ability to specifically identify those pathogens. There have been recent improvements in classical microbiological techniques for both enumeration and identification that have included more rapid plating techniques, automation of these processes, and computerized aids to the interpretation of results. With regard to the detection and enumeration of bacteria, departures from microbiological assays include impedance measurements, bioluminescence assays, and so on. Two methods that are truly a result of genetic engineering applications to problems in the food industry can now be added to the biosensor repertoire: the use of DNA probes and monoclonal antibodies as diagnostic tools for the specific identification of microorganisms. Monoclonal antibodies have provided an additional benefit of being useful reagents for the detection of nonmicrobial components of foods.

MICROBIOLOGICAL ASSAYS FOR MICROORGANISMS

Throughout this discussion, the enteric pathogens provide a good example. In food, *Salmonella* has received substantial attention from the biotechnology arena, partly because it is such a prevalent pathogen, but also because *Salmonella*-testing programs are so well established in the food industry, and governmental guidelines exist for the detection of *Salmonella* in foods.

The microbiological assessment of *Salmonella* spp., for example, in foods requires a well-defined series of culture and biochemical steps that has been carefully outlined by the AOAC (1). Typically, examination of a food sample for this organism begins with a 24 hour pre-enrichment culture that allows the salmonellae that are present an opportunity to recover from their stressed state in the food sample and to grow to numbers that are detectable. The next day, an aliquot from this pre-enrichment culture is used to inoculate a medium that selectively enriches the microbial population for salmonellae. This medium is usually selenite cystine broth or tetrathionate broth. After 24 hours of growth, aliquots of the selective enrichment culture are streaked onto agar media that further selectively enriches for salmonellae and allows presumptive identification of *Salmonella* colonies. Hektoen enteric agar and bismuth sulfite agar are the usual choices for these platings. After 24 to 48 hours, the agar plates are examined for presumptive *Salmonella* colonies, which are picked from the plates and are used to inoculate several media that allow further biochemical characterization. Finally, bacterial isolates that are found to be positive after this regimen are serologically identified for somatic (O) and flagellar (H) antigens.

In the case of *Salmonella*, one should note that because the testing procedure amplifies at every step, it is possible at the end, to state that a food sample contains salmonellae, even if the initial level of contamination was very low. Theoretically, just one *Salmonella* organism in the tested food sample (often about 400 grams) is enough to give at the end of all of the amplification steps a positive result. However, because the flora of food samples is often very complicated, the competition for survival in the food sample and culture steps can be fierce, and any salmonellae present may not outgrow the competitive organisms, despite the selective enrichment steps. Thus, there is probably a threshold of salmonellae numbers at each step that is necessary for detection in this method. A great deal of effort has gone into improving the selection for salmonellae in foods, although in the end the regimen described above is still considered efficacious (2). The reader should note that the time involved from start to finish is approximately five days, and that in the end one has a qualitative, but not quantitative assay.

Aside from changing the selective media involved in *Salmonella* testing, an emphasis has been placed on faster handling methods of samples. For example, food samples can be treated with a variety of enzymes to enhance their filterability, and can then be filtered onto hydrophobic grid membranes for subsequent culture steps (3,4,5,6). An alternative to traditional media for testing fermentation patterns of enteric isolates, such as triple sugar iron agar, is to use prepackaged test media, such as API or Enterotek, along with their respective databases to interpret test results.

NON-MICROBIOLOGICAL METHODS FOR DETECTING THE PRESENCE OF BACTERIA IN FOODS

In many applications, it is of interest to know if any bacteria of any kind are present, without regard to specific identification of organisms that are present. In other words, there are typically two questions that one asks about the microbiological flora of food: how many, and what kinds of microorganisms are present. Traditionally, the first question of "how many" has been done with "total plate counts," where an aliquot of a food culture or emulsion is plated on a non-selective medium, and after a period of incubation, all organisms present are counted. Several alternatives to this methodology exist that are departures from microbiological assays, but still may require a culture step. One such alternative is the use of impedance measurements of food samples, especially milk (7,8).

Impedance is the total opposition of an electrical circuit to the flow of an alternating current of a single frequency. It is a combination of resistance and reactance, and is measured in ohms. If a population of microorganisms is provided with nutrients that are nonelectrolytes, such as lactose, and these microorganisms can utilize that nutrient and thereby convert it to an ionic form, i.e., lactic acid, then the impedance of the culture will change as a result of the metabolic activity of the microbial population. There appears to be a threshold for such measurements of about 10^6 to 10^7 organisms per ml. The time that is required for a culture to grow to that threshold is called the impedance detection time or IDT. The performance of an impedance assay for the presence of microorganisms in a food sample involves the inoculation of special media, and then the measurement of impedance over a period of about 20 hours. This method is fully automated and available commercially (9,10).

Another alternative is the measurement of adenosine triphosphate (ATP) in a sample. Since all known microorganisms contain ATP, an assessment of the microbial content of a sample can be made by measuring the amount of ATP present. Bioluminescence is often used for such as assay. The reaction requires the compound luciferin, which is a heterocyclic phenol, ATP, and the enzyme firefly luciferase. Luciferin is oxidized in this reaction, and thereby emits light. This assay can be easily automated and is very sensitive, and is still under development for use in the food industry (11,12,13).

DNA PROBES

The use of nucleic acid sequences as probes in hybridization assays for the detection of specific microorganisms, such as *Salmonella*, has tremendous versatility in many systems. Virtually any organism of interest can be detected in complex samples: viruses, bacteria, protozoans, and so on, in foods, water, body tissues and fluids, etc. Although this technology is very new, there are numerous examples of how well DNA probes work, and commercial test kits are available and are in use in the field.

The most fundamental requirements for a DNA hybridization assay are an appropriate DNA sequence to use as a probe, a method of labeling that probe molecule, and a method for detecting the label. The principles of DNA hybridization assays have been extensively reviewed elsewhere (14,15,16). The choice of a DNA sequence to use as a probe molecule depends on the properties of the organism one wishes to detect. In some instances, the target organism has some obvious feature that would generate a suitable probe. For example, if one wishes to detect the presence of toxigenic *Escherichia coli*, a good choice would be to clone the toxin gene from that organism and use that as a probe (17). When one wishes to detect virulent forms of an organism, if virulence factors are known for that organism, the gene for those factors can be used as probes (18). In the case of viruses, all or part of the viral genome may be used. Hepatitis B virus, for example, is a straightforward example of using the entire viral genome as a probe (19,20). A larger virus, such as cytomegalovirus, shares DNA sequence homology with other viruses, thus, it is necessary in that case to subclone the viral genome into smaller pieces, and then identify those pieces that do not crosshybridize with other sources of DNA (21).

When a eukaryotic organism is under investigation, organelle DNA may provide good probe sequences. For example, kinetoplast DNA from *Leishmania* has been very useful in diagnosing *Leishmania* infections (22).

When the target organism does not have such a readily identifiable feature to use as a probe, one other possibility has been the use of cryptic plasmids. This approach has provide a probe for *Neisseria gonorrhoeae* (23).

On the other hand, there will be instances where the search for a probe sequence will be much harder. For a *Salmonella* probe, for example, it was necessary to make a library of the *Salmonella* genome by cloning the bacterial DNA into small pieces and then examining these clones for sequence homology to other organisms. Those cloned DNA sequences which did not cross-hybridize to other DNAs could then serve as good probes (24,25). A similar approach has been used to find a probe for *Legionella pneumophilia* (26).

Once a probe sequence has been identified, it is necessary to choose a label or reporter molecule, a method for incorporating that reporter molecule into the probe sequence, and a method for detecting that probe sequence. A radioisotope, such as ^{32}P, is usually incorporated into the probe DNA by nick-translation (27) and is detected either by autoradiography or by counting in a scintillation counter. The vitamin biotin is frequently used as a non-isotopic reporter molecule by incorporation of biotinylated nucleotides into the DNA by nick-translation. After hybridization, an avidin-enzyme conjugate is added to the system. The avidin binds tightly to the biotin in the DNA. Finally, the substrate for the enzyme is added (28,29,30). A further alternative is the use of a hapten, such a dinitrophenol, as a reporter molecule. Haptens can be added to DNA either through nick-translation with substituted nucleotides, or by direct chemical modification of the DNA. After hybridization, the hapten is detected by addition of antihapten antibodies (31). It is also possible to directly attach enzymes, such as horseradish peroxidase or alkaline phosphatase, to DNA, thereby allowing detection through the simple addition of the enzyme substrate (32).

MONOCLONAL ANTIBODIES

Immunoassays for diagnostic purposes have been in use for many years, although those assays have typically used polyclonal antibodies. Monoclonal antibodies are wonderful tools for both researchers and diagnosticians. The work of Kohler and Milstein (33) has made possible the efficient production of these reagents for diagnostic assays and for the analysis of the structure of antigens and antibodies (34).

A monoclonal antibody to a particular antigen, such as *Salmonella* flagella, is obtained by first immunizing a mouse with that antigen. After the mouse has developed a strong antibody response to that antigen, the mouse is sacrificed and its spleen is removed. These spleen cells are then fused to myeloma cells *in vitro*. The hybrids, or hybridomas, that result from this fusion have the antibody-producing ability of the spleen cells and ability to be propagated indefinitely from the myeloma cells. Antibodies can be harvested directly from these hybridomas in culture, or hybridoma cells can be injected into mice to form ascites tumors, and then antibodies can be harvested from the ascites fluid.

There are a variety of assay formats in which these antibodies can be used. There are four basic formats for these assays: direct, indirect, competitive, or sandwich. The choice of format depends greatly on the particular application of the assay. Detection methods in immunoassays are similar to those described above for DNA probes: both isotopic and non-isotopic assays are feasible.

The specific example of a monoclonal-based immunoassay for *Salmonella* is worth discussing, since *Salmonella* detection is one of the few examples where both DNA probes and monoclonal antibodies have been used in the food industry and are commercially available. A monoclonal antibody directed against flagellar antigenic determinants has been developed into a test kit using a sandwich format. In this assay, the anti-*Salmonella* antibody is attached to a polystyrene bead. Extracts of food samples are mixed with these beads to allow *Salmonella* antigens to react with the antibody on the bead. After a washing step, a second antibody reagent specific for the *Salmonella* antigen, but labeled with an enzyme, is introduced to the system. Thus, the *Salmonella* antigen is "sandwiched" between antibodies. The substrate for the enzyme-labeled antibody can be added for the final detection step (35). This procedure is termed an enzyme-linked immunoassay, or ELISA.

DNA PROBES VS. IMMUNOASSAYS

Are there relative advantages of DNA hybridization assays and immunoassays that utilize monoclonal antibodies? Although assays can be devised with either of these reagents that will give a yes or no answer to the question of whether a particular organism is present, they intrinsically provide different information. In a DNA hybridization assay, a positive reaction indicates that the genetic information for some particular trait is there. For example, a probe that detects the presence of a toxin gene can indicate whether an organism has the potential to be toxigenic. On the other hand, an antibody directed against the toxin molecule itself can show that the genetic information for that toxin is actually being expressed. In many instances, one of these facts is more important than the other. Thus some diagnostic tests may be better suited to a DNA hybridization assay than an immunoassay, and vice versa. Both tests together give complementary information that can provide a much clearer picture of the status of a test organism, and thus, rather than compete with each other, these technologies can act in a cooperative fashion.

REFERENCES

1. FDA. "Bacteriological Analytical Manual for Foods". 5th ed. Food and Drug Administration, Washington, D.C. (1978)
2. Andrews, W.H. A review of culture methods and their relation to rapid methods for the detection of *Salmonella* in foods. Food. Tech. 39:77-82 (1985)
3. Brodky, M.H., Entis, P., Entis, M.P., Sharpe, A.N., and Jarvis, G.A. A determination of aerobic and yeast and mold counts in foods using an automated hydrophobic grid-membrane filter technique. J. Food Prot. 45:301-304 (1982)
4. Entis, P. Enumeration of total coliforms, fecal coliforms and *Escherichia coli* in foods by hydrophobic grid membrane filter: collaborative study. J. Assoc. Off. Anal. Chem. 67:812-823 (1984)
5. Entis, P., Brodsky, M.H., and Sharpe, A.N. Effect of prefiltration and enzyme treatment on membrane filtration of foods. J. Food Prot. 45: 8-11 (1982)
6. Sharpe, A.N. and Michaud, G.L. Hydrophobic grid-membrane filters: new approach to microbiological enumeration. Appl. Microbiol. 28: 223-225 (1974)
7. Eden, R., and Eden, G. "Impedance Microbiology". (A.N. Sharpe, ed.) John Wiley and Sons, New York (1984)
8. Bishop, J.R., White, C.H., and Firstenberg-Eden, R. Rapid impedimetric method for determining the potential shelf-life of pasteurized whole milk. J. Food Prot. 47:471-475 (1984)
9. Martins, S.B., Hodapp, S., Dufour, S.W., and Kraeger, S.J. Evaluation of rapid impedimetric method for determining the keeping quality of milk. J. Food Prot. 45:1221-1226 (1982)

10. Waes, G.M. and Bossuyt, R.G. Impedance measurements to detect bacteriophage problems in cheddar cheesemaking. J. Food Prot. 47: 349-351 (1984)
11. Karl, D.M. Cellular nucleotide measurements and applications in microbial ecology. Microbiol. Rev. 44:739-796 (1983)
12. Sharpe, A.N., Woodrow, M.N., and Jackson, A.K. Adenosine triphosphate levels in foods contaminated by bacteria. J. Appl. Bacteriol. 33:758-767 (1970)
13. Stannard, C.J., and Wood, J.M. The rapid estimation of microbial contamination of raw meat by measurement of adenosine triphosphate (ATP). J. Appl. Bacteriol. 55:429-438 (1983)
14. Meinkoth, J. and Wahl, G. Hybridization of nucleic acid immobilized on solid supports. Anal. Biochem. 138:267-284 (1984)
15. Fitts R. In: "Topics in Food Microbiology II: Methods". (T. Montville, ed.) CRC Press, Cleveland, Ohio (1986)
16. Fitts, R. In: "Food Microorganisms and Their Toxins - Developing Methodology". (N. J. Stern and M. Pierson, eds.) AVI Publishing Co., Westport, CT (1986)
17. Moseley S.L., Huq, I., Alim, A.R.M.A., So, M., Samadpour-Motalebi, M., and Falkow, S. Detection of enterotoxigenic *Escherichia coli* by DNA hybridization. J. Infect. Dis. 142:897-898 (1980)
18. Hill, W.E., Payne, W.L., and Aulesio, C.C.G. Detection and enumeration of virulent *Yersinia enterocolitica* in food by colony hybridization. Appl. Env. Microbiol. 46:636-641 (1981)
19. Berninger, M., Hammer, M., Hoyer, B., and Gerin, J.L. An assay for the detection of the DNA genome of hepatitis B virus in serum. J. Med. Virol. 9:57-68 (1982)
20. Scotto J., Hadchouel, M., Henry, S., Alvarez, F., Yvart, J., Tiollais, P., Bernard, O., and Brechot, C. Hepatitis B virus DNA in children's liver diseases.. detection by blot hybridizations in liver and serum. Gut 24:618-624 (1983)
21. Chou, S., and Merigan, T.C. Rapid detection and quantitation of human cytomegalovirus in urine through DNA hybridization. New Engl. J. Med. 380:921-925 (1983)
22. Wirth, D.F., and Pratt, D.M. Rapid identification of *Leishmania* species by specific hybridization of kinetoplast DNA in cutaneous lesions. Proc. Natl. Acad. Sci. USA 79:6999-7003 (1982)
23. Totten, P.A., Holmes, K.K., Handsfield, H.H., Knapp, P.L., Perine, P.L., and Falkow, S. DNA hybridization technique for the detection of *Neisseria gonorrhoeae* in men with urethritis. J. Infect. Dis. 148:462-471 (1983)
24. Fitts, R., Diamond, M., Hamilton, C., and Neri, M. DNA-DNA hybridization assay for the detection of *Salmonella* spp. in foods. App. Env. Micro. 46:1146-1151 (1983)
25. Fitts, R. Development of a DNA-DNA hybridization test for the presence of *Salmonella* in foods. Food Tech. 39:95-102 (1985)
26. Grimont, P.A.D., Grimont, F., Desplaces, N., and Tchen, P. DNA probe specific for *Legionella pneumophilia*. J. Clin. Micro. 21: 431-437 (1985)
27. Rigby, P.W.J., Dieckman, M., Rhodes, C., and Berg, P. Labeling deoxyribonucleic acid to high specific activity *in vitro* by nick-translation with DNA polymerase I. J. Mol. Biol. 113: 237-252 (1977)
28. Singer, R.H., and Ward, D.C. Actin gene expression visualized in chicken muscle tissue culture by using *in situ* hybridization with a biotinylated nucleotide analog. Proc. Natl. Acad. Sci. USA 7: 7331-7335 (1982)
29. Langer, P.R., Waldrop, A.A., and Ward, D.C. Enzymatic synthesis of biotin-labeled nucleotides: novel nucleic acid affinity probes. Proc. Natl. Acad. Sci. USA 78:6633-6637 (1981)

30. Broker, T.R., Angerer, L.M., Yen, P.H., Hershey, N.D., and Davidson, N. Electron microscopic visualization of tRNA genes with ferritinavidin: biotin labels. Nucleic Acids Res. 5:363-384 (1978)
31. Tchen, P., Fuchs, R.P.P., Sage, E., and Leng, M. Chemically modified nucleic acids as immunodetectable probes in hybridization experiments. Proc. Natl. Acad. Sci. USA 81:3466-3470 (1984)
32. Renz, M., and Kurz, C.A. A colorimetric method for DNA hybridization. Nucleic Acids Res. 12:3466-3470 (1984)
33. Kohler, G., and Milstein, C. Continuous cultures of fused cells secreting antibody of predefined specificity. Nature 256: 495-497 (1975)
34. Pollack, R.P., Teilaud, J-L., and Scharff, M.D. Monoclonal antibodies: a powerful tool for selecting and analyzing mutations in antigens and antibodies. Ann. Rev. Microbiol. 38:389-417 (1984)
35. Mattingly, J.A., Robison, B.J., Boehm, A., and Gehle, W.D. Use of monoclonal antibodies for the detection of *Salmonella* in foods. Food Tech. 39:90-94 (1985)

Chapter 19

Strategies for Commercialization of Biotechnology in the Food Industry

David Wheat

INTRODUCTION

In the past few years many applications and potential impacts of biotechnology in the food industry have surfaced. The food processing industry may follow the drug industry in the commitment of substantial resources to application of biotechnology. Many companies, however, have made hasty decisions regarding development of biotechnology-based products and processes in the food area. These companies, and others faced with decisions regarding future funding of food biotechnology programs, are asking themselves what the appropriate response to the promise of biotechnology should be for them. What can one do today to avoid being left behind as this promise becomes reality?

THE APPROPRIATE RESPONSE TO BIOTECHNOLOGY IN THE FOOD INDUSTRY

No participant in the food processing industry can afford to risk ignoring the use of biotechnology, which will have a pervasive impact in coming years. Nonetheless this impact will be incremental, and in many cases there will be situations where biotechnology will play a relatively minor role, rather than a revolutionary one. Many industry participants need not take immediate steps to get into biotechnology.

Applications Now Emphasize Cost Reduction

The applications of biotechnology in the food processing industry will be relatively focused in the intermediate term. Many of the efforts currently under way are appropriately directed toward cost reduction.

"Biotechnology" as used here includes gene splicing, protein engineering, immobilized cells and proteins, cell and tissue culture, biosensors and biotechnology-based tests, and a range of advanced biological techniques for genetic analysis and related research. These techniques can be used with two general commercial objectives: cost reduction and product differentiation. Cost reduction in the food industry might arise from improvements in raw material quality, process improvements based on biotechnology (such as new enzyme processes), use of lower cost ingredients, or upgrading of waste streams. Product differentiation could be based on use of proprietary raw materials or on exclusive processes and could play a significant role in the growing area of branded produce. However, the fruits of the application of biotechnology in this field are some years away.

It is not surprising that the focus of the food industry is on cost reduction, in contrast to the commonly perceived application of biotechnology in the pharmaceutical industry for the development of proprietary products. Table 19-1 demonstrates the basic structural and operational differences between the food and drug industries which reflect these differences in emphasis. One can expect the applications of biotechnology in the food processing industry to be quite different from those in the drug and chemical industries in years to come. Biotechnology firms and research managers will have to take these differences into account.

Table 19-1. Industry Comparison

Food Industry	Drug Industry
Marketing and Cost Driven	Technology Driven
R&D Less than 1% of Sales	R&D More than 10% of Sales
Process Oriented	Product Oriented
Little Patent Protection	Proprietary Products

Current Efforts to Apply Biotechnology

A review of the current programs to apply biotechnology to food-related problems, based on published reports, bears out the emphasis on cost reduction, but suggests some other applications as well. Tables 19-2 through 19-7 site examples of food industry involvement in biotechnology-related projects, illustrating both the diversity of applications of biotechnology and how projects addressing these applications have been structured.

Table 19-2 lists those food companies that have currently entered into collaborative research ventures with biotechnology companies for the improvement of crops as food processing raw materials. A recent development affecting improvement of plant varieties is the decision by the U.S. Board of Patent Appeals granting Molecular Genetics, Inc., the right to patent a new laboratory-derived corn variety high in tryptophan.

Table 19-2. Joint Ventures for Crop Improvement

Food Company	Biotechnology Company	Crop
Campbell Soup	Calgene	High Solids Tomato
Campbell Soup	DNA Plant Technology	High Solids Tomato
Del Monte	Int'l Plant Research Institute	High Solids Tomato
Kraft	DNA Plant Technology	"VegiSnax"
Heinz	ARCO	High Solids Tomato
McCormick	Native Plants Inc	Plant Cell Culture for Spices
United Fruit	DNA Plant Technology	Cloning Techniques for Oil Palm
Nestle	Calgene	Herbicide-Tolerant Soybean

Genetic improvement of microorganisms used in the food industry, either for use in fermentation processess and production of food supplements (Table 19-3), or for production of food ingredients, such as flavors and fragrances (Table 19-4) and sweeteners (Table 19-5), are active areas of investigation.

Table 19-3. Genetic Improvement of Microorganisms Used in Food Processing

Company	Application
RHM/ICI	Mycoprotein
Cyanotech	High Protein Algae
Microalgae International	High Protein Algae
Labatt Brewing Co.	Improved Brewing Yeast
Moet-Hennessy	Immobilized Yeast for Champagne

Table 19-4. Companies Involved in Flavor/Fragrance Technology

Company	Application
BASF/Fritzsche, Dodge & Olcott	Fruit-Based Flavors
Calgene	Mint Oil and Mint
Food Research Institute	Cell Culture, Quinine and Saffron
W.R. Grace/Synergen	rDNA for Flavors and Fragrances
IPRI	Fruit Flavors and Fragrances
Firmenich/DNAP	Cell Culture for Improved Flavor Sources

Table 19-5. Development of Sweeteners

Company	Application
Searle/Monsanto	Aspartame
Genex	Phenylalanine
CPC	Glucose Isomerase
Synthetech	Aspartame
Toyo Soda	Aspartame
Tate & Lyle and other sugar companies	Polypeptide Sweeteners

Several companies have become involved in the production of rennin derived from recombinant organisms (Table 19-6). It is likely that lower cost rennin from recombinant strains, with the properties of natural calf rennin, will be available to the cheese manufacturing industry. Because of the number of research programs in this field there may be an oversupply of this enzyme in the future, leading to price competition if production cost are low enough.

Table 19-6. Companies Involved in Production of Rennin

Celltech
Collaborative Research/Dow
Genencor/Chr. Hansen
Genex
Gist Brocades
Miles

Many of the more interesting programs are in the miscellaneous applications shown in Table 19-7. The Corning Joint Ventures include Nutrisearch with Kroger, and Specialty Dairy Products with the Milk Marketing Board in Britain (see Chapter 17). Dr. Cayle of Kraft has specifically discussed another approach to upgrading whey in Chapter 10. These efforts and others like them suggest the possibility of food companies diversifying more into non-food specialty chemicals derived from traditional food raw materials or by-products.

Table 19-7. Other Companies Involved in Biotechnology

Company	Application
Corning	Whey Products
Litton Bionetics	Monoclonal *Salmonella* Tests
Biocontrol Systems	*Salmonella* and Other Tests
Integrated Genetics	*Salmonella* DNA Probe

One other area of the application of biotechnology of particular interest to the food industry is the development of biotechnology-based sensors for process control. Workers at the Leicester Biocenter and Cambridge Consultants Ltd. in England, together with other researchers in academia and industry, are making substantial advances in biosensor technology.

These devices offer the promise of more accurate process control leading to improved product quality and/or reduced processing cost.

Impact of Biotechnology in Food Processing

There have been several estimates of the impact of biotechnology in the food industry, as shown in Table 19-8. Part of the range of these forecasts can be explained by the various criteria used by the forecasters to delimit their area of application studied. Some estimates, for example, may include the entire seed industry including non-food crops. Others may include the animal feed industry or others.

Table 19-8. Forecast of Worldwide Markets for Biotechnology-Derived Food and Agriculture Products

Source	$ (in millions)	Year
Business Communications	430	1990
Strategic, Inc.	4,500	1990
Policy Research	50,000 - 100,000	2000
Predicasts	101,000	1995
T.A. Sheets	21,300	2000

We must recognize that many of the products which will be coming to market which embody biotechnology will be replacement products. The example of rennin given above may not be unusual. Genetically engineered natural rennin may replace natural calf rennin from extraction, but unless it supplants microbial rennin the total value of sales will not exceed those of calf rennin today. If significant price declines occur due to price competition the rennin market may not grow much in dollar value. Yet a biotechnology-based product will have replaced traditional products. It is not reasonable, however, to count the cheese produced using genetically engineered rennin as a biotechnology product.

Arthur D. Little, Inc., estimates that the total value of biotechnology-based products associated with the U.S. food processing industry, including ingredients supplied to the industry, will be about $1 billion by 1990. This will be substantially dependent on sales of enzymes and sweeteners. By 1995 this total will rise to about $3 billion, as the use of other biotechnology-derived raw materials, ingredients, and fermentation processes become more widespread.

Developing the Appropriate Response

Food processing companies, faced with the situation described above, have choices to make. For many, the result of careful assessment of the role of key technologies in their ability to compete, and the coming impact of biotechnology on these technologies, may be a decision not to invest now in biotechnology. Others may decide to curtail existing commitments which do not offer adequate returns. The effort to assess the specific areas where biotechnology can provide competitive advantage is necessary. This assessment is best undertaken from the point of view of the company's existing product and technology base and strategic plan, not from an evaluation of the attractiveness of this or that interesting idea for the application of biotechnology.

Although some firms, such as Monsanto, Du Pont, and Ciba Geigy, have made substantial commitments to biotechnology, this will rarely be appropriate for a food processing company. The substantial activities of such participants as Tate & Lyle may have a realistic business justification. More appropriate in most cases will be focussed applications designed to enhance, defend, or build technological competitiveness in some key area vital to a company's long term position in its industry. Every company has such key areas of technology, and biotechnology can be applied to some. The key which unlocks the commercial potential of biotechnology for food companies is the discovery of the fit between existing technology needs and the realistic current uses of biotechnology.

Chapter 20

Cost Reductions in Food Processing Using Biotechnology

David Jackson

INTRODUCTION

This chapter will cover some opportunities for cost reduction in the food processing industry which have been occasioned by the development of new technologies with a focus on enzymes. There are several cases that have been developed enough to do a meaningful economic evaluation.

One can begin with a discussion of the advantages of the use of enzymes in the food processing industry. The most obvious reason for use is that they perform useful chemistry on various compounds which are relevant to the food and feed industries. These compounds include vitamins, amino acids, flavoring and fragrance agents and sugars and other sweeteners. In addition, some enzymes operate on biological macromolecules which are major components of food and feed including starches, other polysaccharides, gums and proteins. An advantage in using enzymes is that they operate as highly efficient, active and very selective catalysts under moderate conditions of temperature, pH, ionic strength, and solvent composition such that they are compatible with preserving the chemical integrity of the biochemical substances that they act on.

One of the more important reasons for using enzymes to deal with biological material is that the process preserves or often creates the biologically correct optical isomer (chiral component) necessary from a nutritional standpoint. This includes nearly all of the naturally occurring amino acids, most vitamins, all sugars, etc. This property is vitally important in many cases when enzymes are used in a synthetic mode because most nonenzymatic catalysts usually produce the racemic mixture which then requires separation and/or purification.

ENZYME USE AND COSTS

Enzymes are not used widely in the food processing industry in certain cases because of economic reasons. Of the thousands of enzymes present in the world that might have some relevance to food processing, many are too expensive to isolate in adequate quantities using traditional techniques. The development of rDNA technology, in particular, the ability to transfer a gene coding for a specific enzyme from nonmicrobial sources will help reduce this cost. Also, the introduction of immobilized enzyme engineering technologies should create a fundamentally new economic environment in which a variety of different applications for enzymes ought to be reassessed.

The current total world market for industrial enzymes, including the use of enzymes in laundry detergents is about 500 million dollars per year. Only about 20 enzymes account for the vast bulk of this market. One reason for the low number of enzymes in use is the inability to produce them in large enough quantities at low cost. There is a need to invest a great deal of time, effort and expense to improve the sourcing organism to the point that it could produce the enzyme with an economically reasonable yield. A second limitation is that enzymes, in general, have evolved to function optimally under conditions of moderate temperature and ionic strength in an aqueous solution and not necessarily under the conditions that are demanded by the economic constraints of whatever process they will be employed in. This particular constraint is not so important in the food processing industry as it is in some other industries, particularly the specialty chemical industry. Nonetheless these are real constraints. One obvious but expensive solution to this second problem is through recombinant DNA techniques and engineering as was discussed in Chapter 5. A key issue would be to get the cell to secrete the enzyme into the medium from which it is much easier and therefore more cost effective to recover. This secretion process also has some other advantages, in particular, the stability and activity of the enzyme that is secreted is often more desireable than the one produced in organisms where it is accumulated inside the cell.

CHYMOSIN AS AN EXAMPLE

Once produced, the next step is to transform these technological achievements into real products which make real profits. A good case study is that of the use of rennin in cheese making, a product under development at Genex as well as elsewhere. Rennin is used to initiate the process of curd formation in cheese making by coagulation of kappa-casein. There are two classes of enzymes used for this: 1) authentic calf rennin or chymosin, which is extracted from the fore-stomach of calves; and, 2) various enzymes isolated from a variety of different microorganisms. The authentic chymosin is typically used for the higher quality, long cured cheeses, while the microbial rennins are used for short cured and processed cheese products. The chymosin gene has been cloned and expressed in a variety of different microorganisms including *E. coli*, *Saccharomyces cerevisiae*, *Bacillus subtilis*, and *Aspergillus* species.

E. coli produce the chymosin in large inclusion bodies inside the cell. An appropriate recovery process to isolate the enzyme with a purity that compares with the authentic calf chymosin had to be developed. SDS polyacrylamide gel patterns for an industry standard chymosin are used for evaluation of the chymosins produced by genetic engineering. In general the recombinant products look quite good with respect to purity. Genex has conducted fairly extensive cheese testing for over three years and the results have been quite satisfactory.

PRICE OF GENETICALLY ENGINEERED PROTEINS

Proteins from genetic engineering can be classified in terms of their selling price. There are certain very high value proteins like the interferons, urokinases and lymphokines which are used in clinical medicine. In 1979 interferon was purchased by the American Cancer Society at a value which was in the billions of dollars per pound. These products and the clinical evaluations necessary can really be supported only by the pharmaceutical industry. The proteins which are of relevance to the food processing industry such as rennins, proteases, and amylases sell in the range of several hundred to a few thousand dollars per pound. Thus, for a recombinant enzyme product to be successful in the food market place, the economics of the production of the protein including regulatory clearance, should not exceed these lower values. In fact, the cost will probably have to be less by some modest margin. The major costs include the value of the fermentation medium, the fermentation yield of the enzyme per liter of medium, the scale of production, the cycle time for the fermentation, and the cost and yield of the recovery and purification process for the adequately pure final product. The most important factors are the yield of the protein in grams per liter of medium and whether the product is accumulated in the cell or is secreted by the cell into the growth medium.

COMPARISON OF PRODUCT COSTS

Table 20-1 shows the yield of some products (g/L) as a function of both the fermentation efficiency (yield of cells per liter of the fermentation medium), and as a function of the percent of the total cell protein represented by the protein of interest. These data are for a protein that is accumulated intracellularly. To meet the cost constraints for the food processing industry one needs to have about 3 to 5 grams/liter. This also construes the fact that one must achieve fermentation yields in the range of 100 grams of wet weight of cells per liter with total cell proteins being around 20%. Both of these have been achieved in some cases (for example see Chapter 16). They represent relatively difficult parameters to meet routinely.

Table 20-1. Yield of Protein X in g/L of Medium as Function of Fermentation Efficiency and Percent of Total Cell Protein as Protein X*

		GRAMS WET CELLS PER LITER OF MEDIUM			
		10	**20**	**50**	**100**
PERCENT	5	0.13	0.25	0.63	1.25
TOTAL CELL	10	0.25	0.50	1.25	2.50
PROTEIN AS	15	0.38	0.75	1.88	3.75
PROTEIN x	20	0.50	1.00	2.50	5.00

*Assumptions:
1. 50% of wet weight of cells is water
2. 50% of dry weight of cells is protein

The yields represented in Table 20-1 are for the unpurified state. Losses suffered during purification of the intracellularly accumulated protein is not a trivial problem, since many proteins that are produced at high yield in *E. coli* are produced in a non-active form so that the product recovery process has to involve a denaturation-renaturation step that has some inherent kinetic and thermodynamic limitations.

The contrasting situation is one in which the enzyme is produced as an extracellular protein. An economic analysis at two different scales, 5,000 kilos and 50,000 kilos per year of a protein is shown in Figure 20-1. As noted when the yield of enzyme increases, the cost decreases significantly. However, there is a much lower cost for a protein that is secreted into the extracellular medium. The assumptions in Figure 20-1 are for the intracellular process at 40 grams per liter dry cell weight, which corresponds to 80 to 100 grams per liter of wet cell weight, depending upon the microorganism and an assumption of a 40% yield in the product recovery and purification step in both cases. If one goes to larger scale production, the unit costs will obviously be lower in each case, but the dramatic difference between the unit cost of an intracellular protein and extracellular protein remains. Figure 20-2 further illustrates these points. The expected impact of increasing scale is observed, but again, the unit cost for an intracellular protein is significantly higher than unit cost of an extracellularly secreted protein.

The organisms most frequently used for the secretion of recombinant proteins are yeast, other fungi, and *Bacillus subtilis.* In such systems one typically gets higher titers and lower costs with equivalent cellular protein synthetic capacity. Because the protein is secreted, product recovery processes are much simpler and tend to be less costly. The secreted protein is generally soluble and is typically secreted in its native form. Yeast have the capability of glycosylating secreted proteins, which is important for some pharmaceutical applications. In addition, the knowledge base of large scale fermentations with bacilli and yeast is quite extensive.

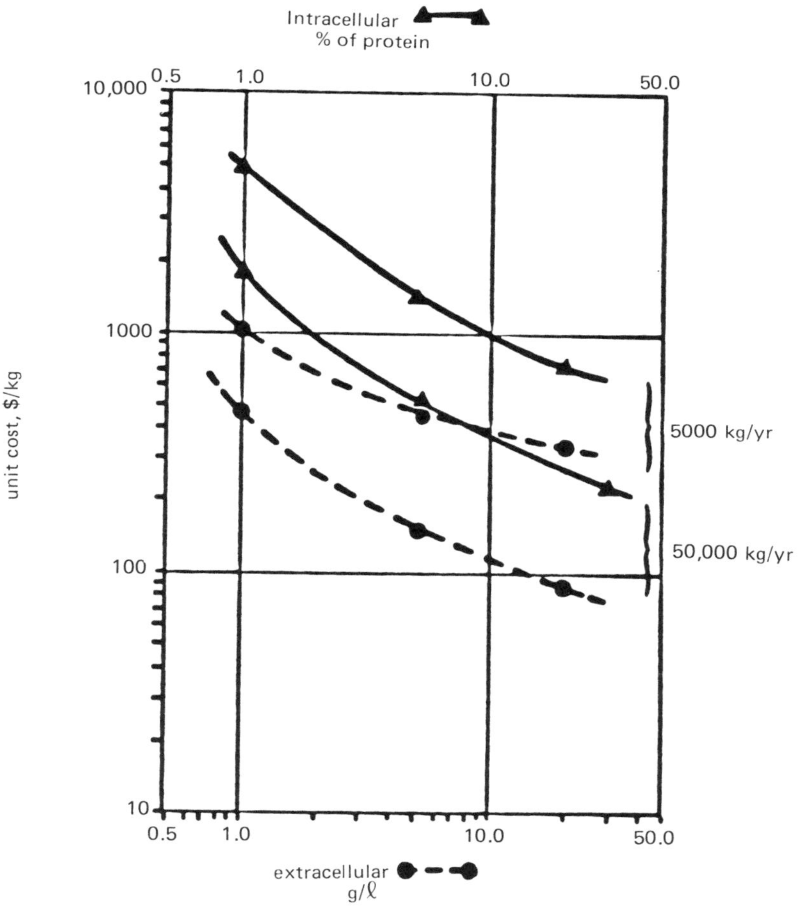

Figure 20-1. Relationship of Unit Cost to Protein Expression Level

There are also some disadvantages to the secretion processes. Secreted proteins tend to be relatively unstable in the presence of extracellular proteases produced by yeast and bacilli. There are a variety of successful genetic strategies that can deal with this. The other nontrivial problem is that bacilli and yeast are still not as easy to manipulate as *E. coli*. This tends to increase the cost of the research and development required to construct suitable organisms. Table 20-2 summarizes the advantages and disadvantages of secretion in protein production.

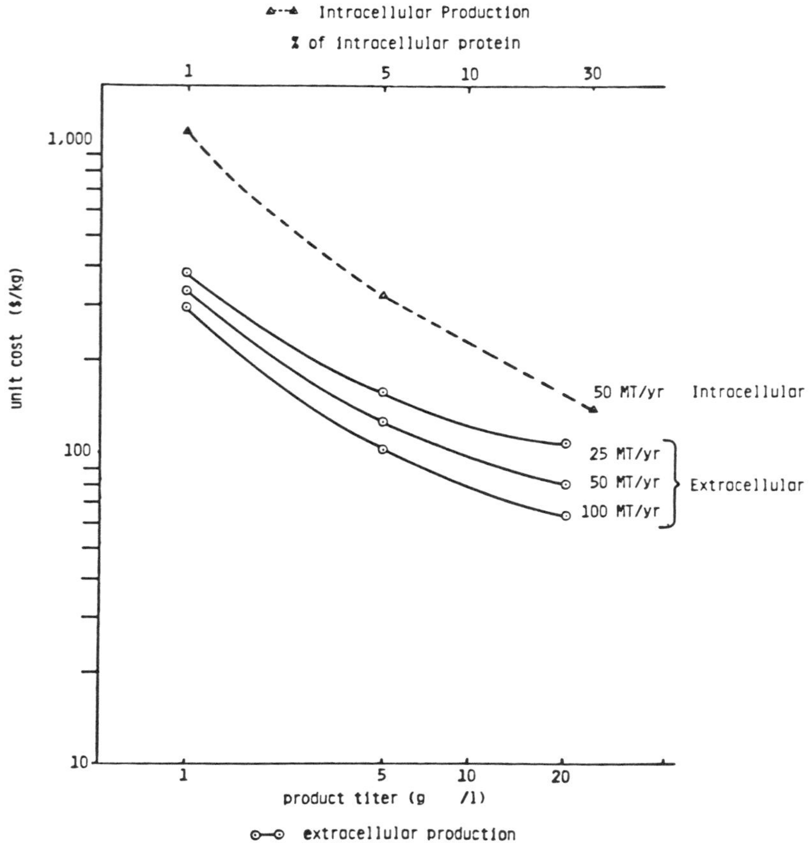

Figure 20-2. Production Costs of Intracellular vs. Extracellular Proteins

Table 20-2. Potential Advantages and Disadvantages of Secretion in Protein Production

ADVANTAGES:

1. Higher titer, lower cost
2. Simpler, lower cost recovery processes
3. Processing can give mature protein without leader sequences or N-terminal methionine
4. Secreted protein is generally soluble and in a native conformation
5. With yeast some secreted proteins can also be glycosylated
6. Freedom from endotoxins
7. Knowledge base with bacilli, yeast in industrial fermentations

DISADVANTAGES:

1. Instability of secreted protein to extracellular proteases
2. Bacilli and yeast are still not as easy to manipulate genetically as *E. coli*

IMMOBILIZED ENZYME TECHNOLOGY

The food industry has seen some success in the immobilization of enzymes in bioreactors. The most successful process is the use of glucose isomerase for the final step in the production of high fructose corn syrup. There are a number of companies and academic laboratories working on immobilizing the second enzyme used in this process, glucoamylase, in order to increase yield from corn starch.

The goals in any immobilization process are to increase the stability of the enzyme, increase the ability to recycle the enzyme, and easily separate the enzyme from the product, all of which have economic benefits. A typical configuration for a bioreactor is that of non-viable cells or enzymes immobilized to some sort of solid matrix as a support, with the immobilized biocatalyst then being packed into a column. A process stream of reactants is then pumped through the column and the immobilized biocatalyst converts the feedstock to the product. Genex has developed such a system for the conversion of fumaric acid plus ammonia to L-aspartic acid, using the enzyme aspartase from *E. coli* as the biocatalyst. This bioreactor is both very stable and very productive. A column of 12 liters bed volume packed with this biocatalyst can make up to 70 tons of L-aspartic acid per year under manufacturing conditions. Moreover, the process stream is very clean, the concentration of reactants in the process stream is high, and the recovery process is consequently very simple. All of these characteristics lower costs. In addition, the biocatalyst is very stable in its immobilized form. After more than half a year of continuous operation, over 90% of the initial biocatalyst activity remained. The example of aspartase is almost certainly an exceptional one. Not all immobilized enzymes have half lives in excess of a year. However, increases in stability upon suitable immobilization are not at all unusual, and research strategies that anticipate them are appropriate.

The economic impact of an immobilization as successful as the one cited here can be substantial. Fermentation, product recovery, waste disposal, and capital costs are all lowered significantly. Table 20-3 compares this bioreactor process for the production of aspartic acid with a typical batch fermentation process to produce the same amino acid. The bottom line is that the unit cost of production is about 40% lower with the bioreactor process.

Figure 20-3 shows data demonstrating the economic impact of enzyme immobilization. These two processes have been running commercially with a free enzyme and then were converted to immobilized enzyme processes. One is for the production of an amino acid and the other is for the deacylation of an amino acid. In each case the actual savings as reflected in unit costs were in excess of 40%.

Table 20-3. Process Economics: Amino Acid Production by Batch Fermentation and by an Immobilized Cell System

	Production Costs ($ Millions)	
	Batch Fermentation	Immobilized Cell Systems
Raw Materials	3.8	3.0
Labor	1.5	1.1
Utilities	0.2	0.1
Equipment	1.6	0.7
Buildings	0.3	0.1
Direct Costs	7.4	5.0
Overhead (60% of labor)	0.9	0.7
Total costs	8.3	5.7
Annual Production	2,000,000 lb	2,000,000 lb
Unit Cost	$4.15/lb	$2.85/lb

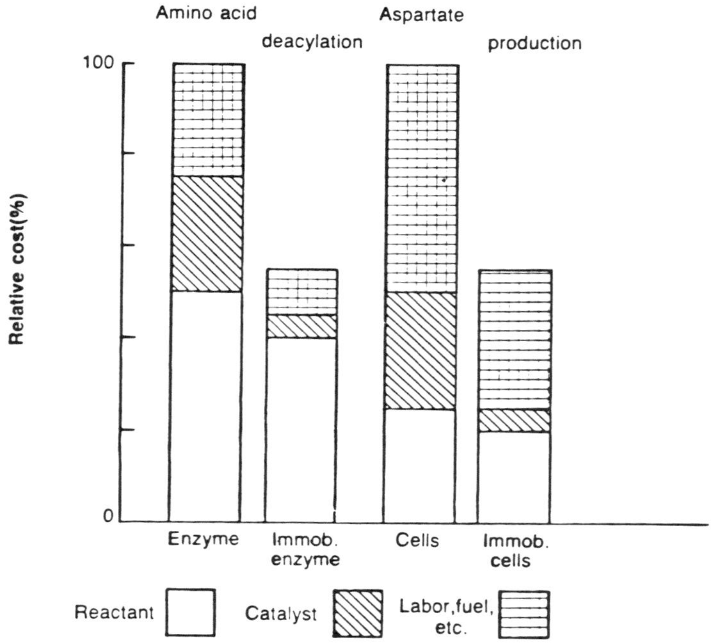

Figure 20-3. Relative Operating Cost for Immobilized vs. Non-Immobilized Systems

One of the major advantages of enzyme based bioreactors as opposed to fermentations involving viable cells is that the concentration of reactants and products in the process stream can often be substantially higher than cells can tolerate. Table 20-4 lists the most efficient fermentation processes known in terms of final titer of product in the fermentation broth. Concentrations of product in excess of 10% are very rare. In contrast, many enzyme based bioreactors operate at concentrations of 20% or higher (and do so with process streams that are much cleaner and less complex than a fermentation broth). As an example, Table 20-5 shows data from a bioreactor developed by Genex to convert glycine plus formaldehyde to the amino acid L-serine. L-serine is used as a precursor in another bioreactor process to make L-tryptophan, an amino acid with a potential market in excess of $100 million as an animal feed supplement. Table 20-5 shows that it is possible to perform this conversion in a bioreactor using the enzyme serine-hydroxy-methyl-transferase at reactant concentrations of 45% and achieve molar yields on the order of 90% (based on glycine). This is true even though formaldehyde is a highly reactive compound with a substantial capacity to inactivate enzymes.

Table 20-4. Titers for Selected Industrial Fermentations

Product	Microorganism	Titer (grams per liter)
Citric acid	*Aspergillus niger, Candida*	150
MSG	Coryneforms	>100
Ethanol	Yeasts	100
L-lysine	Coryneforms	75
Inosine	*Brevibacterium ammoniagenes*	30
IMP	*Brevibacterium ammoniagenes*	19
Penicillin	*Penicillin chrysogenum*	>30
Cellulase	*Trichoderma reesei*	>20
Alkaline protease	*Bacillus subtilis*	14

Table 20-5. Conversion of Glycine Plus Formaldehyde to L-Serine Using Serine-hydroxy-methyl-transferase in a Bioreactor

Initial glycine concentration (g/L)	Final glycine concentration (g/L)	Final L-serine concentration (g/L)	Molar conversion (per cent)
225	19.5	280	90
300	23.5	339	91
375	43.5	402	87
450	43.5	453	88

The L-serine made in this bioreactor is subsequently combined with indole in another bioreactor to produce L-tryptophan, using the enzyme tryptophanase. This bioreactor illustrates another advantage of using enzymes as opposed to viable cells. Both indole and tryptophan are relatively insoluble compounds and indole is quite toxic to cells. Genex has produced tryptophan at concentrations of 190 g/L in its bioreactor, about 15 times the solubility limit. The bioreactor in fact converts a slurry of indole into a slurry of tryptophan, conditions under which few cells could survive but which have many economic benefits.

SUMMARY OF BIOREACTOR ADVANTAGES/DISADVANTAGES

The advantages of bioreactors using enzymes relative to fermenters are: 1) higher concentrations of reactants in the process stream, which lead to lower fixed capital costs; 2) cleaner, less complex process streams, which leads to much simpler product recovery processes and lower costs; 3) lower capital costs per unit product; 4) bioreactors tend to lend themselves to continuous rather than batch processing approaches resulting in significant savings due to automation; and, 5) the simplicity of the process stream in terms of biological oxygen demand in particular tends to mean that the waste treatment requirements are less stringent for a bioreactor process stream than they are for the same product produced in a batch fermentation. As waste treatment costs become an increasingly larger part of the overall economic evaluation this factor will assume greater and greater importance.

There are disadvantages of bioreactors compared to fermenters as well. The most significant disadvantage is that the raw materials cost for bioreactors is typically higher than for batch fermentations which run on glucose. The economic reality is that in spite of all the other disadvantages of a batch fermentation it may be the process of choice. One must do that analysis on a case by case basis.

Another fact in the fermentation industry is that there is substantial installed and depreciated fermentation capacity available. The overall economics from an individual company's point of view have to take that into account. Typically installed and depreciated capacity available for large scale bioreactor approaches do not exist. Finally, there is always the problem of the stability of the biocatalyst but as indicated that is a problem where technological solutions will occur through both enzyme immobilization approaches and protein engineering. Table 20-6 summarizes all these factors.

Table 20-6. Advantages and Disadvantages of Bioreactors Relative to Batch Fermentations

ADVANTAGES:
1. Higher concentrations of reactants in process stream
2. Cleaner, less complex process stream
3. Simpler product recovery processes required
4. Lower capital cost per unit product
5. Continuous (rather than batch) processing utilizes equipment, facilities, and labor more efficiently
6. Bioreactor processes lend themselves to automation
7. Smaller and simpler requirements for waste disposal

DISADVANTAGES:
1. Raw materials costs for bioreactors are typically significantly higher than for fermentations
2. Substantial installed and depreciated fermentation capacity currently exists
3. Stability of biocatalyst has traditionally been a problem

CONCLUSIONS

We are in the midst of an economic transformation with respect to the feasibility of using enzymes in the food processing industry as well as for other applications. The driving force is the ability to produce enzymes in high yield in genetically engineered microorganisms. For many, if not most of the enzymes that are of interest to the food processing industry, the economic realities will demand that they be produced in organisms that can secrete the enzyme into the fermentation medium at high yields, greater than 5 grams per liter. It should be pointed out that even *E. coli* processes can be made competitive. For example, the Genex process for rennin results in high quality at a price that can actually compete with other microbial rennins which are selling at half the cost of the authentic chymosin. Finally, for production of many food and feed additive ingredients, bioreactors will increasingly be used in preference to traditional batch fermentation approaches.

Chapter 21

Profit Opportunities in Biotechnology for the Food Processing Industry

Nanette Newell and Susan Gordon

INTRODUCTION

The many advances in biological research at the molecular and cellular level in the 1960's and 1970's led to greatly intensified biological research efforts for the development of new commercial products. New biotechnology has been touted as the most important industrial advance since semiconductors, and the predictions of sales of biotechnology products by the year 2000 range from tens to hundreds of billions of dollars. New biotechnology is defined as the use of *novel* commercial techniques that use living organisms, or substances from those organisms, to make or modify a product and techniques used for the improvement of the characteristics of economically important plants and animals and for the development of microorganisms to act on the environment (1). One reason biotechnology has caught the interest of so many people is its broad applicability across many industrial sectors, from pharmaceuticals to chemicals to electronics. The food and food processing industries have found that biotechnology can be used for a variety of products and processes, including improved processibility of crops, food additives, and improved processing methods. Here we shall concentrate on the application of biotechnology to food processing and not on agricultural applications except as they relate to food processing.

This chapter will first discuss the opportunities for biotechnology in the industry, and then concentrate on the various industrial players beginning to incorporate biotechnology into the food processing industry. Biotechnology will be important in the food processing industry, but it will be several years before it has a substantial impact due to the present investment in current processes and capital equipment. The development of biotechnological products and processes will be driven by consumer demand, and later by more efficient processes as they are developed. As long as the major U.S. food processing companies keep in touch with the developing technology and are ready to apply it when the market dictates, they should remain competitive in the world marketplace.

THE PRODUCTS

There is no dearth of potential biotechnology products in the food processing industry as seen in Table 21-1. Some of these products, such as amino acids and vitamins, are already made biologically, and new biotechnology is being used to improve production processes and lower costs. New products, such as the non-nutritive sweeteners, will appear because biotechnology can be used to produce them cost-effectively for the first time. Additionally, biotechnology, in the form of improved enzymes, will change and improve production processes.

There have been many estimates of future biotechnology product markets. Table 21-2 presents some estimates for the food and agriculture industries. The estimates vary widely, and there are several reasons for this variation. First and foremost, it is difficult to predict the extent to which biotechnology will affect the industry, especially by 1990. The results of the rapid pace of biological research since 1970 show that we are only scratching the surface of the potential that exists. Additionally, because biotechnology can affect so many other industries, the major trends in the world marketplace will influence the rapidity with which biotechnology is developed. For instance, the cost of petroleum will help determine whether biological processes will replace chemical ones. And finally, in many instances, the product development cycle is so much in its infancy that we just don't know exactly what's going to work. Biological systems are extremely complex, and we are just beginning to have some real understanding of their molecular basis. Biotechnology will be important, but how quickly and to what extent is difficult to predict.

PROCESSING ECONOMICS

The food processor will benefit from biotechnology in a multitude of ways. The benefits could be improved product margins, unique product advantages, and improved processing economics, to name just a few. Two examples are discussed here: development of tomatoes with higher solids content, and increasing the use of enzymatic production processes.

Tomato Solids

The tomato industry could significantly reduce overall processing costs by using tomatoes with higher than the average solids content. The average solids content of a tomato is approximately five percent, and the rest of the tomato is water. This water must be removed to give processed tomato products that can have solids as high as 32%. Biotechnology has the capability of increasing the solids content by several percentage points, thereby reducing processing costs.

Table 21-1. Biotechnology Products for the Food Industry

Product	Use
Amino Acids	
glutamic acid	Food additive
aspartic acid	Aspartame
phenylalanine	
methionine	Animal feed
lysine	
tryptophan	
Vitamins	
	Human and animal diet supplements
B2	
B12	
C	
E	
Enzymes	
alpha-amylase	High fructose corn syrup
glucoamylase	
glucose isomerase	
amyloglucosidase	Light beer
pullulanase	
proteases	Food treatment
rennin, lysozyme	Cheese
Improved Organisms	
yeast	Wine and beer
lactic acid bacteria	Milk fermentation
Low Calorie Products	
aspartame	Non-nutritive sweeteners
thaumatin	
monellin	
stevioside	
modified triglycerides and fatty acids	Cooking oil and food additives
Microbial Polysaccarides	Thickeners and gelling agents
Flavors, Fragrances, and Colorants	
Food Testing	
monoclonal antibody kits	*Salmonella*, aflatoxin
Single Cell Protein	Animal and human food supplement

Table 21-2. Forecasts on Size of Worldwide Market for Biotechnology—Agriculture and Food Processing Products

Source	Year	$ (in millions)
Arthur D. Little	1990	2,000-4,000
Business Communications Co.	1990	430
Policy Research Corp.	2000	50,000-100,000
Predicasts, Inc.	1985	6,200
	1995	101,000
Strategic, Inc.	1990	4,500
	2000	9,500
T. A. Sheets & Company	2000	21,300

It has been estimated that for every one percent increase in tomato solids, the tomato processing industry would save $80 million per year (2). This value was obtained by estimating the savings a tomato processor would realize from reduced raw product volumes, raw material transportation costs, and processing energy costs. Savings will also come from increased productivity without facilities expansion. Specifically, the higher the tomato solids (to a point), the shorter the time required to move raw product through a processing plant. This shorter time corresponds to a reduction in capital equipment requirements for a determined amount of tomato product, resulting in an improvement in processing margins.

Enzymes

The food processing industry is currently the largest consumer of industrial enzymes, making up about 40% of a $400 million market (3). The enzyme market, in contrast to other areas of food processing, is in a growth phase. A recent report suggested that the food-related enzyme market will grow at a 6% annual rate to approach $200 million by 1988 (4). Currently, the enzyme industry is dominated by two European companies, Novo and Gist-Brocades, who command 65% of the world market share (5). Because of the potential growth in this industry sector, several large companies, including such food processing companies as CPC International, ADM, and Miles, have entered the enzyme business. Newer biotechnology companies, such as Collaborative Research and Genencor, have also targeted the food-related enzyme market.

Only a handful of enzymes are now used in the food industry. In many cases, they present enough disadvantages to outweigh their major advantage: to carry out complex chemical reactions with a high degree of specificity. Enzymes used industrially must be able to withstand the rigors of production processes, such as heat. With the advent of new biotechnology and an increased understanding of protein structure-function relationships, the use of enzymes industrially should increase.

For instance, glucose isomerase is used for the final step in the production of high fructose corn syrup. Recombinant DNA technology could potentially be used to increase glucose isomerase production in microorganisms and to improve the enzyme's properties. An improved glucose isomerase would have the following properties: 1) a lower pH optimum to decrease the browning reaction caused by the alkaline pH now required; 2) thermostability so that the reaction temperature can be raised, thus pushing the equilibrium of isomerization to a higher percentage fructose; and, 3) improved reaction rates to decrease production time. Improvements in glucose isomerase will first come from the cloning of its gene into microorganisms that have been developed for high production. It is also possible that screening a broad range of microorganisms will yield enzymes with some improved properties. Finally, it will be possible in the future to identify the regions of the enzyme that are responsible for its various properties, such as pH optimum, and to direct changes in the gene structure to modify these properties.

As more is learned about how to stabilize enzymes for industrial production, the number of enzymes used is likely to increase. Their use for the production of new products could increase dramatically over the next decade. It is unlikely, however, that we will see a large increase in enzymatic synthesis of products replacing chemical synthesis. The food processing/chemical industry has large existing investments in capital equipment for production using petrochemical feedstocks. Feedstock prices would have to greatly increase to justify a complete change in production methodology in the near future.

CONSUMER TRENDS

Biotechnology can be used to answer market demands by consumers. Here we discuss two trends: natural products and low calorie foods.

Natural Products

The current trend toward a more health and nutrition conscious lifestyle has encouraged the development of natural food products. One way the food processors have addressed this trend is by increasing their use of natural food colorants and flavorants found in plants. However, this is not without its problems. There is a shortage of natural food additives available, and those that are used are often in short supply and noncompetitively priced in comparison to synthetics as seen in Table 21-3. Generally, natural additives elicit a higher price than their synthetic counterparts due to the difficulty in growth of the plants and extraction of the products. The industry, if it is to meet consumer demand effectively, needs competitive new sources of natural food products, and biotechnology provides one solution.

Enhancing levels of high valued specialty chemicals within the source plant or production of these chemicals in plant cell culture systems, are two ways to meet consumer demands. Economics dictate, however, that biotechnology as a tool must be used selectively. Only compounds with relatively large market volumes and value, and without existing satisfactory means of production, will find biotechnology the method to use.

Table 21-3. Secondary Plant Products of Economic Significance and Their Synthetic Counterparts

Ingredient	Price ($/lb.) Natural	Synthetic
Flavoring and Fragrances:		
Vanilla extract (in gallons)	50	5
Cocoa distillate	10	None
Rose Oil	1,000	None
Menthol	13	5
Jasmine	2,275	
Jasmone*		175
Colorings and Pigments:		
Peruvian Cochineal	230	
Red Dye No. 40*		28
Peruvian Annatto	25	
Yellow Dye No. 6*		10
Shikonin	2,045	
Fats and Oils:		
Cocoa butter	2	1

* Synthetic compound is not directly exchangeable with the natural counterpart.

Vanilla is the most widely used food flavoring in the world. Natural vanilla extract (40% vanillin) sells for $50 per gallon while its synthetic counterpart sells for much less, $5 per gallon. However, synthetic vanilla is generally not as attractive to the food processor as natural vanilla because of its lower quality and "artificial" FDA status. Vanilla is not grown domestically for a variety of reasons, the most important of which is cost related, and supply and price can fluctuate dramatically. Engineering high concentrations of vanilla in a domestic "Mexican" vanilla variety or producing vanillin via tissue culture could offer a solution to these U.S. food processors.

Another example, shikonin, a napthoquinone, is one of only a few chemical compounds produced via plant cell culture. Shikonin is a deep purple pigment used largely in Japan as a dye and pharmaceutical with sales of $0.6 million per year. Traditionally extracted from the root of the shikon plant, Japan's only source of the plant has been China. The plant, harvested once every seven years, contains less than two percent shikonin. Correspondingly, the price for the natural compound is very high and the supply is inconsistent (6). Mitsui Chemical Company has developed a plant cell culture system to produce high concentrations of this chemical, which is harvested every 23 days. While the price has remained as high as that of the natural chemical, supply has been expanded and new market opportunities have arisen.

The food processor is playing an expanding role in the genetic development of new and better natural compounds to meet the "natural" food consumer trend. This is evident when looking at those flavor, color, and fragrance companies investing in biotechnology research either in-house or out of house. McCormick, the world's leading producer of spice and seasonings, has recently invested in a program at Native Plants Inc. to develop modified and improved seasonings. Firmenich, a large flavor and fragrance supplier, has similarly invested in DNA Plant Technology to develop improved methods of production for plant derived flavor compounds. W. R. Grace has signed an agreement with Synergen to modify microbes for the more economic manufacturing of flavoring agents. In the future, we will be seeing many more such tie-ups.

Low Calorie Foods

Hand in hand with the materializing consumer demand for natural food products is the demand for low calorie processed foods. This, in part, can be detected in the existing food processing product lines as low calorie convenience foods, texturizing and bulking agents, non-nutritive sweeteners, and low calorie beers. With the advent of biotechnology, the potential for production of new and unique low calorie foods is tremendous.

Perhaps the greatest opportunity for biotechnology in low calorie food production is the development of a low calorie fats and oils (CALO) market. One approach to engineering CALO fats may be to induce the production of shorter chain fatty acids in commonly used vegetable oils like soybean or rapeseed. It has been predicted that the CALO market could reach $2 billion a year by the end of the next decade (7).

Biotechnology has and will continue to play a major role in developing the non-nutritive sweetener market which, according to consumer consumption trends, will have a significant place in the food industry. Reach Associates has predicted that the market will be $500 million by the year 2000. A new class of compounds called taste active proteins will be targets of biotechnology research (7). These compounds can act as either or both sweeteners and flavor modifiers and include compounds such as aspartame, thaumatin, monellin, and stevioside. Thaumatin, a sweetener known under the trade name as Talin, is the sweetest compound in the world (2,500 times as sweet as sucrose). It is capable of enhancing flavors associated with sweet products like peppermint and menthol as well as savory flavors (Table 21-4) (8). Tate & Lyle, located in the United Kingdom, has isolated the gene that codes for this protein.

Table 21-4. Examples of Flavor Potentiation with Thaumatin (18)

Flavor	Normal Perception Threshold	Reduction of Threshold with Thaumatin ($5x10^{-5}$%) (times lower)
Peppermint	0.00006	10
Menthol	0.00006	3-5
Coffee essence	0.0096	2
Coffee extract	0.01	3-4
Chocolate flavor	0.0048	2
Milk base	0.048	2
Vanilla essence	0.0048	2
Orange essence	0.0048	2
Strawberry essence	0.00024	1.5-2
Lemon essence	0.02	2
Apple flavor	0.0024	2
Shiso (perilla)	0.0024	4
Mustard base	0.0006	2
Beef extract	0.0045	2
Chicken extract	0.072	1.5-2
Tuna fish (Bonita)	0.0048	1.5-2

This gene can be used in the production of Talin itself or engineered into plants to give new and unique foods.

Low calorie beer is currently produced by the addition of certain enzymes designed to cleave dextrin, a complex starch structure, into units that can be acted upon by amyloglucosidase and degraded into alcohol. Biotechnology is being used to develop specialized yeasts and enhanced enzymes that will aid in the production of low calorie beer. The desire for a high quality low calorie beer within the brewing industry is perhaps the prime motivational force behind the application of biotechnology to brewing.

THE U.S. PLAYERS

In the United States, the industrial application of new biotechnology has been led by two types of firms. Large established companies have invested in biotechnology as a means of diversifying either their product lines or processing methods. Small entrepreneurial companies began to open their doors in the mid 1970's in response to the advances in molecular biology and genetics that made the commercial development of biotechnology possible. The established firms either made a commitment to an in-house research program, or, more often, invested in the new biotechnology companies in the form of equity or research contracts. The food processing sector was no exception, although the level of commitment was not nearly as large as that of pharmaceuticals and animal health. Much of the investment that did occur in food processing was in agriculture and novel bioprocessing, especially enzymes. In terms of R&D, applications in the food processing industry are longer term and potentially riskier than those in pharmaceuticals.

There still is a hesitation on the part of large food processing firms in making a major commitment to biotechnology. Table 21-5 shows the largest U.S. food companies and their commitment to biotechnology. Only one-third of these companies have in-house R&D programs, and some have chosen not to invest in the new technology at all. Table 21-6 shows other major U.S. food processors who are actively investigating biotechnology. Most of their effort is in the form of contracts to smaller biotechnology companies. In general, the U.S. food processing industry does not have a large commitment to biotechnology.

Table 21-5. The Top 15 Food Processors and Their Stances on Biotechnology

Food Company	In-house Program	Biotechnology Company Tie-Up	Agreement
General Foods	yes	DNA Plant Technology	Cocoa bean improvement
		Ergenics	Process improvements
Dart and Kraft	no	—	—
Anheuser-Busch	no	Interferon Sciences	Bioprocessing products
Beatrice	no	Ingene	Enzymes, protein, sweeteners
Nabisco	yes	Cetus	Many food related technologies, including enzymes
Coca Cola	no	—	—
PepsiCo	no	—	—
R.J. Reynolds	yes	—	—
H.J. Heinz	no	ARCO, Biotechnica	Tomato improvement, low cost amino acid production
Phillip Morris, Inc.	yes	—	—
Campbell Soup Co.	yes	Calgene, DNA Plant Technology	Tomato and carrot improvement
Sara Lee Corp.	no	—	—
Carnation	no	—	—
General Mills	no	—	—

Table 21-6. Other Major Food Processing Companies with Biotechnology Efforts

Food Concern	In-house	Biotechnology Company	Agreement
Food Processing			
Archer Daniels Midland	yes	DNA Plant Technology	Enzymes
CPC International	yes	Enzyme Biosystems	Food enzymes
Hershey	no	DNA Plant Technology	Cocoa bean improvement
Kellogg	no	Agrigenetics	Equity
Seagram Co., Ltd.	no	Biotechnica	Equity; genetic engineering of yeast and bioprocessing systems
Flavor and Fragrance			
Firmenich	no	DNA Plant Technology	Improved production methods for desired flavorings
W.R. Grace	yes	Synergen	Development of microbiological systems to produce flavors and fragrances
Spice and Seasoning			
American Basic	yes		Onion, garlic improvement
McCormick	no	Native Plants Inc.	Improved production of seasonings

In many instances, especially in Japan, large companies have used biotechnology as a means of diversification into other industrial sectors. This has not been the case for the US food processing companies. To a very limited extent, other large U.S. companies are using biotechnology to diversify into food. Examples include ARCO's subsidiary, The Plant Cell Research Institute, Dow's investment in Collaborative Research, Phillips Petroleum's single cell protein production process, and G.D. Searle's NutraSweet^{R}.

As can be seen from Table 21-7, the number of major start up companies applying biotechnology to food is not large compared to other industrial sectors. For the most part, these companies are the major players in plant agriculture, and some of their R&D is devoted to improving processing characteristics of crops. For these companies, it is thought that a greater rate of return can be achieved through processing improvements than through improvements to farmers.

Table 21-7. Some U.S. Biotechnology Companies in Food Processing

Company	Affiliations with the Food Industry
Advanced Genetic Sciences	Hilleshog
Agrigenetics	Kellogg
Biotechnica International	Heinz
Calgene	Campbell
	Nestle
DNA Plant Technology	Arthur Daniels Midland
	Campbell
	Firmenich
	General Foods
	Hershey
	Kraft
Engenics	General Foods
Genencor	Corning Glass
	A.E. Staley
	Chr. Hansen
Native Plants, Inc.	McCormick Spice
Plant Genetics, Inc.	Kirin
Synergen	W.R. Grace

Therefore, at this time there are not a large number of players in food processing biotechnology, but the further development of the technology should attract more in the future.

COMPETITION FROM JAPAN

The food and beverage companies of Japan began to enter the biotechnology field aggressively in 1980. In marked contrast to the food companies in the U.S., those in Japan have extensive experience in bioprocessing, one of the key underlying technologies important in commercializing biotechnology products. Bioprocessing technology can be used to make food additives such as amino acids and vitamins, and fermented food products such as tempeh. Additionally, bioprocessing is critical to the production of beer and whiskey. Japan leads the world in production of amino acids and fermented food products and has many successful companies using bioprocessing. In the U.S., on the other hand, it is the pharmaceutical industry that has the necessary expertise in bioprocessing, while very little biological production occurs in the food industry.

The food processing industry in Japan is a mature industry with very little growth prospects ahead. Biotechnology is giving those companies a method of diversification, by allowing them to exploit the new technology using their bioprocessing base. Diversification, in the majority of cases, has meant entering the pharmaceutical field (Table 21-8). The commercialization targets, such as antibiotics, immune regulators, and heart disease drugs, are the same as those of U.S. pharmaceutical companies. A few other examples include: 1) Suntory, best known for its whiskey, has isolated a novel cerebroside from a species of sponge that is effective in lowering blood pressure in mice (9); 2) Yaegaki, a major producer of sake, has developed a sake that contains monakolin K, a fungal metabolite that inhibits cholesterol biosynthesis (10); and, 3) Yamase Shoyu, the second largest soy sauce company in Japan, is diversifying into monoclonal antibody diagnostics (11). These Japanese food companies, however, are not ignoring biotechnology as applied to their main business. They intend to use biotechnology to further their development of food and food additives. Two large food additive companies, Ajinomoto and Kyowa Hakko, have in-house biotechnology programs and have developed more efficient production methods for phenylalanine, a component of aspartame (12,13). Suntory, in a joint venture with two other Japanese companies, is developing a bioreactor system to develop new food additives and preservatives (14).

Table 21-8. Japan's Food Processing Industry: Companies Entering Biotechnology (19)

Field of Entry: Pharmaceuticals	
Ajinomoto	Morinaga & Co.
	Nishin Flour Milling
Fuji Oil	Sanraku-Ocean
Kikkoman Shoyu	Sapporo Breweries
Kirin Brewery	Suntory
Kyowa Hakko Kogyo	Takara Shuzo
Meiji Milk Products	Toyo Jozo
Meiji Seika Kaisha	Yakult Honsha
Field of Entry: Food	
Ajinomoto	Morinaga Milk Industry
Japan Maize Products	Snow Brand Milk Products
Meiji Seika Kaisha	Yakult Honsha
Meito Sangyo	
Field of Entry: Environmental Applications	
Ajinomoto	Snow Brand Milk Products
Takara Shuzo	

Thus, the Japanese food processing companies see the application of biotechnology to their business as a natural extension of their current technological capabilities. They have aggressively incorporated new biotechnology into their corporate strategy both for diversification and for development of novel processes to remain competitive. Often, these commitments have not required large sums of money. The research expenditure of the Japanese food processing industry in FY 83 was $10.5 million (not including amino acids) (15). Yet their success in biologically derived products continues and certainly presents a competitive threat to the U.S. Interestingly, a recent Japanese industry report predicts that biotechnology products from the food and beverage sector will top the list in sales by the year 2000 (Table 21-9).

Table 21-9. Estimates of Output of Biotechnology Products for Japan's 14 Key Bio-industry Groups in the Year 2000 (20)

SECTOR	YEN (billions)	DOLLARS (millions)
food, beverages	4,247.4	17,196
pharmaceuticals	3,151.4	12,759
commodity chemicals	1,522.2	6,163
agricultural produce	1,401.4	5,674
industrial wastewater	1,370.2	5,547
specialty chemicals	1,076.1	4,357
electronics	603.5	2,443
cattle, dairy products	475.7	1,926
petroleum	462.8	1,874
metals	334.5	1,354
agricultural chemicals	141.6	573
fisheries	118.1	478
wood and dissolved	89.9	364
forestry products	8.5	34
TOTAL	¥15,003.3	$60,742

A LOOK TO THE FUTURE: THE PHENYLALANINE STORY

The recent history of phenylalanine production can be used to exemplify some aspects of competition between U.S. and Japanese companies and between established firms and smaller biotechnology companies. Phenylalanine is one of two amino acids that make up the sweetener aspartame marketed by G.D. Searle under the tradename NutraSweet^{R}. Phenylalanine was never an important product target until its use in a sweetener was discovered, and at that time, the cost of production was high. In 1983, it was reported that the cost of making phenylalanine was too high to make the aspartame cost competitive with saccharin (16). At that time, several companies were pursuing the development of efficient production methods. Searle gave Genex, one of the top rated biotechnology companies at the time, a contract for phenylalanine production. Genex, seeing a rosy future, invested $25 million in a production facility in Kentucky. Then, in June 1985, Searle informed Genex that they were cancelling the contract (17). Genex does not have any other well developed product lines, and the loss of the Searle contract may well cause

Genex not to survive as an independent entity. The novel bioprocessing methods, developed for phenylalanine by Genex were cost effective three years ago, but advances in biotechnology have been so dramatic that the Genex process was no longer able to compete on the world market. It appears that Genex was not able to continue to refine its production methodology enough to remain competitive.

Searle has retained one supplier, the Japanese company Ajinomoto. Ajinomoto was able to develop its technology to be competitive in the world marketplace. Ajinomoto is sharing its technology with Searle, and Searle intends to become an in-house producer (17).

Several questions arise from this story, which could be generally applicable to the emerging biotechnology industry: 1) Are Japanese companies really more innovative than U.S. companies in the development stages of a product?, 2) Will the small U.S. biotechnology companies survive into the production stages, or is their usefulness at the early innovation stages coming to an end? and, 3) Are the established U.S. food processing companies positioning themselves to take advantage of novel processes and products?

The large U.S. companies have been and will continue to be conservative as far as setting up major in-house programs. The U.S. food processing companies, for the most part, do not see biotechnological applications as a natural outgrowth of their current expertise. Entering a new technological area will take a longer time. A fairly major amount of proof of the technology will have to exist before the established firms make a commitment. It may be that the new biotechnology firms and Japanese firms will do the R&D that proves the technology. With the amount of money that can be drawn upon by large U.S. companies, they should be able to compete successfully in biotechnology if they are successful at realizing biotechnology's potential.

REFERENCES

1. U.S. Congress, Office of Technological Assessment. Commercial Biotechnology: An International Analysis. (1984)
2. Agricultural Genetics Report, p 7, Nov./Dec. (1983)
3. Layman, P. Chemical and Engineering News. pp 11-13, Sept. 12 (1983)
4. Biotechnology News, p 8, May 22 (1985)
5. Eveleigh, D.E. Scientific American 245:155-177 (1981)
6. Biotechnology Newswatch, p 8, Oct. 15 (1984)
7. Chemical Week, pp 6-10, June 19 (1985)
8. Stephens, P. Food Ingredients, March (1983)
9. Biotechnology in Japan Newsservice, p 3, March (1985)
10. Biotechnology in Japan Newsservice, p 5, April (1985)
11. Biotechnology in Japan Newsservice, p 2, April (1985)
12. Biotechnology in Japan Newsservice, p 7, February (1985)
13. Business Week, p 37, July 15 (1985)
14. Biotechnology in Japan Newsservice, p 5, February (1985)
15. Biotechnology News, p 4, March 25 (1985)

16. Chemical and Engineering News, p 8, June 13 (1983)
17. Business Week, p 37, July 15 (1985)
18. Stephens, P. Food Ingredients, March (1983)
19. Saxonhouse, G. Biotechnology in Japan, contract report prepared for the Office of Technology Assessment, U.S. Congress (1983)
20. Japanese Association of Industrial Fermentation, reprinted in Biotechnology Newswatch, p 16, January 7 (1985)

Index